Lecture Notes
in Control and Information Sciences
192

Editor: M. Thoma

Lecture Notes
in Control and Information Sciences   197

Roy S. Smith and Mohammed Dahleh (Eds.)

The Modeling of Uncertainty in Control Systems

Proceedings of the 1992 Santa Barbara Workshop

Springer-Verlag London Ltd.

Editors

Roy S. Smith, PhD
Electrical and Computer Engineering Department, University of California
at Santa Barbara, Santa Barbara, California 93106-9560, USA

Mohammed Dahleh, PhD
Mechanical and Environmental Engineering Department, University of California
at Santa Barbara, Santa Barbara, California 93106-9560, USA

This book was processed using the LaTeX macro package with the LMAMULT style
and additional modifications by Roy S. Smith

ISBN 978-3-540-19870-3   ISBN 978-3-540-39327-6 (eBook)
DOI 10.1007/978-3-540-39327-6

British Library Cataloguing in Publication Data
A catalogue record for this book is available from the British Library

Typesetting: Camera ready by editors

69/3830-543210 Printed on acid-free paper

Preface

This volume collects together a series of essays and papers arising from a workshop held at the University of California, Santa Barbara, 18–20th June, 1992. The official title was "The Modeling of Uncertainty in Control Systems," and the sponsorship was provided by the National Science Foundation (NSF) and Air Force Office of Scientific Research (AFOSR).[1]

The format of the workshop was somewhat novel and is worthy of comment here. For logistical reasons, it was held the week prior to the 1992 American Control Conference (ACC) in Chicago. As almost all workshop attendees would also be at the ACC, there was little point in having a program of formal presentations — only to have it repeated the following week. The format chosen was suggested by members of the advisory committee with the emphasis being placed on discussion groups rather than formal presentations. A tour of several identification and control experiments at the Jet Propulsion Laboratory was conducted for the attendees prior to the workshop.

Attendees typically had research experience in either traditional identification or robust control. In each area, some progress has been made towards the common goal of developing a unified identification and design paradigm. Researchers in the identification area are now considering models with uncertainty expressed as confidence intervals. Several groups of robust control researchers have begun considering the identification in a worst case setting which attempts to obtain models compatible with the existing robust control design approaches. Others are involved in studying the properties and application of iterative identification/design schemes. Research in these areas is only just beginning and the objective of this workshop was to broaden the perspectives of those working in these areas, initiate collaborations, and outline common research goals. Each academic participant was asked to bring a graduate student interested in the area. Most did and these students often played an active role in the discussions, contributing their own ideas as well as benefiting from the close involvement with experienced researchers.

The participants were divided into four groups for detailed discussions on

---

[1] The workshop was funded by NSF and AFOSR under NSF grant number: ECS-9203908

a particular topic. The topics were essentially defined by the participants and are loosely categorized as:

Probablistic uncertainty modeling. This group covered the probablistic bounding approaches to the problem. This is in the line of the more traditional identification theory and addressed such things as the effects of undermodeling.

Worst case identification. The theory of worst case deterministic identification, in the presence of bounded noise, was discussed here. This area has seen a recent burst of activity.

Iterative identification/control design. Several of these schemes have been proposed, both in probabilistic and H_∞ settings. The discussion focused on whether a closed loop approach was necessary and what benefits might be gained from it.

"What is it that we really want from control models?" A long term view was taken by the participants in this discussion group in an effort to find some unifying ideas or approaches.

Each discussion group was given the task of summarizing their area and identifying open problems and research directions. After a series of such discussions each group presented the results of their discussion to the wider audience and lead a general discussion on the overall context of this work. This format proved to be very successful with most participants gaining an understanding of the issues involved in the other related areas.

An effort has been made to maintain some of the informal discussion aspects in this proceedings volume. After the workshop, participants were invited to submit papers in either or both of two categories: speculative essays and formal papers. The essays are intended to provide an overview of the area and outline future research directions. This perspective is often difficult to obtain purely from research papers. This research area is in its infancy and the essays allow others to see the motivating issues and potential research directions. The papers, on the other hand, give a concrete indication of where the field currently stands. Both the essays and the papers were written after the workshop, giving the authors an opportunity to present their ideas and work in the context of the wider issues.

Part I contains twelve essays. Several give general comments and opinions on the field and others address specific aspects in more detail. Pramod Khargonekar, in *"Identification and Robust Control,"* provides a broad overview of the identification for robust control field. The extensive references will give a reader new to the area an excellent starting point.

"An Essay on Identification of Feedback Systems," by Pertti Mäkilä and Jonathon Partington outlines some of the issues that arise in deciding between probabilistic or deterministic frameworks.

In *"Thoughts on Identification for Control,"* Michael Safonov discusses the achievements of modern robust and robust adaptive control and how they are limited by the lack of appropriate identification for control techniques. The mathematical as well as the philosophical implications of a priori knowledge of the systems' structure, in addition to the separation between identification and control paradigm, are discussed.

A similarly broad view of the problem is taken by Hidenori Kimura in *"An Essay on Robust Control."* This essay analyzes the role of models in control systems design, the difficulty of arriving at such models, and contrasts control design based on models with control design based on mere insight and experience. The impact of modeling of uncertainty on control design is discussed.

Robert Kosut, in *"On the Character of Uncertainty for System Identification and Robust Control Design,"* points out that most traditional identification methods do not provide information for error estimates. Robust control calls for a nominal model and an error estimate. This essays contrasts the deterministic and probabilistic paradigms for uncertainty modeling.

A parametric approach is discussed by John Norton in *"Extensions of Parametric Bounding Formulation of Identification for Robust Control Design."* In this note the parametric bounding methods are discussed with the aim of motivating the incorporation of more a priori information, such as joint bounds on transfer function parameters. It also includes a discussion of matching of identification and control design methods, performance guarantees for modeling techniques, and performance guarantees for controllers.

"Connecting Identification and Robust Control," by Michel Gevers discusses the gap between the estimation of uncertain bounds and the current identification for robust control design. Iterative identification/control design schemes are presented as a means of bridging this gap.

Iterative methods are also considered by Paul Van den Hof, Ruud Schrama and Peter Bongers in *"On Nominal Models, Model Uncertainty and Iterative Methods in Identification and Control Design."* Their paper discusses the issues related to the role of nominal models and model uncertainty in the problem of joint identification and control design. General viewpoints on the construction of iterative schemes and their related criteria are discussed.

Model validation takes a somewhat different approach to the problem. Roy Smith discusses this area in *"An Informal Review of Model Validation."* The problem is discussed in the context of identification for robust control. Future research prospects, as well as the current state of the art, are outlined for both the H_∞ and l_1 cases.

Munther Dahleh and John Doyle, in *"From Data to Control,"* discuss an iterative validation approach. The idea involves characterizing and determining the set of models that are unfalsified by the data, designing controllers for these models, and collecting further data to reduce the set of unfalsified models. The controller performance is improved as the set of unfalsified models shrinks.

The applicability of worst case design approaches is called into question by Jeffrey Kantor and Billie Spencer, Jr. in their essay, *"Is Robust Control*

Reliable?" They suggest methods based on estimating structural reliability as a reliable alternative.

A significant proportion of the researchers came from application specific backgrounds. Daniel Rivera, in *"Modeling Uncertainty in Control Systems: A Process Control Perspective,"* discusses the requirements of any identification/design theory from the chemical process control point of view. A distillation column is considered as an example of a benchmark problem.

Nineteen technical papers are included in Part II. They can be considered as falling within several broad categories. The first of these is the problem of identification in a worst case setting. The probabilistic framework is considered next and then the iterative identification and design approaches are covered. The volume closes with several papers discussing practical applications and benchmark problems.

"A Note on H_∞ System Identification with Probabilistic A Priori Information," by Clas Jacobson and Gilead Tadmor, deals with control-oriented identification in the H_∞ setting. They consider a probabilistic setting of a priori information and show how, under certain specifications of a priori information, the probabilistic problem setting can be converted to a related worst case problem. An algorithm to solve the related worst case problem is given.

Jie Chen, Carl Nett and Michael Fan, present work in the identification for H_∞ area in *"A Worst Case Identification Method Using Time Series Data."* A discrete-time, SISO, linear time-invariant system is considered. Noise is assumed to be additive, at the system output, and l_∞ bounded. It is also assumed that system has an impulse response bounded by an a priori specified decaying exponential. An algorithm for generating a nominal system, with a calculable H_∞ error bound, is presented. The exponentially bounded system and l_∞ bounded noise are assumptions common to much of the work in this area. The concepts of algorithm convergence and optimality under these assumptions are also discussed.

The paper, *"Identification in H_∞ using Time-Domain Measurement Data,"* by Guoxiang Gu, looks at a similar problem for systems under feedback. Sine-dwell experiments are proposed, leading to an assumed frequency domain noise bound. The identification error is quantified in terms of the directed gap metric.

A time domain approach is taken by Pertti Mäkilä and Jonathon Partington in *"Identification of Feedback Systems from Time Series."* They discuss closed loop identification methods with a gap metric. The focus is on ARX models where the AR and X parts are coprime.

A different approach to this problem is taken by Tong Zhou and Hidenori Kimura in *"Input-Output Extrapolation-Minimization Theorem and its Application to Model Validation and Robust Identification."* They characterize the smallest H_∞ norm transfer function matching input-output data in the time domain. Details are provided for the SISO noiseless case although the work can be extended to include noise and MIMO problems. This result is

applied to model validation for additive, multiplicative or co-prime perturbation descriptions. The results are also applied to the identification in H_∞
problem in the case where the system poles are specified a priori. Each of
these applications leads to a convex optimization problem.

Richard Hakvoort and Paul Van den Hof consider a similar problem in
"Identification of Model Error Bounds in l_1 and H_∞-Norm." A discrete-
time formulation is used and the system pulse response is assumed to satisfy
an a priori exponential bound. The additive perturbation can be a priori
frequency weighted and the noise is assumed to satisfy a time-domain l_∞
bound. In the case where an l_1 norm bound is applied to the perturbation,
the approach results in a linear programming problem. In the H_∞ bound
case, the perturbation is overbounded and a linear programming approach is
again used to find the minimum overbound.

In *"Asymptotic Worst-Case Identification with Bounded Noise,"*, Munther
A. Dahleh presents an overview of the work of the author and co-workers in
the area of worst-case identification with bounded noise. The objective of the
study is to address the fundamental limitations and capabilities of worst-case
identification along with their impact on input design and algorithms. The
approach is applied to a variety of model sets such as stable LTI systems and
nonlinear systems with fading memory.

An adaptive approach is taken in *"Sequential Approximation of Uncertainty Sets via Parallelotopes,"* by Antonio Vicino and Giovanni Zappa. This
paper adopts a deterministic setting for disturbances through hard bounds.
An adaptive algorithm is presented for constructing recursively an outer
bounding parallelotopic approximation of the parameter uncertainty set. This
can be employed in a purely parametric or in mixed parametric/non-parametric
settings.

Tung-Ching Tsao and Michael Safonov also consider an adaptive approach
in *"A Robust Ellipsoidal-Bound Approach to Direct Adaptive Control."* Their
direct robust adaptive control scheme overcomes some of the conservatism in
ellipsoidal-bounding indirect schemes. A key idea is the introduction of a fictitious reference signal which facilitates direct identification of the controller
to meet performance specifications using measurements of plant input and
output. No use is made of the separation of identification and control which
is essential for indirect methods.

Er-Wei Bai and Sundar Raman propose an on-line approach in *"On Line
Model Uncertainty Quantification: Hard Upper Bounds and Convergence."*
They present a recursive, on-line, adaptive identification scheme to determine
bounds in either an H_2, H_∞ or pointwise sense. The models are of restricted
complexity and the order selection issues are discussed.

Douwe de Vries and Paul Van den Hof consider a mix of both frameworks in *"A Mixed Deterministic-Probabilistic Approach for Quantifying Uncertainty in Transfer Function Estimation."* They present a procedure to obtain an estimate of the transfer function of a linear system together with
an upper bound on the error, using only limited a priori information on the

data generating process. The model error consists of both probabilistic and deterministic parts.

"Estimation for Robust Control," by Brett Ninness and Graham Goodwin, considers the estimation of the frequency response of LTI systems from observed noise corrupted data. In contrast to related work it examines the extension of the stochastic paradigm to problems of undermodeling. This approach requires certain a priori information to be imposed on the class of undermodelings which in turn makes the problem amenable to analysis by stochastic estimation techniques.

The consequences of time variation are considered by Håkan Hjalmarsson in *"Non-Vanishing Model Errors."* This paper considers time-varying systems that are essentially time-invariant but with small, arbitrary fast, fluctuations of the dynamics. The approach aims at providing a method for detecting such time-varying terms, and to obtain a methodology to determine whether the assumption that with infinitely many data it is possible to get a perfect model is valid.

A more classical identification approach is taken by Wallace Larimore in *"Accuracy Confidence Bands Including the Bias of Model Under-Fitting."* A parametric model is obtained from an underlying linear system in the presence of stochastic noise and disturbances. Confidence bands are obtained as well as a nominal model. Model order selection issues are also discussed.

The iterative identification and design approach is addressed in several papers. In the first of these, *"Iterative Identification and Control Design: A Worked Out Example,"* by Ruud Schrama and Paul Van den Hof, the H_∞ norm of a particular transfer matrix is applied as the performance criterion. A weighted identification problem is used to minimize the error between this transfer matrix for the nominal model and the actual system. The tradeoff between nominal performance and robust stability is also adjusted, via a user specified parameter, during the iteration. A simulation example illustrates that a good open-loop fit is not necessary for the design of a good robust control system.

An iterative framework is outlined by David Bayard and Yeung Yam in *"Frequency Domain Identification for Robust Control Design."* The approach given here involves iterative reweighting of the identification scheme. This contrasts with the previous approach in that new input-output data is not obtained with the controller designed at each stage of the iteration. The underlying system is assumed to satisfy an a priori exponential decay on the impulse response. The noise and disturbances are assumed to be Gaussian and the approach gives a statistical confidence bound on the H_∞ error of the identified model. A simulation example illustrates the application of the approach.

Hamid Ajbar and Jeffrey Kantor present an integrated design and diagnostic approach in *"Time Domain Approach to the Design of Integrated Control and Diagnosis Systems."* They consider the l_1 design framework and develop a design framework which allows fault detection in the presence of unknown bounded perturbations.

One of the major motivations in this field is the difficulties that arise when applying standard robust control techniques to certain physical systems. It is therefore fitting to close the volume with some benchmark problem applications. Elling Jacobsen and Sigurd Skogestad give a detailed description of a heat exchanger in *"Identification of Ill-Conditioned Plants — A Benchmark Problem."* A standard identification procedure is shown to give a very poor model in the multivariable sense. A nonlinear model, linearized model and Matlab based simulation are provided for this system to allow other researchers to investigate and apply their own techniques to this problem.

Peter Bongers and Gregor van Baars describe a wind turbine system in *"Control Design and Implementation Based on Experimental Wind Turbine Models."* The physical aspects of this problem are used to illustrate some of the issues that typically arise in the practical design of robust controllers. Multiple input-output data is taken and multiple models are obtained. The gap metric is used to pick a suitable nominal with the criteria being the smallest gap between the nominal and all other identified models. The design is done in order to obtain robustness with respect to this size gap.

As these outlines illustrate, many viewpoints are represented in the following pages. This research area is fundamental to the application of robust control techniques to practical problems. The work is only just beginning and we hope that volume serves as an interesting introduction of the field to the wider control community.

Roy Smith
Mohammed Dahleh

University of California, Santa Barbara
January, 1993.

Acknowledgements

There are many people to thank for their efforts in producing this work; our co-organizer, John Doyle; the sponsors: Radhakisan Baheti of NSF and Marc Jacobs of AFOSR; the support of the Center for Control Engineering & Computation (CCEC) at UCSB; David Bayard and other JPL personnel for their help in arranging the tour; the international advisory committee (Graham Goodwin, Pramod Khargonekar, Hidenori Kimura, Robert Kosut, James Krause, Lennart Ljung and Carl Nett); and all the participants of the workshop, particularly those that took the time to document their viewpoints.

Table of Contents

Part I Essays   **1**

Identification and Robust Control
Pramod P. Khargonekar   3

An Essay on Identification of Feedback Systems
Pertti Mäkilä and Jonathan Partington 11

Thoughts on Identification for Control
Michael G. Safonov 15

An Essay on Robust Control
Hidenori Kimura 19

On the Character of Uncertainty for System Identification and
Robust Control Design
Robert L. Kosut 25

Extensions of Parametric Bounding Formulation of Identification
for Robust Control Design
J. P. Norton 29

Connecting Identification and Robust Control
Michel Gevers 35

On Nominal Models, Model Uncertainty and Iterative Methods
in Identification and Control Design
Paul Van den Hof, Ruud Schrama and Peter Bongers 39

An Informal Review of Model Validation
Roy Smith 51

From Data to Control
Munther A. Dahleh and John C. Doyle 61

Is Robust Control Reliable?
Jeffrey C. Kantor and Billie F. Spencer, Jr. 65

Modeling Uncertainty in Control Systems: A Process Control
Perspective
Daniel E. Rivera 69

Part II   Technical Papers   77

A Note on H_∞ System Identification with Probabilistic
A Priori Information
Clas A. Jacobson and Gilead Tadmor   79

A Worst Case Identification Method Using Time Series Data
Jie Chen, Carl N. Nett, and Michael K.H. Fan   93

Identification in $\mathcal{H}^\infty$ Using Time-Domain Measurement Data
Guoxiang Gu 105

Identification of Feedback Systems from Time Series
Pertti Mäkilä and Jonathan Partington 117

Input-Output Extrapolation-Minimization Theorem and Its
Applications to Model Validation and Robust Identification
Tong Zhou and Hidenori Kimura 127

Identification of Model Error Bounds in ℓ_1- and $\mathcal{H}_\infty$-Norm
Richard G. Hakvoort and Paul M.J. Van den Hof 139

Asymptotic Worst-Case Identification with Bounded Noise
Munther A. Dahleh 157

Sequential Approximation of Uncertainty Sets via Parallelotopes
Antonio Vicino and Giovanni Zappa 171

A Robust Ellipsoidal-Bound Approach to Direct Adaptive Control
Tung-Ching Tsao and Michael G. Safonov 181

On Line Model Uncertainty Quantification: Hard Upper Bounds
and Convergence
Er-Wei Bai and Sundar Raman 197

A Mixed Deterministic-Probabilistic Approach for Quantifying
Uncertainty in Transfer Function Estimation
Douwe K. de Vries and Paul M.J. Van den Hof 221

Estimation for Robust Control
Brett M. Ninness and Graham G. Goodwin 235

Non-Vanishing Model Errors
Håkan Hjalmarsson 261

Accuracy Confidence Bands Including the Bias of Model
Under-Fitting
Wallace E. Larimore 275

Iterative Identification and Control Design: A Worked Out
Example
Ruud Schrama and Paul Van den Hof 289

Frequency Domain Identification for Robust Control Design
    David S. Bayard and Yeung Yam 303

Time Domain Approach to the Design of Integrated Control and
    Diagnosis Systems
    Hamid Ajbar and Jeffrey C. Kantor 337

Identification of Ill-Conditioned Plants — A Benchmark Problem
    Elling W. Jacobsen and Sigurd Skogestad 367

Control Design and Implementation based on Experimental Wind
    Turbine Models
    Peter Bongers, Gregor van Baars 377

Part I

Essays

Identification and Robust Control

Pramod P. Khargonekar

Department of Electrical Engineering and Computer Science, The University of Michigan, Ann Arbor, MI 48109-2122, USA.

System identification is a well established area of research. Indeed, there is a large body of literature on system identification. Several impressive results have been obtained and effective algorithms have been developed for system identification. There are several excellent books on this subject which describe the accomplishments during the last decades. See, for example, the book by Ljung [28].

A common traditional approach to system identification has been to take a stochastic problem formulation. This has led to a fairly well developed theory of system identification and identification algorithms. Until recently, the issue of obtaining bounds on the resulting model error had not been a focal point of research. However, in some recent papers, results on the size of the model error from using classical identification algorithms have been obtained by [9, 15, 5, 21, 29, 52, 53]. This is a subject of much current research interest. The survey paper by Gevers [9] gives a nice and readable account of certain aspects of this line of research.

In modern robust control, the starting point for control system analysis and design is a nominal plant model and (norm) bounds on model uncertainty. This has fueled a renewed interest in worst-case deterministic formulations of the system identification problem. These worst-case deterministic formulations of the identification problem are motivated by a perceived need to develop a theory of system identification that is compatible with modern robust control. As a result, system identification techniques should be required to provide guaranteed error bounds in addition to a candidate system model, or more generally a set of systems as the model. Recently, some papers that take a deterministic approach to the identification problem have appeared, see for example, [3, 5, 7, 15, 17, 23, 24, 16, 26, 34, 35, 36, 39, 21, 42, 22, 48, 24, 51, 52, 55, 27] and the references cited there. Many of these papers deal with what may be broadly termed as "robust identification". Another approach to the interaction between robust control and identification is via iterative identification and control designs. This approach is being investigated in [16, 22, 27, 45, 57]. Another direction of research activity is the so-called identification in $\mathcal{H}_\infty$ problem which is *one* particular formulation of an identification problem for robust control. The problem of identification in $\mathcal{H}_\infty$ was first formulated by Helmicki et al. [17]. Since then there has been a spurt of activity on this problem [1, 2, 4, 5, 7, 8, 6, 17, 18, 19, 20, 31, 17, 18, 19, 40, 44] and the references cited there. Worst case identification theory is also related to certain aspects of complexity theory [30, 47, 49, 56].

Any theory of identification has (at least) three main ingredients:

- A priori knowledge or assumptions about the unknown system, the noise, and model set from which the identified model is chosen by a potential identification algorithm;
- Measured data; and
- Measure of performance of the identification algorithm.

There are many different choices for these three ingredients, and each choice gives rise to a different identification theory. They also determine the scope and nature of results one can logically expect to obtain. Thus, it is important to fully understand the various issues which should be considered in making these choices.

- The unknown system is most often taken to be a linear time-invariant system. One may take the noise to be a stochastic process with some particular characteristics. In this case, one can not expect to derive hard deterministic, bounds on the modeling error, if that is taken to be a measure of performance of the identification algorithm. However, it may be possible to obtain probabilistic error bounds for a given confidence level. The other choice is take the noise to be a bounded signal with a known bound. This can be followed by either a worst-case or an average-case analysis. In the worst-case analysis, one may reasonably expect to arrive at hard bounds on the modeling error.
- Measured data may be taken to be either time-domain input-output measurements or frequency domain measurements. It should be recognized that one can not get frequency domain measurements directly without a time-domain experiment. However, in some cases, for example if the inputs are sinusoidal, it may be more efficient to represent the measured data to be frequency domain measurements. The model set may be a parametrized set of systems. On the other hand, one may let the model set be the set of all stable systems as in nonparametric identification. The former choice leads to restricted complexity identification.
- The performance of an identification algorithm depends on the intended use of the identified model. Thus, if a model is sought for the purpose of prediction, then prediction error is a natural performance measure. On the other hand, if the purpose of the model is control design and analysis, then the performance measure must reflect this objective.
  Generally speaking, performance of an identification algorithm often involves the following aspects: asymptotic convergence properties, rate of convergence, hard (deterministic) or soft (probabilistic) bounds on the modeling error, intrinsic and algorithmic complexity, etc. Each of these

is an important issue and leads to different technical problems. At this point the choice between the hard bounds and the soft (probabilistic) bounds remains largely open. It seems that hard deterministic bounds may be overly conservative. On the other hand, from a theoretical point of view, it is important to analyze the worst case behavior of any given identification algorithm. Also, robust control theory is almost exclusively a "hard bounds – worst case" theory.

Model (in)validation is a complement to identification. An identification algorithm leads to a model and bounds on model-plant mismatch. Now one can take this as a postulated model for the system to be identified. Then one can ask the question whether this postulated model correctly describes the system. This question can be addressed by performing new experiments on the system or using old experimental data which was not used for constructing the model. The problem then is to decide whether this new data contradicts the postulated model. If it does contradict the postulated model then the one should declare that the postulated model is invalidated; otherwise, the postulated model is not an invalidated model of the system under consideration. Model invalidation has been an important component of classical identification theory and is usually performed by residual analysis as described in the book by Ljung [28]. It is also an important aspect of the model building in the "behavior framework" in the work of Willems [54]. Recently, model invalidation problems in the setting of robust control models have been investigated in [6, 21, 22].

Model validation is one potential bridge that may connect stochastic identification theory with deterministic robust control. One may take the model that results from the traditional identification techniques with uncertainty bounds obtained by confidence interval methods as a robust control model. A good model validation test then may give the robust control designer the confidence needed to use this model for robust control design.

Despite the flurry of activity in this research area, at this time there is no broad consensus on a paradigm for the interaction between robust control and identification. Many of the reasons for this lack of agreement touch upon the very foundations of *theoretical engineering*. (Is this, perhaps, a contradiction in terms?) What are appropriate models for uncertainty in signals and systems? How do we choose between stochastic vs deterministic formulations? How, if at all, can control or identification theories be compared? What is the precise sense in which the term "theory" is used in systems and control? For example, does it make sense to *falsify* [43] or validate a control or an identification theory? Or are control and identification theories essentially mathematical so that the concept of falsification does not make sense? In the context of physics and other natural sciences, the notion of a "scientific theory" has been analyzed in depth by many eminent philosophers. However, it is not at all clear that the various analyses of scientific theories are applicable in the context of engineering. One way to obtain a deeper understanding of some of these issues would be to explore the relations between synthesis theories and the engineering design process. It seems that the value of theory in engineering is in providing conceptual frameworks, synthesis procedures, and computational tools useful in solving engineering problems. A thorough analysis of the concept of "theory" in the engineering context is a very challenging and worthwhile objective.

As far as the interaction between identification and robust control is concerned, it is very important to develop *simple paradigm problems* that capture at least some of the important features of the identification for robust control problem. These simple paradigm problems would focus on specific topics on the interface between robust control and identification. These paradigms might also emerge from an analysis of the current identification and robust control practice. The current focus seems to be directed towards developing identification theory so that there is a better match with the existing robust control methodologies. It may be that new robust control paradigms should be developed which are more compatible with identification techniques. For example, it seems quite appropriate to develop a stochastic robust control framework. It is clear that the combination of identification and control provides a potential for significant gains in design of engineering control systems as well as for further development of control and system theory.

References

1. H. Akçay, G. Gu, and P. P. Khargonekar, "A class of algorithms for identification in $\mathcal{H}_\infty$: continuous-time case," accepted for publication in *IEEE Transactions on Automatic Control*.

2. H. Akçay, G. Gu, and P. P. Khargonekar, "Identification in $\mathcal{H}_\infty$ with nonuniformly space frequency response measurements," *Proc. 1992 American Control Conference*, pp. 246–250.

3. H. Akçay and P. P. Khargonekar, "The least squares algorithm, parametric system identification, and bounded noise," submitted for publication in *Automatica*.

4. J. Chen, G. Gu, and C.N. Nett, "Worst case identification of continuous time systems via interpolation," submitted to *Automatica*.

5. J. Chen, C.N. Nett, and M.K.H. Fan, "Optimal non-parametric system identification from arbitrary corrupt finite time series: a worst case /deterministic approach," *Proc. 1992 American Control Conference*, pp. 279–285.

6. J. Chen, C.N. Nett, and M.K.H. Fan, "Worst-case system identification in H_∞: validation of a priori information, essentially optimal algorithms, and error bounds," *Proc. 1992 American Control Conference*, pp. 251–257, also submitted to *IEEE Transactions on Automatic Control*.

7. M. Dahleh, E. Sontag, D. Tse, and J. Tsitsiklis, "Worst case identification of nonlinear fading memory systems," *Proc. 1992 American Control Conference*, pp. 241–245.

8. T. T. Georgiou, C. Shankwitz, and M. C. Smith, "Identification of linear systems: a graph point of view," *Proc. 1992 American Control Conference*, pp. 307–311.

9. M. Gevers, "Connecting identification and robust control: a new challenge," Technical Report 91.58, CESAME, Universite Catholique de Louvain, Belgium, 1991.

10. G. C. Goodwin, G. Gevers, and B. Ninnes, "Quantifying the error in estimated transfer functions with applications to model order selection," *IEEE Transactions on Automatic Control*, vol. 37, pp. 913–928, 1992.

11. G. C. Goodwin, B. Ninnes, and M. E. Salgado, "Quantification of uncertainty in estimation," *Proc. 1990 American Control Conference*, pp. 2400–2405.

12. G. Gu and P. P. Khargonekar, "Linear and nonlinear algorithms for identification in $\mathcal{H}_\infty$ with error bounds," *IEEE Transactions on Automatic Control*, vol. 37, pp. 953-963, 1992.

13. G. Gu and P. P. Khargonekar, "A class of algorithms for identification in $\mathcal{H}_\infty$," *Automatica*, vol. 28, pp. 299–312, 1992.

14. G. Gu, P. P. Khargonekar, and Y. Li, "Robust convergence of two-stage nonlinear algorithms for identification in $\mathcal{H}_\infty$," *Systems and Control Letters*, vol. 18, pp. 253–263, 1992.

15. R. G. Hakvoort, "Worst-case system identification in $\mathcal{H}_\infty$: Error bounds, interpolation, and optimal models," Internal Report, Delft University of Technology, The Netherlands, 1992.

16. F. Hansen, G. F. Franklin, and R.L.Kosut, "Closed-loop identification via the fractional representation: experiment design," *Proc. 1989 American Control Conference*, pp. 1422–1427.

17. A. J. Helmicki, C. A. Jacobson and C. N. Nett, "Control-oriented system identification: a worst-case/deterministic approach in $\mathcal{H}_\infty$ " *IEEE Trans. Automat. Contr.*, vol. 36, pp. 1163–1176, 1991.

18. A. J. Helmicki, C. A. Jacobson and C. N. Nett, "Identification in $\mathcal{H}_\infty$: linear algorithms," *Proc. 1990 American Control Conference*, pp. 2418–2423.

19. A. J. Helmicki, C. A. Jacobson and C. N. Nett, "Identification in $\mathcal{H}_\infty$: the continuous-time case," *IEEE Trans. Automat. Contr.*, vol. 37, pp. 604-610, 1992.

20. A. J. Helmicki, C. A. Jacobson, and C. N. Nett, "Fundamentals of control oriented system identification and their application for identification in $\mathcal{H}_\infty$," *Proc. 1991 American Control Conference*, pp. 89–99.

21. R. L. Kosut, "Adaptive uncertainty modeling: on-line robust control design," *Proc. 1987 American Control Conference*, pp. 245–250.

22. R. Kosut and H. Ailing, "Worst case control design from batch least squares," *Proc. 1992 American Control Conference*, pp. 318–322.

23. R. L. Kosut, M. Lau and S. Boyd, "Identification of systems with parametric and nonparametric uncertainty," *Proc. of the 1990 American Control Conference*, pp. 2412–2417.

24. J. M. Krause and P. P. Khargonekar, "Robust parameter adjustment with nonparametric weighted-ball-in-$\mathcal{H}_\infty$ uncertainty," *IEEE Transactions on Automatic Control*, vol. 35, pp. 225–229, 1990.

25. J. M. Krause and P. P. Khargonekar, "Parameter identification in the presence of non-parametric dynamic uncertainty," *Automatica*, vol. 26, pp. 113–124, 1990.

26. M. Lau, R. L. Kosut and S. Boyd, "Parameter set identification of systems with uncertain nonparametric dynamics and disturbances," *Proc. 29th IEEE Conference on Decision and Control*, pp. 3162–3167, 1990.

27. W. Lee, B. D. O. Anderson, and R. Kosut, "On adaptive robust control and control relevant identification," *Proc. 1992 American Control Conference.*

28. L. Ljung, *System Identification, Theory for the User*, Prentice-Hall, Inc., Englewood Cliffs, NJ, 1987.

29. L. Ljung and Z.-D. Yuan, "Asymptotic properties of black-box identification of transfer functions," *IEEE Transactions on Automatic Control*, vol. 30, pp. 514–530, 1985.

30. L. Lin, L. Wang, and G. Zames,"Uncertainty principles and identification n-widths for LTI and slowly varying systems," Proc. 1992 American Control Conference, pp. 296–300.

31. P. M. Mäkilä, "Robust discrete approximation and worst case identification", Preprint, Department of Chemical Engineering, Åbo Akademi, Finland, 1991.

32. P. M. Mäkilä and J. R. Partington, "Robust approximation and identification in $\mathcal{H}_\infty$," *Proc. 1991 American Control Conference*, pp. 70–76.

33. P. M. Mäkilä and J. R. Partington, "Worst-case identification from closed-loop time series," *Proc. 1992 American Control Conference*, pp. 301–306.

34. S. H. Mo and J. P. Norton, "Parameter-bounding identification algorithms for bounded-noise records," *IEE Proceedings. Part D, Control Theory and Applications*, vol. 135, pp. 127–132, 1988

35. P. J. Parker and R. R. Bitmead, "Adaptive frequency response identification," *Proc. 28th IEEE Conference on Decision and Control*, pp. 348–353, 1987.

36. P. J. Parker and R. R. Bitmead, "Approximation of stable and unstable systems via frequency response identification," *Proc. 10th IFAC World Congress*, Munich, Germany, pp. 358–363, 1987.

37. J. R. Partington, "Robust identification in $\mathcal{H}_\infty$," presented at the *Int. Symp. on the Mathematical Theory of Network and Systems*, Kobe, Japan, 1991. To appear in *J. Mathematical Analysis and Applications*, 1992.

38. J. R. Partington, "Robust identification and interpolation in $\mathcal{H}_\infty$," *International J. Control*, vol. 54, pp. 1281–1290, 1991.

39. J. Partington, "Worst-case identification in Banach spaces," *Systems and Control Letters*, pp. 423–429, 1992.

40. J. R. Partington and P. M. Mäkilä, "Robust identification of stabilizable systems," preprint, University of Leeds, 1991, *Proc. 30th IEEE Conference on Decision and Control*, pp. 629–633, 1991.

41. K. Poolla, P. Khargonekar, A. Tikku, J. Krause, and K. Nagpal, "A time approach to model validation," *Proc. 1992 American Control Conference*, pp. 313–317.

42. K. Poolla and A. Tikku, "Time complexity of worst-case identification," submitted for publication, Dept. of Mechanical Engineering, University of California, Berkeley, CA, 1992.

43. K. R. Popper, *Conjectures and Refutations: The Growth of Scientific Knowledge*, Routledge and Paul, London, 1969.

44. S. Raman and E. W. Bai, "A linear, robust and convergent interpolatory algorithm for quantifying model uncertainties," *Systems and Control Letters*, vol. 18, pp. 173-178, 1992.

45. R. J. P. Schrama, "Accurate models for control design: the necessity of an iterative scheme," *IEEE Transactions on Automatic Control*, vol. 37, pp. 991-993, 1992.

46. R. Smith and J. C. Doyle, "Towards a methodology for robust parameter identification," *IEEE Transactions on Automatic Control*, vol. 37, pp. 942-952, 1992.

47. R. Tempo, "IBC: a working tool for robust parameter estimation," *Proc. 1992 American Control Conference*, pp. 237-240.

48. R. Tempo and G. Wasilkowski, "Maximum likelihood estimators and worst case optimal algorithms for system identification," *Systems and Control Letters*, vol. 10, pp. 265–270, 1988.

49. J. F. Traub, G. W. Wasilkowski, and H. Wozniakowski, *Information-Based Complexity*, Academic Press, New York, NY, 1988.

50. D. N. C. Tse, M. A. Dahleh, and J. N. Tsitsiklis, "Optimal asymptotic identification under bounded disturbances", Preprint, LIDS, MIT, Cambridge, MA. An abridged version is in *Proc. 1991 American Control Conference*, pp. 1786–1787.

51. T. van den Boom, "MIMO-systems identification for $\mathcal{H}_\infty$ robust control: A frequency domain approach with minimum error bounds," Ph. D. Thesis, Eindhoven University, The NETHERLANDS.

52. B. Wahlberg and L. Ljung, "Hard frequency domain bounds from least squares like identification techniques," *IEEE Transactions on Automatic Control*, vol. 37, pp. 900–912, 1992.

53. W. Wang and M. G. Safonov, "Relative-error $\mathcal{H}_\infty$ identification from autocorrelation data — a stochastic realization method," *IEEE Transactions on Automatic Control*, vol. 37, pp. 1000-1003, 1992.

54. J. C. Willems, "From time series to linear system — Parts 1, 2, 3," *Automatica*, vol. 22-23, 1986-87.

55. R. C. Younce and C. E. Rohrs, "Identification with non–parametric uncertainty," *IEEE Transactions on Automatic Control*, vol. 37, pp. 715-128, 1992.

56. G. Zames, "On the metric complexity of causal linear systems: ϵ-entropy and ϵ-dimension for continuous time" *IEEE Transactions on Automatic Control*,, vol. AC-24, pp. 222–230, 1979.

57. Z. Zang, R. Bitmead, and M. Gevers, "$\mathcal{H}_2$ Iterative model refinement and control robustness enhancement," *Proc. 30th IEEE Conference on Decision and Control*, pp. 279-284, 1991.

58. T. Zhou and H. Kimura, "Identification for robust control in time-domain," submitted for publication, Dept. of Mechanical Engineering, Osaka University, Osaka, JAPAN, 1992.

An Essay on Identification of Feedback Systems

Pertti Mäkilä[1] and Jonathan Partington[2]

[1] Åbo Akademi University, Department of Engineering, SF-20500 Åbo, Finland

[2] School of Mathematics, University of Leeds, Leeds LS2 9JT, UK

> '*Mathematics has often been used as a tool for defusing contro-versy and for allowing ideas running counter to some dominant doc-trine to be developed in neutral form*'— Jacob T. Schwartz.

This statement fits very well in the present controversial situation in the field of identification for robust control design. As the situation was just two or three years ago it was generally acknowledged that mainstream probabilis-tic system identification methodologies do not address the issue of estimation of the size of the unmodelled nonparametric dynamics in any rigorous sense. This started an exciting race which has, however, turned out to be as tough as the marathon run in the 1992 Olympics at Barcelona.

A consequence of this race has been a renewed interest in set-valued non-probabilistic identification methodologies and in worst-case analysis of iden-tification problems (for earlier work see e.g. the surveys [23, 25]). We would like to say next a few words on this latter topic.

Worst-Case Analysis of Identification Problems

It is often said that the first task of theory is to delineate what is possible. Keeping this in mind several researchers have started to analyse nonparamet-ric identification problems in nonprobabilistic settings (see [10, 14, 16, 5, 8, 11, 46] and the special issue of IEEE Trans. Automatic Control, July 1992, on system identification for robust control design).

Worst-case analysis is well-suited for studying what is possible under var-ious mathematical assumptions. Obviously, the analysis is done with math-ematical models, so that the validity of the statements in the real (phys-ical) world is another matter. Even mainstream probabilistic identification methodologies make numerous assumptions in their mathematical modelling which would be hard to verify.

If you have a mathematical theory which is to be used to do engineering you need to make sure that the "engineering" statements of the theory (i.e.

those that relate to manipulations of the real world and to real measurements) are insensitive or robust to purely mathematical constructs such as whether a quantity whose value must be obtained through measurement is a rational number or a irrational number! Since, in modern robust control theory, methods have been developed which can deal with system uncertainty (of a certain type) in a quantitative way, it is natural to try to see whether the functional analytic and approximation theoretic philosophy so successful in robust control theory could be extended to system identification.

Some worst-case identification methods are criticized for requiring unrealistic experimental data in order to yield various information about the system. However there is here a trade-off between the information one might like, and the information one needs for robust control design, which are not the same things.

It is sometimes implicitly assumed that the Principle of Least Squares and the Principle of Least Prediction Error are the fundamental concepts in system identification. But in identification for robust control design, this clearly need not be so. We should be much more concerned about the robust closed-loop performance of the control system designed with help of experimental data. This gives us many possible measures of identification algorithm performance which so far have not been studied in a systematic way. By this we do not mean to imply that the Principle of Least Squares and related Principles will suddenly become obsolete. There is a considerable possibility that it is quite difficult to find something better and more practical (after all the Principle of Least Squares, due to Gauss and others, has withstood the test of time for 200 years).

Future ?

'Prediction is difficult, especially of the future'– Mark Twain.

Some members of the mainstream probabilistic identification community have made the interesting suggestion that rather than develop a theory of system identification based on the needs of robust control design, it could be more realistic to develop a probabilistic robust control theory based on the needs of probabilistic system identification.

The importance of robustness considerations suggest yet another possibility. Develop a robust probabilistic identification theory (robust in the probabilistic sense [1]).

Whatever modelling approach is chosen, it is possible to consider using recently developed tools on uncertainty model validation [32, 37].

There are many open issues in identification for robust control design. In worst-case identification we expect more studies to come out on other error (noise) structures than the bounded error structure. It is hoped that the future efforts on both nonprobabilistic and probabilistic system identification, as well on related modelling topics, will establish this field as an important bridge between feedback control theory and applications.

References

1. Andrews, D.F., P.J. Bickel, F.R. Hampel, P.J. Huber, W.H. Rogers and J.W. Tukey : Robust Estimates of Location. Princeton University Press, Princeton (1972).
2. Helmicki, A.J., C.A. Jacobson and C.N. Nett : Control oriented system identification : A worst-case/deterministic approach in H^∞. IEEE Trans. Automat. Control 36 (1991) 1163–1176.
3. Krause, J.M. and P.P. Khargonekar : Parameter identification in the presence of nonparametric uncertainty. Automatica 26 (1990) 113-124.
4. Mäkilä, P.M. : Approximation and identification of continuous-time systems. Int. J. Control 52 (1990) 669–687.
5. Mäkilä, P.M. and J.R. Partington : Robust identification of strongly stabilizable systems. IEEE Trans. Automat. Control. 37 (1992) 1709–1716.
6. Milanese, M. and A. Vicino : Optimal estimation theory for dynamic systems with set membership uncertainty : An overview. Automatica 27 (1991) 997–1009.
7. Norton, J.P. : Identification and application of bounded-parameter models. Automatica 23 (1987) 497–507.
8. Partington, J.R. : Robust identification in H^∞. J. Math. Anal. Appl. 166 (1992) 428–441.
9. Poolla, K., P. Khargonekar, A. Tikku, J. Krause and K. Nagpal : A time-domain approach to model validation. Proc. 1992 American Control Conf., Chicago.
10. Smith, R.S. and J.C. Doyle : Model validation : a connection between robust control design and identification. IEEE Trans. Automat. Control. IEEE Trans. Automat. Control 37 (1992) 942–952.
11. Tse, D.N.C, M.A. Dahleh and J.N. Tsitsiklis : Optimal asymptotic identification under bounded disturbances. IEEE Trans. Automat. Control. To appear (1992).
12. Zames, G. : On the metric complexity of causal linear systems : ϵ-entropy and ϵ-dimension for continuous time. IEEE Trans. Automat. Control AC-24 (1979) 222–230.

Thoughts on Identification for Control

Michael G. Safonov

Department of Electrical Engineering — Systems,
University of Southern California, Los Angeles, CA 90089-2563, U.S.A

For most of the past fifteen years, it seemed that robust control theorists (and, more recently, robust adaptive control theorists) had earned the right to be smug about the fact that they were among the few in the system theory world who are privy to a mathematically sound theory for analyzing and optimizing the performance of systems whose dynamics are, to within certain quantitative bounds, uncertain. It is now clear that we have been too quick to be smug about our achievements. The issue of identification for control has become a stumbling block for the advancement of control and decision theory, including robust control and robust adaptive control. The problem is that to effectively design controllers within the framework of robust control theory, one needs a theoretical framework that enables one to identify best estimates of bounds on modeling accuracy along with best estimates of nominal model parameters. Traditional system identification techniques do not extract information about estimate accuracy from the measurement data. It is, of course, naive to suppose that the covariances which accompany extended Kalman filter parameter identifiers can be used to obtain estimates of model uncertainty bounds since Kalman filter covariance matrices come from Riccati equations whose coefficients do not even depend on the measurement data except in a secondary fashion. It is therefore vital that new methods of system identification be developed, or that old methods be radically altered, so that model accuracy can be estimated from measurement data. Initial efforts by participants in this workshop have focused on defining sets of assumptions about a priori information under which it is possible to develop identification techniques which produce bounds on the H^∞ norm, or the L_2-gain, of the error between estimated models and true models since it is information of precisely this sort that is required by the favored methods for robust control synthesis, viz. H^∞ synthesis and μ-synthesis.

At this juncture in time, it may be sobering to consider the fact that the question of how to formulate the problem of identification for control is as much philosophical as mathematical. It is clear that it will be necessary to make some assumptions about a priori knowledge about system structure, since nothing can be estimated from a finite amount of measurement data in a complete vacuum of a priori information. But there remains the difficult philosophical question of just what sort of assumptions to make. Is it realistic to assume bounds on slopes of frequency responses and the accuracy of frequency response measurements as proposed by the work of Helmicki, Jacobson and Nett? Is it sensible to impose a parametric structure on the identification problem as some researchers are doing? If so, how are we to ascertain the accuracy of sensors as required by unknown-but-bounded parametric estimation methods involving ellipsoids (e.g., Schweppe, Bertsekas, Krause, Kosut, Younce, etc.)? Or, are non-parametric time-domain approaches (e.g., Hankel SVD, balanced truncation or BST) more realistic? The observations about the modeling accuracy needed for control in [1] provide a compelling case for the use of non-parametric identification methods based on an H^∞ multiplicative error criterion [2].

Still more sobering is the possibility that it may be necessary to re-assess the essentially time-invariant nature of the frequency-domain methods which form the basis for much of robust control theory in order to accommodate the slowly-time-varying uncertainty-bound estimates and the intrinsically finite time-interval nature of real-time estimation and control. Moreover, even within the infinite-time-interval frequency-domain framework of H^∞ robust control, there are still questions as to how best to quantify the "minimal" modeling accuracy required for control as we are forced to use intuition and trial-and-error methods in choosing between additive, multiplicative, gap and more general LFT representations of modeling error. Inevitably, the "right" answer to the difficult philosophical question of how best to formulate the problem of identification for control will depend to a great extent on what proves to be the most mathematically tractable, in addition to depending on the control performance specifications.

Matters are further complicated by the fact that there seems to be a complex interplay between the requirements for accuracy in identification and achievable control performance. Because of this, it may make more sense to bypass identification of the plant for control and instead focus on approaches in which one directly identifies the controller gains which produce the desired closed-loop performance. That is, maybe we would do better to stop attempting to artificially impose the concept of separation between plant identification and controller design on a problem which seems to be unreceptive. This is the approach that we take in [3].

References

1. Michael G. Safonov. Quantifying the modeling accuracy needed for control. In S.P. Bhattacharyya and L. H. Keel, editors, *Control of Uncertain Dynamic Systems*, Boca Raton, FL, 1991. CRC Press.
2. W. Wang and M. G. Safonov. Relative-error H_∞ identification from autocorrelation data — a stochastic realization method. *IEEE Trans. on Automatic Control*, AC-33(7), 1992.

3. T. C. Tsao and M. G. Safonov. A robust ellipsoidal-bound approach to direct adaptive control. In this proceedings, 1992.

An Essay on Robust Control

Hidenori Kimura

Dept. of Mechanical Engineering for Computer-Controlled Machinery,
Osaka University, 2-1, Yamada-oka, Suita, Osaka 565, Japan.

Prologue

The recent success of robust control theory is making a significant contribution to eliminate a long-standing issue of theory-practice gap in control engineering. It has established a deterministic way of representing model uncertainty and a method of worst-case performance expression. This new framework would give a significant impact on the science of design in the future.

The Role of Model in Control System Design

Control systems are always composed of the two parts: the fixed part called a *plant* and the changeable part called a *controller*. The performance of a control system is measured in terms of the overall behaviors of the controller and the plant which are inseparably integrated to form a closed loop. The design of a control system is a systematic procedure of selecting, assembling and adjusting a set of devices to build a controller for a given plant. It crucially depends on the characteristics of the plant. Hence, it is necessary to have enough prior knowledge of the plant on which the design is based.

The role of control theory in the design is to tell the designer what is the relevant information on the plant and how it is used to achieve a good design. Control theory asks the designer to represent his knowledge on the plant in a specially structured form, which we call a *model*. Without a model of the plant, no designer can use theory in his design. Once a reliable model is available, theory can access the real plant through the model in a variety of ways. The model is the unique meeting point of theory and real plants. The model is *an interface* between the abstract control theory and the real world of plants (Fig. 1).

Why Modeling is so Difficult?

To build a model of a given plant is a difficult task in general due to a multitude of reasons. The difficulty of obtaining sufficient prior knowledge on the object is perhaps the most salient feature that distinguishes control system design from other engineering designs.

The most fundamental reason lies in the fact that the plant is a highly man-made object in which a number of physical processes are combined temporally and spatially in a sophisticated way to attain its goal. The object to be controlled belongs to the *artificial world* (Simon [1]), not to the *natural world*.

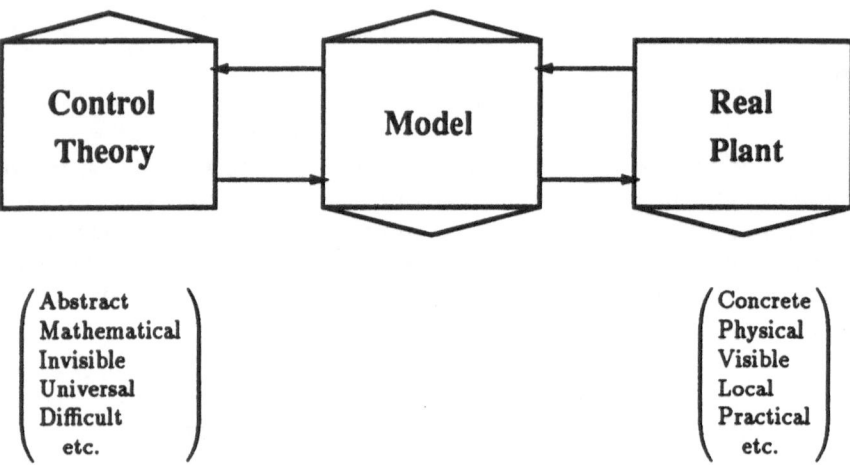

$$
\begin{pmatrix}
\text{Abstract} \\
\text{Mathematical} \\
\text{Invisible} \\
\text{Universal} \\
\text{Difficult} \\
\text{etc.}
\end{pmatrix}
\qquad\qquad
\begin{pmatrix}
\text{Concrete} \\
\text{Physical} \\
\text{Visible} \\
\text{Local} \\
\text{Practical} \\
\text{etc.}
\end{pmatrix}
$$

Fig. 1. The model is an interface between theory and plants

Perhaps, physicists believe that *Nature is simple*, governed by a small number of fundamental equations. For them, the macroscopic complexity is just a mask for the intrinsic simplicity. For engineers, however, *Nature is complex*, at least rich enough to create a vast diversity of technical products. Even if they agree with physicists' belief, they are very interested in the way the complexity is woven out of simplicity.

For instance, the steel strip rolling is a relatively simple mechanical process based on plastic deformation. However, an actual rolling process is far from simple due to the use of a tandem mill system. It generates interstand tensions which delicately affect the thickness. The rotating speed of the mill crucially affects the friction and the lubrication between the mill and the material. If we take the deformation profile of the strip into account, the problem becomes three dimensional. No mechanical theory of deformation is available to deal with such three dimensional phenomena.

We must be aware of the fact that the physics which deals with idealized and pure processes is far from sufficient to describe industrial processes which belong to the artificial world. Though the plant must obey physical laws, the artificial aspects are frequently more important than the physical aspects. We must admit that we have currently no systematic way of coping with the overwhelming complexity that is intrinsic in man-made objects. Some sort of "metaphysics" that belongs to "the science of the artificial" would solve the problem in future. The complexity of the plant is the main source of difficulty of modeling.

The complexity of engineering systems is the problem for engineering science in general, not particular in control system design. If you are interested only in the analysis of your plant and your purpose is to obtain a model that simulates your plants as precisely as possible, you might be able to enhance the precision of your model by incorporating as many phenomenological equations as you wish at the cost of increasing effort in validating them. This is exactly the way of research taken by most engineering scientists.

Control system design must go further beyond the pure analysis. It must combine the knowledge of the plant with the control strategy. Irrelevant information should be discarded to obtain a simple, workable model for control. The model simplification is actually the task of distinguishing relevant information from irrelevant information and is the most difficult and important step of control system design. Thus, the need for model simplification is another reason of the difficulty of modeling.

Impact of Robust Control in Design Science

Now, the fundamental issue of modeling becomes clear. On the one hand, in order to design a control system based on theory, we need a model of the plant which is simple but captures the relevant information on the characteristics of the plant. On the other hand, it is difficult to obtain such a workable model of the plant due to the fundamental structure of control system design. The issue of modeling is really the most fundamental and serious reason of theory-practice gap in control engineering.

One remedy for this issue is to make the design *model-free*. On-spot tuning of the PID controller is a typical example of model-free control. Fuzzy control and adaptive control can also be categorized as model-free control. This way of control abandons the rationality and reduces the design to a matter of insight and experience.

The other remedy is to establish a model-based design method that tolerates the uncertainty of the model because the difficulty of modeling results in the increasing uncertainty of the model. This is exactly the goal of robust control. It is the unique solution to the modeling issue, provided that we wish to preserve the rationality of design.

The probabilistic framework has been the usual way of representing uncertainty. If the modern science was originated as a systematic effort of human beings who wanted to eliminate uncertainties surrounding them, the probability theory was born as a means of coping with uncertainties by dealing with them properly and rationally. It is an interesting coincidence that probability theory was originated at the same time as the birth of modern science.

As is well known, robust control theory does not use a probabilistic framework. It takes the deterministic framework for representing uncertainty, and as a result, it measures the performance at the worst case uncertainty. This framework, which dates back to Lufe's absolute stability (Lufe et al. [2]), has been very fruitful. A number of innovative results have been obtained in various fields of robust control (Zames [3], Doyle [4]). We just mention that the robust stabilizability results obtained by Kimura [5] and Glover [6] combine the stability theory and the interpolation theory to give a fundamental contribution in the long history of stability theory.

Conventionally, the factors which limit the control performance have been ascribed to the quality of control devices such as the power and the dynamic range of actuators, noise level of sensors, the computational capability of CPU etc. Robust control theory has shown that there is another crucial factor that limits the control performance. This is *the quality of our knowledge* of the plant. Robust control theory asserts that the ultimate control performance is a function of the two variables : one is the quality of devices which is denoted by D and the other is the quality of knowledge (or the level of uncertainty) which is denoted by K (Fig.2). If we can calculate the function $P = f(D, K)$ properly, it would be possible to make a tradeoff between D and K.

Epilogue

In many fields of research in engineering and economics, the theoretical model plays a fundamental role in design, decision making, simulation, prediction etc. It is also common sense that a model is always an idealized abstraction of the real world and some sort of gap between the model and the real world is inevitable. However, it seems that the question of how this obvious gap affects the design, the decision making, the simulation and the prediction concerning the real world has been largely ignored, or at least not discussed seriously.

Since the control system involves the plant which exchanges large amount of materials and energy with its environment in real time, it is affected by the gap between the model and the real plant in a very serious way. That is why control theory first became aware of the effects of model uncertainty. It introduced a new design philosophy to tolerate the model uncertainty. The success of robust control will establish a new design philosophy incorporating the knowledge quality as a limiting factor.

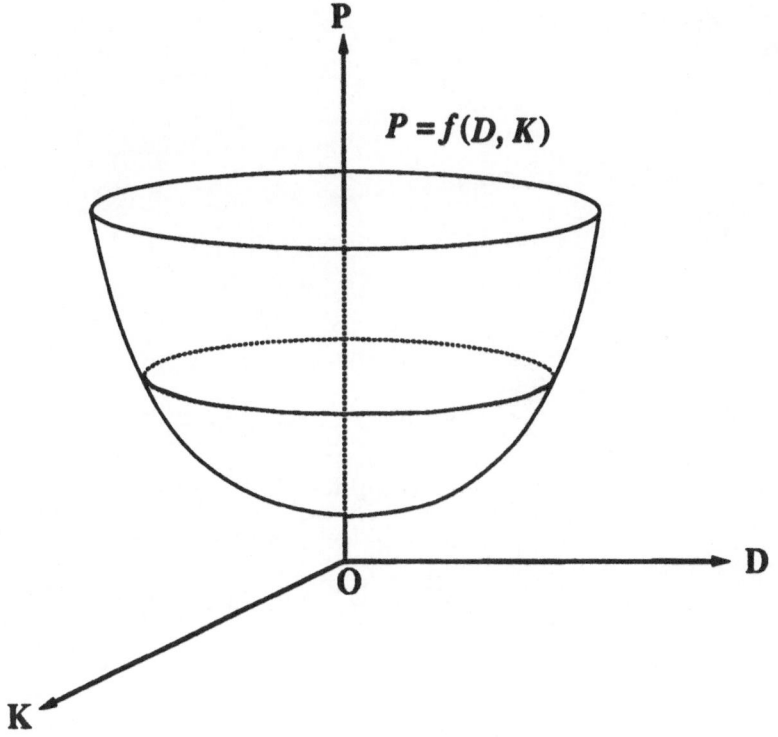

$$P = f(D, K)$$

Fig. 2. Performance depends on both device quality and knowledge quality

References

1. H.A. Simon: The Science of the Artificial. MIT Press, (1969).
2. A.I. Lufe and V.N. Postnikov: On the theory of stability of control systems. Prikl. Mat.i.Mehk., VIII, 3, (1944).
3. G. Zames: Feedback and optimal sensitivity: Model reference transformations, multiplicative seminorms, and approximate inverses. IEEE Trans. Automat. Contr, vol.AC-26, pp.585-601, (1981).
4. J. Doyle: An analysis of feedback systems with structured uncertainties. IEE Proc., vol.129, Pt.D, pp.242-250, (1982).
5. H. Kimura: Robust stabilizability for a class of transfer functions. IEEE Trans. Automat. Contr., vol.AC-29, pp.788-793, (1984).
6. K. Glover: Robust stabilization of linear multivariable systems: Relations to approximation. Int.J. Control, vol.43, pp.741-766, (1986).

On the Character of Uncertainty for System Identification and Robust Control Design

Robert L. Kosut

Integrated Systems, Inc., 3260 Jay St., Santa Clara, CA, 95054 and Department of Electrical Engineering, Stanford University, Stanford, CA.

> *"It ain't the things you don't know what gets you in deep trouble. It's the things you know for sure, but what ain't so."* – Uncle Remus.

Nothing could more aptly describe the predicament when faced with the problem of designing a controller from accumulated sensed input-output data. The identification, or estimation, of a system's transfer function from input-output data has a long history and there are many excellent survey articles and textbooks that can be referenced, e.g., [4, 7, 8, 2, 15]. The problem with all the methods discussed in these references, insofar as robust control design is concerned, is that model error estimates are usually not available, and if available, cannot be trusted. The principal reason for this difficulty is that the identification algorithms are developed under the false assumption that the true system is in the model set. As a result, the model estimate, loosely speaking, is "biased", and hence, a controller designed using the estimate may result in unacceptable closed-loop behavior, a phenomenon which is well documented, e.g., [1, 3, 2]. To paraphrase the above aphorism, "Trouble is bound to follow if the identified model is *known for sure* to be the true system."

This situation is unfortunate, because all the standing assumptions made in current robust control design methods require a model set description which typically consists of a nominal model and an error estimate, usually a norm bound, where both together are *guaranteed* to encompass the true system. To fulfill the needs of robust control design will therefore require a new approach to system identification which provides both a nominal model and a measure of its uncertainty. Such schemes have been referred to by various names, e.g., set-membership identification, set-estimation, uncertainty modeling, as well as other self-canceling phrases – how does one model an uncertainty? This research topic has received strong interest recently as evidenced by this workshop, the recent special issue [10], and the many conference sessions planned at the next ACC and CDC.

Formulating the Problem

"If what is said is not meant, then what ought to be done, remains undone."
– Confucious.

Sometimes solving a problem means finding a simple or direct statement of the problem in the first place. In attempting to distill the problem formulation to its essence, perhaps it is this: given a finite collection of sensed sampled input/output data from an unknown system, what level of confidence can be assigned to a feedback controller design or modification? If, other than the measured data, there is no additional knowledge about the system, then the problem is solved: there is no safe controller. Anything can happen, because there is no means for inferring the future from the past. Therefore, to make the problem meaningful, it is necessary to make a priori assumptions about the system. These assumptions can be either qualitative or quantitative. For example, assuming that the unknown system is linear-time-invariant is *qualitative* a priori knowledge. Knowing that it is stable can still be classified as qualitative, but assigning a region for pole locations or knowing a bound on the impulse response is *quantitative*. A similar classification can be made regarding signal characteristics. Knowing that a signal is white is qualitative; but knowing a precise value for the variance is quantitative.

Although a priori quantitative information may be readily available, e.g., from the underlying physics, I think that it is first necessary to resolve the more pristine problem of specifying a minimal amount of qualitative a priori data so as to assign a high degree of confidence to a controller design.

Is Nature Good, Evil, or Indifferent?

The phrase "high degree of confidence" needs clarification. Do we mean worst-case or high probability?

Current robust control formats are based on worst-case scenarios. Nature is perceived as Evil, and hence, does the *wrong* thing, from our perspective. However, if this is not the case, and Nature is at worst Indifferent or Neutral, then the problem should be posed in reverse: to fulfill the needs of system identification, long resting on a probabilistic (neutral) foundation, may require a new approach to robust control which allows for a probabilistic description of uncertainty! This latter possibility invokes the current debate on the intrinsic nature or character of the uncertainty set. Is it probabilistic or worst-case deterministic? Clearly both can be used to quantify uncertainty in either disturbances and transfer functions. However, searching for the worst-case may be a hopeless task. If the worst-case has not yet occurred, it might in the future, and hence, the search never ends. Fitting a probablistic model is more sensible in this regard, but a 99.99% confidence level does not preclude the remaining .01% from occurring.

A probabilistic, or stochastic, description of a disturbance is common practice and forms the basis for $\mathcal{H}_2$-filtering and control design, i.e., optimal filtering and LQG control design [2]. A power bounded set of disturbances and/or a worst-case deterministic description of transfer function uncertainty leads to $\mathcal{H}_\infty$ methods of control design, e.g., [5]. These sets can be combined leading to mixed $\mathcal{H}_2/\mathcal{H}_\infty$ control design, e.g., [9]. The above examples by no means exhaust the possible deterministic and probabilistic sets. For example, sequences can be uncertain but have a bounded spectrum or a bounded magnitude. Transfer functions can be uncertain but with (time) bounded impulse responses, and so on. The choice of which uncertainty characterization to use depends upon prior knowledge about the true system. Clearly different assumptions ought to lead to set estimators with differing forms and mixtures of probabilistic and/or worst-case deterministic uncertainty types.

As a case in point, if we begin with a stochastic description of the exogenous inputs to a system, then the standard least-squares based identification method with a high-order ARX model structure leads naturally to a *purely parametric* uncertainty which, depending on further assumptions, is either probabilistic (normally distributed) or worst-case deterministic (ellipsoid bounded), e.g., [6, 11, 12]. To conform to current robust control paradigms, the parametric characterization of uncertainty must be transformed to a nonparametric worst-case deterministic frequency domain bound, a transformation that is not without a considerable loss of information. Dealing directly with the worst-case deterministic (ellipsoid bounded) parameter uncertainty leads to some new insights into robust control design e.g., [13]. For the probabilistic form of parameter uncertainty, it is my view that it would be better to develop a compatible theory of "probabilistic" robust control.

Going in this direction, however, immediately raises the question: what does a robust control mean in the context of probabilities? We tend to think of a robust controller as providing an absolute guaranty against instability and/or certain levels of performance degradation given a deterministic, or "hard " bound on plant uncertainty. With a probabilistic description, or "soft" bound, we must decide if 99.99% is safe enough. To turn the question the other way, the deterministic bounds necessitates guarding against the worst-case, which may be extreme, i.e., unlikely, thereby leading to a conservative controller. But this brings us back to exactly the question of probabilities and outcomes, and finally to a more fundamental question: is Nature neutral or conspiratorial?

Towards a New Paradigm, or Paradigm Lost

Attempting an answer may not be necessary, nor very fruitful. I think that a better attitude at this point is to follow the consequences, without prejudice, of developing a theory of set-membership identification and corresponding "robust" control design methods compatible with probabilistic plant set descriptions. This to me seems the more sensible engineering oriented character of uncertainty.

Hopefully, as a result of research efforts in many different directions, new paradigms will arise which combine system identification and robust control design. With the wide availability and use of CACSD packages, the benefits of this research could be widely utilized in many engineering fields. Hence, it becomes imperative that the resulting methodologies are comprehensible and useful for the engineering community at large; not just understandable to a few experts. The onus is on us!

References

1. B. D. O. Anderson, R. R. Bitmead, C. R. Johnson, Jr., P. V. Kokotović, R. L. Kosut, I. M. Y. Mareels, L. Praly, and B. D. Riedle, *Stability of Adaptive Systems: Passivity and Averaging Analysis*, MIT Press, Cambridge, MA, 1986.

2. B.D.O. Anderson and J. Moore, *Optimal Control: Linear Quadratic Methods*, Prentice-Hall 1990.

3. K.J. Astrom and B. Wittenmark, *Adaptive Control*, Addison-Wesley, 1989.

4. H. Cramér, *Mathematical Methods of Statistics*, Princeton University Press, 1946.

5. J.C. Doyle, B.A. Francis, and A.R. Tannenbaum, *Feedback Control Theory*, Macmillan, 1992.

6. G.C.Goodwin, M. Gevers, and B. Ninness, "Quantifying the error in estimated transfer functions with application to model order selection", "Special Issue on System Identification for Robust Control Design," *IEEE Trans. Aut. Contr.*, vol. 37, no. 7, July, 1992.

7. G.C. Goodwin and R.L. Payne, *Dynamic System Identification: Experiment Design and Data Analysis*, Academic Press, 1977.

8. G.M. Jenkins, D.G. Watts, *Spectral Analysis and its Applications*, Holden-Day, San Francisco.

9. P.P. Khargonekar and M.A. Rotea, "Mixed $\mathcal{H}_2/\mathcal{H}_\infty$ control: a convex optimization approach", *IEEE Trans. Aut. Contr.*, vol. 36, pp.824-837, 1991.

10. R.L. Kosut, G.C. Goodwin, and M. Polis, Guest Editors, "Special Issue on System Identification for Robust Control Design," *IEEE Trans. Aut. Contr.*, vol. 37, no. 7, July, 1992.

11. R.L. Kosut and H. Aling, "Worst-case control design from batch-least-squares identification", *Proc. 1992 ACC*, Chicago, IL, June 1992.

12. R. L. Kosut, "Adaptive control via parameter set estimation", *Int. Journal of Adapt. Contr. and Sig. Proc.,*" vol. 2, pp. 371-399, 1988.

13. M. Lau, S. Boyd, and R.L. Kosut (1991), "Robust control design for ellipsoidal plant set," *Proc. 30th CDC*, Brighton, U.K., Dec. 1991.

14. L. Ljung, *System Identification: Theory for the User*, Prentice-Hall, 1987.

15. J.P. Norton, *An Introduction to Identification*, Academic Press, 1986.

Extensions of Parametric Bounding Formulation of Identification for Robust Control Design

J. P. Norton

School of Electronic & Electrical Engineering, University of Birmingham, Edgbaston, Birmingham B15 2TT, UK

Introduction

The purpose of this note is to point to some areas where the requisites for a practical, integrated identification and control design technology for uncertain systems do not yet exist or are in a rudimentary state. The possible role of parametric bounding [1, 2, 3] will be discussed.

Let's start with a truism or two. The form of a model for control design has to match the knowledge of the plant and the nature of the controller performance requirements. Nevertheless, most modelling techniques do not take into account control design as an ultimate aim. Conspicuous exceptions are Ziegler-Nichols tuning and Bode plots for classical compensator design. The enduring popularity of these methods is at least partly due to their good matching of model form to control-design technique. By contrast, more recent identification and control-design methods tend to be viewed in isolation. Also, a striving for generality often results in the omission of messy practical features from the model, paradoxically restricting its scope. The identification problem is usually regarded as a two-stage procedure: model-structure selection from a family of tidy, well behaved models (typically linear, time-invariant or slowly varying, low-order, and in transfer-function or state-variable form), plus parameter estimation according to a statistical criterion. Most commonly, both stages are based entirely on a given set of input-output records; prior (collateral) knowledge of the plant is not considered, except perhaps in providing prior parameter estimates. Control design often omits features such as structured disturbances, state and actuator constraints, or systematic measurement errors. The design procedure may take parameter uncertainty into account, but it is unusual to find that the uncertainty is so structured as to constrain the plant to behave realistically.

Developments in two areas could improve this state of affairs: incorporation of more collateral knowledge into models, and better matching of identification and control-design methods to take account of model quality in control synthesis and to implement plant-behaviour, state and control constraints readily. Both areas are considered below in a framework in which model parameters, noise, disturbances and control performance are all bounded.

Incorporation of Collateral Knowledge as Bounds

Model uncertainty allows a range of possible behaviour. To exclude unrealistic behaviour, it may be necessary to impose extra restrictions in the form of joint bounds on model parameters. As a simple example, consider a SISO system with z-transform transfer function,

$$\frac{b_1 z^{-1} + b_2 z^{-2}}{1 + a_1 z^{-1} + a_2 z^{-2}} = \frac{0.87 z^{-1} - 0.67 z^{-2}}{1 - 1.7 z^{-1} + 0.72 z^{-2}}$$
$$= \frac{1.13 z^{-1}}{1 - 0.9 z^{-1}} - \frac{0.26 z^{-1}}{1 - 0.8 z^{-1}} .$$

Even with a_1 and a_2 accurate, independent errors up to $\pm 1\%$ in b_1 and b_2 allow the amplitudes of the modal components (partial fractions making up the unit-pulse response) to range from 0.9847 to 1.2753 and -0.3966 to -0.1234 about their actual values, 1.13 and -0.26; their ratio ranges from -7.98 to -3.22. Restriction of either or both modal amplitudes (or their ratio) to a modest range, on the basis of prior knowledge, would add joint bounds on the model parameters; their intersection with the original bounds on b_1 and b_2 would define b_1 and b_2 much more sharply.

Some commonly available collateral information can be expressed as convenient bounds on the parameters of a transfer-function model. For example, a bound on a modal amplitude can be shown to give a hyperplane bound on the parameters of a z-transform transfer function with real, distinct poles. A bound on steady-state gain trivially yields another hyperplane parameter bound. Some other items of information translate into parameter bounds which are less convenient (given present technology for computing parameter bounds) although still quite simple. For instance, an amplitude bound on a mode corresponding to a complex-conjugate pole pair bounds a quadratic form in the transfer-function parameters (containing no cross-product terms between numerator and denominator parameters). Another example is a bound on the phase change of the transfer function at a specific frequency, which yields a bound on a bilinear form in the numerator and denominator parameters. Other parameterizations may simplify collateral bounds; state-variable parameterizations, in particular, would be worth examining.

A practically worthwhile research aim is to compile a catalogue of the transfer-function or state-space model-parameter bounds given by bounds on quantities which are often known a priori to be in known ranges, then to devise efficient ways of incorporating such extra bounds into parameter-bounding algorithms. Further examples of quantities which may be bounded are rate constants, feed composition in process plant, mass, density, inertia,

instrument drift rate, rise time or bandwidth, damping or overshoot, and gain
at specified frequencies. Incorporation of such bounds is likely to require significant extension of existing parameter-bounding algorithms, most of which
allow only ellipsoidal or hyperplane bounds.

Similar comments apply to bounds on functions of state variables. Examples are non-negativity restrictions, bounds on flows and bounds on stored
energy. Bounds on forcing and disturbances are already catered for in existing
state-bounding methods [4, 5, 6], so long as they are very simple.

Matching of Identification and Control-Design Methods

For reasons of academic history not well supported by engineering practice,
control design has come to be viewed primarily as an optimization problem.
One consequence is that a common reaction, on being offered computed parameter or state bounds, is to ask for optimal estimates to be derived from
them. If there is no information about what happens between the bounds, the
optimality criterion has to be minimax. Minimax-of-parameter-error/state-
error estimates are the means of the extrema of the feasible set in the coordi-
nate directions, and the minimax-of-model-output-error estimate is the singleton to which the parameter feasible set is reduced as the specified bounds
on model-output error are reduced. In fact, many applications do not require
unique parameter or state estimates and need not be viewed as optimization problems. Examples are prediction of the range of future behaviour and
experiment review/redesign, where the thing of interest is the extent of uncertainty in the items being bounded. In addition, control problems are often
in practice a matter of meeting a specification expressible as a set of inequalities. This is explicit in classical control design but not in most modern
design techniques. A bounding framework is a natural way to handle such
requirements.

Identification and control design should match in that the information
provided by the model (including, for robust control, its uncertainty) is what
is required to evaluate the control-performance criterion. A bounding approach to identification which embodies bounds on future model-output error
(not assumed in H_∞ control) yields worst-case control-performance guarantees for a single realization, in advance, to the extent that the model bounds
can be predicted. If a fuller description of the uncertain items is essential,
to permit a more detailed measure of controller performance, information
about what happens between the bounds can be included without resorting
to a probabilistic description. For instance, *histogram bounds* on the numbers of samples within each of a number of ranges could be quoted for errors
(instrument noise, disturbances and/or structural model-output error). For
model parameters, a family of feasible sets, each consistent with one pair from
a range of successively smaller specified model-output-error bounds, or with
one of a range of specified proportions of the observations for fixed model-
output-error bounds, may be computed instead of just a single feasible set

for a single pair of bounds on all the model-output errors. In a basic form, the latter idea has been adopted by Walter et al. in the OMNE scheme [1].

Performance Guarantees for Modelling Techniques

Performance guarantees for probabilistic modelling and/or control apply to ensembles and require assumptions about ensemble behaviour. Most are also asymptotic. It's worth asking how the situation differs for bound-based modelling. Convergence of parameter bounding can be analyzed under probabilistic assumptions [7], but this is a bit perverse as the deterministic, finite-sample setting of bounding is one of its most appealing features. So long as the bound computation can be regarded as accurate, it simply re-expresses all the available information in a more convenient form and *the question of performance analysis does not arise*. Of course, approximations or the desire for experiment-design guidance analogous to persistency-of-excitation conditions, guaranteeing well defined parameter bounds, may make an analysis necessary. The natural setting for such an analysis is deterministic and finite-sample. However, there is some point in looking for a more detailed description of the model-output error (noise plus modelling error) and input, in order to provide sharper results than can be obtained from bare bounds on model-output error. *Such a description need not be probabilistic.* For instance, extra bounds can be imposed on noise (sample) correlations [8], on spanning properties of explanatory-variable sequences [9] or, as outlined above, on the (deterministic) distribution of the noise samples within the record. An example might be an assumption that a certain proportion of samples have errors within a specified distance of the overall bounds, the counterpart of a probabilistic assumption about sufficient density at the ends of the support of the distribution.

So far no analysis has appeared of deterministic, finite-sample, performance (smallness of uncertainty) guarantees for parametric models when detailed bounding assumptions are made, such as histogram bounds on the model-output error and lower bounds on the separation in direction of successive "regressor" (explanatory variable) vectors in a linear model. Such an analysis would be valuable.

Performance Guarantees for Controllers

The combination of bounding of model parameters and worst-case control design has been described in detail, in an adaptive-control setting, by Veres and Norton [9]. One topic considered there is the relaxation of controller worst-case-optimal performance so as to give more freedom for optimizing the control input with regard to identification accuracy. Another is the imposition of control constraints to account for actuator limitations and to maintain robust stability. A third issue examined in bound-based adaptive control was joint adaptation of bounds on model-output error and parameter variation to give good tracking of time-varying plant.

The use of histogram bounds could provide an alternative to minimax control performance criteria. The aim would be to guarantee, in a deterministic, finite-sample setting, that control performance (output error, perhaps combined with control effort) is within successively tighter bounds at successively smaller proportions of the sample instants, related to those at which the model-output error is within successively tighter bounds.

Why Go At It This Way?

The motive for elaborating a deterministic, parametric, bound-based problem formulation is that it fits traditional engineering toleranced-design practice and the availability of much prior information as bounds. It has several bonuses: conceptual simplicity, the ability to make definite statements (conditional on the prior assumptions) about performance in a specific run, and parallels with probabilistic treatment which are helpful in suggesting new algorithms and diagnostics. The approach also has some drawbacks which need further attention: potentially poor performance when bounds become loose through having to be approximated, vulnerability to outliers due to the "all or nothing" weighting of observations, and heavy computational load if the parameter bounds are complicated.

References

1. E. Walter and H. Piet-Lahanier, "Estimation of parameter bounds from bounded-error data: a survey," *Maths. & Computers in Simulation*, 32, pp. 449-468, 1990.
2. M. Milanese and A. Vicino, "Estimation theory for dynamic systems with unknown but bounded uncertainty: an overview," *9th IFAC/IFORS Symp. on Identification & Process Parameter Estimation*, Budapest, pp. 859-867, 1991.
3. J. P. Norton, "Identification and application of bounded-parameter models," *Automatica*, 23, pp. 497-507, 1987.
4. F. C. Schweppe, "Recursive state estimation: unknown but bounded errors and system inputs," *IEEE Trans. on Autom. Control*, AC-13, pp. 22-28, 1968.
5. V. Broman and M. Shensa, "A compact algorithm for the intersection and approximation of N-dimensional polytopes," *Maths. & Computers in Simulation*, 32, pp. 469-480, 1990.
6. W. K. Tsai, A. G. Parlos and G. C. Verghese, "Bounding the states of systems with unknown-but-bounded disturbances," *Int. J. Control*, 52, pp. 881-915, 1990.
7. S. M. Veres and J. P. Norton, "Structure selection for bounded-parameter models: consistency conditions and selection criterion," *IEEE Trans. on Autom. Control*, AC-36, pp. 474-481, 1991.

8. S. M. Veres and J. P. Norton, "Structure identification of parameter-bounding models by use of noise-structure bounds," *Int. J. Control*, 50, 2, pp. 639-649, 1989.

9. S. M. Veres and J. P. Norton, "Bound-based worst-case self-tuning controllers," *9th IFAC/IFORS Symp. on Identification & Process Parameter Estimation*, Budapest, pp. 773-778, 1991. [*see also* "Predictive self-tuning control by parameter bounding and worst-case design," *Automatica, to appear*].

Connecting Identification and Robust Control

Michel Gevers

CESAME, Louvain University, Louvain la Neuve, Belgium

The intensive work that is presently going on in the general area of identification in connection with robust control finds its origin in the awareness, among people from both the identification and the robust control community, that there is a wide gap between the premises on which robust control design is built and the tools and results that 'classical' identification theory is able to deliver. (By 'classical' I mean the theory as it existed three years ago.)

The most obvious manifestation of this gap, and the one that has triggered most of the present research activity, was the realization that robust control theory requires a priori *hard bounds* on the model error, whereas classical identification theory delivers at best *soft bounds* in the case where the system is in the model set and no bounds at all in the case of undermodeling.

Nowadays, one has come to realize that the great 'hard-versus-soft' bound debate is not the real issue, but the main focus of research is still on trying to produce identification methods that allow for the computation of uncertainty bounds, whether hard or soft. While this is certainly a most pertinent scientific pursuit, I would like to argue that if the objective of the identification exercise is to design a robust controller, then the most important issue is probably not the estimation of uncertainty bounds on the identified model, but the design of a control-oriented identification or, even better, of a synergistic identification and control design.

To make things clear, let me subdivide our field of research (connecting identification and robust control) into three areas, that correspond to three aspects of the identification and robust control design problem:

1. Estimation of uncertainty bounds on identified models;
2. Identification *for* robust control design;
3. The combined (synergistic) design of the identification and control.

For the moment, the mainstream approach seems to be 'Perform the identification with a method that allows the computation of error bounds on the estimated model, then design a robust controller using that model and its bounds'. The main focus of research is therefore on the estimation of uncertainty bounds and I believe that the methodology developed in [15] is a useful contribution to this line of research. The problem is that identification methods (using restricted complexity model sets), whose sole merit is to deliver accurate error bounds, may well produce nominal models that are ill-suited for robust control design: that is, the frequency distribution of the model uncertainty is such that it may lead to poor closed loop performance.

The idea of the second line of research above is to develop identification methods that will produce models whose uncertainty distribution, over the frequency range, allows for high performance robust control design. Thus these models should have low uncertainty where closed loop control specifications require this, but they may have large uncertainty in frequency bands where this does not imperil closed loop stability or penalize closed loop control performance. Results are now available for the tuning of the identification method towards such objectives: the identification must be performed in closed loop with the appropriate data filter. These results rely on an understanding of the interactions between identification, robust stability and robust performance. In [1] the robust stability and robust performance criteria of H_∞ control design were used as the key ingredient for an understanding of these interactions. On that basis, a design was proposed in an adaptive control framework using a combination of Least Squares identification and LQG/LTR control design. In the scheme of [1], the identification design takes account of the robust control requirements through the data filters; however, the LQG/LTR control design does not explicitly take account of the frequency distribution of the model uncertainty. Thus, this scheme fits in the framework of identification *for* control.

The third line of research is to combine the identification and the control design in a mutually supportive way, from the point of view of robust stability and/or robust performance. Even though the objective might seem overly ambitious and elusive, some preliminary results are available. They take the form of iterative schemes in which a succession of identification and control design steps are performed, leading to more and more performant control systems. The identification steps are performed in closed loop using data obtained with the last controller operating on the actual plant. The control design steps use the most recently identified plant model. The scheme of [3, 4] combines frequency weighted Least Squares identification in closed loop (where the frequency weights take account of the previously identified model and the presently acting controller), with frequency weighted LQG control (where the frequency weights take account of the plant-model mismatch - including the mismatch of the noise models - using signal information). Other such iterative design schemes are emerging: see e.g. Schrama, and Lee, Anderson, Kosut, Mareels. Even though the specific identification and control design techniques vary between these schemes, they all have in common a succession of performance enhancement designs. The idea of redesigning controllers using closed loop data collected on the plant in order to improve performance is what process control engineers naturally tend to do. The merit of the recent research is to develop systematic procedures to achieve this performance enhancement in a theoretically sound way.

To summarize, most of the present focus of research is on the estimation of error bounds. This problem is not only of independent interest, but it is also an important step towards the design of robust controllers based on identified models. However, I strongly believe that the key ingredient for the successful application of robust control design methods to identified models is not so much the computation of error bounds, but it is to *let the global control performance criterion dictate what the identification criterion should be*. The iterative scheme of [3] is an application of this precept. By tuning the identification criterion towards the control design objective, the model error automatically becomes small in the frequency bands where it needs to be small. To achieve this objective requires a better understanding of the interconnections between closed loop identification and control design.

Finally, the ultimate goal is the combined design of the identifier and the controller in the case of plant undermodelling, and this raises some deep but challenging open questions.

References

1. R.R. Bitmead, M. Gevers, V. Wertz, "Adaptive Optimal Control - The Thinking Man's GPC", Prentice Hall International, Series in Systems and Control Engineering, 1990.
2. G.C. Goodwin, M. Gevers, B. Ninness, "Quantifying the error in estimated transfer functions with application to model order selection", *IEEE Transactions on Automatic Control*, Vol. 37, No 7, pp. 913-929, July 1992.
3. Z. Zang, R.R. Bitmead and M. Gevers, "Iterative Model Refinement and Control Robustness Enhancement", submitted for publication to *IEEE Transactions on Automatic Control*.
4. Z. Zang, R.R. Bitmead and M. Gevers, "Disturbance Rejection: On-Line Refinement of Controllers by Closed Loop Modelling", *Proc. 1992 ACC*, Vol.4, pp. 1829-1833, Chicago, Illinois, June 1992.

On Nominal Models, Model Uncertainty and Iterative Methods in Identification and Control Design

Paul Van den Hof, Ruud Schrama and Peter Bongers

Mechanical Engineering Systems and Control Group,
Delft University of Technology, Mekelweg 2, 2628 CD Delft, The Netherlands.

Introduction

Recent years have seen a growing interest in the use of system identification as a means to build models for control design. The traditional identification methods deliver a model in the form of a linear time-invariant finite dimensional system. Such a model, called a nominal model, is bound to only approximately describe the dynamics of the plant of concern. In order to cope with this approximation we have to call on methods for robust model-based control design, in order to reach the goal of high performance plant control. It has been widely recognized that not every controller suited for the nominal model will perform equally well with the plant. The general principle is that robustness must be traded off versus nominal performance.

We take as a starting point that the problem of constructing a high performance controller for a plant with unknown dynamics, is going to be tackled through identification and model-based control design. This involves two complementary prerequisites:

- The controller must be robust for the imperfections of the nominal model, and
- these imperfections must allow the design of a high performance controller.

The first prerequisite shows the need of a quantified (bound on the) model uncertainty, whereas the second one refers to the construction of a sufficiently accurate or suitable nominal model, as a basis for the control design. Accordingly, the field of control-relevant system identification branches into two directions as depicted in Fig. 1.

Attention for the left branch has motivated the development of identification techniques for the estimation of an upper bound on the deviations between a plant and some nominal model. Depending on the kind of a priori assumptions, these techniques yield probabilistic bounds [3, 7, 17] or deterministic bounds, see e.g. [1, 4, 6, 9] and many others, or even a combination

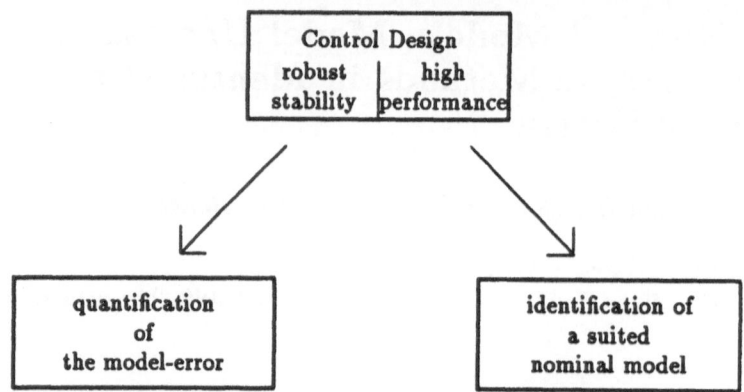

Fig. 1. The two branches of control-relevant system identification

of both [2]. Ideally, an upper bound can be used to design a controller that achieves some guaranteed performance for the plant of concern. However, by itself, a tight estimated upper bound is not sufficient for *high* performance control design: the achievable performance is truly limited by the actual model-error rather than by some upper bound. This is not a matter of only estimating a tight upper bound, but that of selecting a suitable nominal model. This motivates the presence of the right branch. Methods for identification of control relevant nominal models has been subject of research in [6, 7, 10, 11, 12, 13, 16].

We could argue what is meant by a "control-relevant" or "suitable" nominal model. Definitely our aim is to select a nominal model that enables the design of a high performance controller for the plant. Whereas in robust control theory a common observation is that the achievable robust performance is limited for a *given* nominal model and a (bound on the) model mismatch, we make the converse statement that

The requirement of a high performance imposes limitations on the allowed shape and extent of the mismatch between a plant and its nominal model.

For instance it is well understood that a reasonable fit around the crossover frequency of the control system is needed for robust performance, see e.g. [15]. Larger deviations are allowed at other frequencies, as long as they do not impair the control design.

In this setting it is clear that the quality of each candidate nominal model depends on its induced controller and, the other way around, the controller to be designed will definitely be dependent on the nominal model. Hence the problem of constructing a high performance controller for a plant with unknown dynamics boils down to a *joint problem* of identification and control design. A natural way to cope with this problem is to embed the separate procedures in an iterative scheme of consecutive identification, control design and renewed experimentation. This approach is supported in [7, 11, 13, 16].

In this essay-like paper, we are going to discuss some matters related to the role of nominal models and model uncertainty in this joint problem of identification and control design, and we will formulate some general viewpoints on the construction of iterative schemes that intend to deal with the problem. We will also formulate a number of criteria, that according to us should be taken into account when evaluating either existing or to be developed iterative schemes. A worked out example of an iterative scheme is presented in a companion paper [14].

The following discussion will be rather fragmentary. Not in the least this is due to the fact that in this very interesting and challenging area, we believe we have only yet begun to formulate and understand the essential questions. The ultimate goal, a generally applicable, data supported high performance (adaptive) (robust) control design method, is still beyond our reach.

What Do We Require from Models and Controllers

Let us first give some very brief notation. We will denote with P a linear, time-invariant plant, represented by its transfer function; $\hat{P}$ is a nominal model of that plant, and $\mathcal{P}_\Delta(\hat{P}, b)$ refers to any set of models induced by the nominal model $\hat{P}$ and an unspecified uncertainty set where the scalar b is a measure for the "size" of this set. Note that the shape of this perturbative family $\mathcal{P}_\Delta(\hat{P}, b)$ is not specified. The uncertainty set can for instance represent unstructured weighted additive uncertainty, as

$$\mathcal{P}_A(\hat{P}, b) := \{\tilde{P} \mid \sigma_{max}(|\tilde{P}(j\omega) - \hat{P}(j\omega)|)g(\omega)^{-1} \leq b\} \ . \tag{1}$$

We could also think of uncertainties in a multiplicative or structured form. The "size" of the uncertainty set is determined by b. C will denote a linear time-invariant controller, and (P, C) represents the closed loop system composed of plant P and controller C. We will employ the notion of performance of a controlled system in an abstract way, without having it specified in detail at this moment.

In model-based control design it is straightforward that a controller C is designed on the basis of $\hat{P}$ and (in a robust design scheme) on $\mathcal{P}_\Delta(\hat{P}, b)$. Naturally a fundamental requirement that is laid upon C is that it yields a high or even optimal performing control system (P, C).

Apart from this, there is a second requirement that is apparent when we consider how model-based control design generally is put into practice. Based on $\hat{P}$ (and possibly $\mathcal{P}_\Delta(\hat{P}, b)$), the designer chooses a control criterion (performance measure) and weighting functions, and he calculates a controller by numerical optimization. Then the system $(\hat{P}, C)$ is analyzed by e.g. examining its sensitivity, step response, robustness analysis and the like. Only if this nominal performance is satisfactory, then the designed controller will be applied to the plant under consideration. The controller C is required to perform satisfactorily with the nominal model $\hat{P}$. If this would not be true, we would have no confidence in it. We stress this point since it plays a role in discussing some of the iterative identification and control design schemes in this paper.

Should Models be "Realistic"?

In many situations of modelling for control design, the designer will have a fair idea about how the dynamics of the plant look like. Sometimes he might even know about which part of the dynamics he is really sure (e.g. steady state gain), and about which part he is rather uncertain (e.g. exact position of resonant frequencies). In that situation it seems logical to think that the designer requires the model to be matching with this a priori information. The same thing holds for situations in which we really want to have a (possibly highly structured) model that is based on first principles with possibly a number of unknown coefficients to be specified through experimental verification / identification. One would require the coefficients to be referring to physical quantities, like masses being positive etc.

The requirement of a model being "realistic" in this sense, generally will refer to an accurate open loop description of the plant. Now this seems to be a fairly moderate requirement if one is aiming at high performance control. However this is not as straightforward as it seems. It has been shown in a number of situations that one easily can arrive at high performance control systems with only a moderate, or even a bad, open loop performance of the nominal model, see. e.g. [6, 11, 12, 16]. Moreover, also the converse holds true: an accurate "looking" open loop model may give rise to bad performing controlled plants. For example, minor model inaccuracies at the bandwidth of the feedback system to be designed, might deteriorate the obtainable performance. The message here is that when arriving at a model that is suited for high performance control, one is not automatically assured of an overall accurate open loop model. And when there is a need for a "realistic" model (which in many situations definitely will exist) it has to be considered as a separate model requirement that is additional to the previous ones, and for which has to be paid, generally in terms of a higher model order and/or more experiments.

Nominal Models and Model Uncertainty

In this section we are going to discuss some aspects related to the role of nominal models and model uncertainty in the combined identification and control design problem. As a start we will consider the following approach which is more or less standard in dealing with the problem. As a result we will refer to it as the "standard approach".

Standard Approach

(1) Given a model set $\mathcal{M}$ and an uncertainty structure (Δ), identify a nominal model $\hat{P} \in \mathcal{M}$ and a bound b such that the measured data and a priori information allows an expression that (it is very likely that) $P \in \mathcal{P}_\Delta(\hat{P}, b)$;

(2) Design a controller C based on $\mathcal{P}_\Delta(\hat{P}, b)$;

(3) There is a resulting guaranteed performance for $(\tilde{P}, C)$ for all $\tilde{P} \in \mathcal{P}_\Delta(\hat{P}, b)$, and thus for (P, C).

In this approach the link between the identification part and the control part of the problem is made through an uncertainty set $\mathcal{P}_\Delta(\hat{P}, b)$. Everybody will agree with the statement that this uncertainty set is far from unique. It heavily relies on the process data that is measured and on which the identification procedure is based, as well as on the a priori assumptions on the plant, the generation of the plant data, and the characteristics of the noise contribution (deterministic, stochastic, hard-bounded, soft-bounded, in time- or frequency domain, choice of bounded norm etc.).

It is very often suggested that whenever a solution is available for problem (1), an appropriate situation is created for successful control of the unknown plant under consideration. As such, this statement is definitely not true, especially when "successful control" is interpreted as "high-performance control". Note that the resulting guaranteed performance, mentioned in step (3), can easily mean a guaranteed low performance, and we will definitely not be satisfied with that if higher performance is necessary and easily achievable. What are the key points here? In the above formulation of the standard approach, there has not been added any optimality requirement. Indeed choosing a specific uncertainty set $\mathcal{P}_\Delta(\hat{P}, b)$ with a very large b, will increase the chance of being able to find evidence in the data that $P \in \mathcal{P}_\Delta(\hat{P}, b)$. However the larger the uncertainty set $\mathcal{P}_\Delta(\hat{P}, b)$, the poorer will be the worst-case performance of $(\tilde{P}, C)$ over $\tilde{P} \in \mathcal{P}_\Delta(\hat{P}, b)$ for any robustly designed C, the poorer will be the guaranteed performance of (P, C), and very likely the poorer will be the actual performance of the controlled plant.

How can we incorporate an optimality requirement in the problem formulation, in order to circumvent the situation as described before. Let us take a look at some different options.

In an alternative formulation for step (1) we could state:

Problem 1α *Given a model set $\mathcal{M}$ and an uncertainty structure Δ, identify a nominal model $\hat{P} \in \mathcal{M}$ and a bound b such that the measured data allows an expression $P \in \mathcal{P}_\Delta(\hat{P}, b)$, and b is the minimal radius over all nominal models in the model set.*

This is actually the formulation of a worst-case identification problem. In (1α) the nominal model can no longer be chosen freely; it becomes the result of an optimization of the minimal radius of the uncertainty set.

Note that the current work on H_∞ identification, as reported e.g. in [6, 4, 9], is of the type as formulated in point (1). There have not been developed any methods for constructing H_∞-optimal models yet, only providing nominal models for which an - minimal- H_∞-bound can be given. In this way the choice of a nominal model is made subordinate to the possibility of constructing H_∞-bounds.

Another formulation introducing an optimality requirement can be formulated as follows:

Problem 1β *Given a model set $\mathcal{M}$ and an uncertainty structure Δ, identify a nominal model $\hat{P} \in \mathcal{M}$ and a bound b such that*

- *the measured data allows an expression $P \in \mathcal{P}_\Delta(\hat{P}, b)$, and*
- *the resulting guaranteed performance in step (3) is optimized.*

Whereas problem (1α) is formulated from an identification point of view, problem (1β) is more control-oriented. It takes account of the fact that control design is our objective. Consequently it relates the requirements on the set $\mathcal{P}_\Delta(\hat{P}, b)$, which acts as an intermediate between the identification and the control part of the problem, to the objectives of the control design.

Note that we already made a compromise in treating step (3) as a control objective. As a matter of fact the real control objective is not the guaranteed performance over a set of models in $\mathcal{P}_\Delta(\hat{P}, b)$, but an optimal performance of the controlled plant (P, C).

Note also that the model set $\mathcal{M}$ acts as an a priori uncertainty set, whereas $\mathcal{P}_\Delta(\hat{P}, b)$ is an a posteriori uncertainty set that can be chosen to have a different structure. It is not necessary at all that $\mathcal{P}_\Delta(\hat{P}, b) \subset \mathcal{M}$. $\mathcal{M}$ can be a set of black box models, but it also can refer to a specific class of structured (parameter) perturbations of some a priori assumed model. It reflects the a priori information on the system dynamics that one likes to provide. An additional requirement $\mathcal{P}_\Delta(\hat{P}, b) \subset \mathcal{M}$, refers to the situation that one is definitely sure about the a priori information, and that one considers it to be impossible that $P \notin \mathcal{M}$.

In both problem formulations (1α) and (1β) the uncertainty structure Δ is fixed beforehand. It has to be stressed that measured data alone will never provide evidence for a specific uncertainty structure. Consequently, besides the problems formulated above, a very relevant question becomes:

What can be a sensible criterion for choosing the uncertainty structure Δ?

Now we again return to the objective of our design: a high-performance controller for the plant P. The result of the standard approach is a guaranteed performance for the set $\mathcal{P}_\Delta(\hat{P}, b)$. Choosing different uncertainty structures Δ will lead to different uncertainty sets $\mathcal{P}_\Delta(\hat{P}, b)$, either through ($1\alpha$) or ($1\beta$), and consequently it will both lead to a different level of guaranteed performance for $(\tilde{P}, C)$ over $\mathcal{P}_\Delta(\hat{P}, b)$ (and thus for P) and to a different actual performance of (P, C). Apparently not only the "size" and "position" of the uncertainty set (reflected by b and $\hat{P}$) is of importance, but definitely also its shape (reflected by Δ). As a result the uncertainty structure Δ acts as a design variable in our combined identification and control problem.

In this line of thought sensible criteria for choosing Δ are: an optimal - actual- achieved performance of (P, C); or an optimal worst-case performance of $(\tilde{P}, C)$ over $\mathcal{P}_\Delta(\hat{P}, b)$.

Note that in either of the two cases the optimal Δ is dependent on the controller C to be designed. This again stresses the interdependence of the identification part (1) and the control part (2) of the problem. The link between the two parts is a two way dependence between the uncertainty set and the controller. When this joint identification and control problem has to be solved by separate identification and model based control, an iterative procedure of separate identification and control is unavoidable, see e.g. [13]. We will discuss iterative procedures further in the next section. First some attention will be given to the nominal model.

What is the role of the nominal model in all this? Does the nominal model itself have specific importance, or is all model information sufficiently represented in the uncertainty set $\mathcal{P}_\Delta(\hat{P}, b)$?
Concerning the role of the nominal model, it is of key importance which step we consider in the identification problem.

In the standard formulation (1), the nominal model is free to choose, and for any $\hat{P}$ and $\mathcal{P}_\Delta$ there seems to exist a bound that certifies that $P \in \mathcal{P}_\Delta(\hat{P}, b)$. As a result, the nominal model plays a key role in determining the uncertainty set $\mathcal{P}_\Delta(\hat{P}, b)$.

In the formulation of problem (1α), the nominal model $\hat{P}$ is the model that should yield a minimum radius b of the uncertainty set. It is determined by the procedure. The question is whether this nominal model in itself plays a role in the problems discussed. In the control design step (2) the real objective is to design a controller that optimizes the performance of the controlled plant. As a tool for approaching a solution to this problem, one generally aims at optimizing the worst-case performance over $\mathcal{P}_\Delta(\hat{P}, b)$. If control design methods would be available that construct controllers on the basis of any given set of models, the choice of a nominal model within this set would not be essential. The control design would be invariant for the choice of nominal model, but determined by the set of models. However in practice for common performance measures and design methodologies, such worst-case synthesis procedures are not available yet. The worst-case performance can be analysed a posteriori, as suggested in step (3), for specific uncertainty sets $\mathcal{P}_\Delta(\hat{P}, b)$ in which the nominal model plays a central role. A similar situation holds on the identification side. In the identification, the set of models that is "constructed from" data, is still strongly determined by the chosen nominal model, even in guaranteed set estimation (overbounding). This is due to the fact that uncertainty sets $\mathcal{P}_\Delta(\hat{P}, b)$ are constructed through simply structured bounds, like $\|\Delta\|_\infty$, or $\|\Delta\|_{\ell_1}$ with Δ an additive or multiplicative model mismatch.

In the approach with step (1β), the choice of the uncertainty set and, as a part of that, the choice of the nominal model is dependent on both the identification part of the problem (i.e. the data and a priori information), as well as on the controller to be designed. In this setting the role of the nominal model becomes even more pronounced.

A Sketch of Iterative Schemes

Currently there have been made some first attempts to solve a combined identification/control problem as specified in formulation (1β) of the standard problem in the previous section. In the same setting as before, these methods take the following approach:

Given a real-valued performance measure $\|T(P, C)\|$, with $\|\cdot\|$ a nonspecified (semi-)norm or distance function and T a nonspecified transfer function, such that optimization of performance refers to minimization of $\|T(P, C)\|$; find Δ, $\hat{P}$ and C such that simultaneously:

(a) The uncertainty set is determined by

$$\mathcal{P}_\Delta(\hat{P}, b) = \{\tilde{P} \mid \|T(\tilde{P}, C) - T(\hat{P}, C)\| \le b\} .$$

(b) The identified model $\hat{P}$ satisfies:

$$\hat{P} = \arg_{\tilde{P} \in \mathcal{M}} \min \{b \mid P \in \mathcal{P}_\Delta(\tilde{P}, b)\} .$$

(c) The controller C satisfies:

$$C = \arg \min_{\tilde{C}} \|T(\hat{P}, \tilde{C})\| .$$

Actually the uncertainty structure Δ has been adapted to the performance measure $\|T(P, C)\|$ of the controlled system. Once the performance has been defined, the uncertainty structure is fixed. As a result, when two models are "close" in terms of the uncertainty "metric", the controller applied to both models will lead to a small performance degradation.

A combination of the three steps shows that

$$\|T(P, C) - T(\hat{P}, C)\| \le \hat{b} \tag{2}$$

with $\hat{b}$ the minimum value of b obtained in part (b). Applying the triangle inequality shows that

$$\|T(\hat{P}, C)\| - \hat{b} \le \|T(P, C)\| \le \|T(\hat{P}, C)\| + \hat{b} . \tag{3}$$

With $\hat{b}$ being small enough, the performance of the controlled plant tends to the (optimized) designed performance. In this setting there is no real need for a robust control design scheme, in the sense of a worst case performance optimization. The control design is a nominal performance design, and robustness is obtained through the identification stage. In order to obtain this robustness, it is required that $\hat{b} << \|T(\hat{P}, C)\|$. In that case the bounds (3) become tight. The accuracy of the identification determines to what extent the identified model $\hat{P}$ can predict the performance of the controlled plant.

When aiming at solutions along the lines described above, with separate stages of identification and control design, one meets the problem that all steps $(a) - (c)$ actually are connected. In an iterative scheme one typically fixes the controller in the identification step to be known, and one fixes the uncertainty set $\mathcal{P}_\Delta(\hat{P}, b)$ in the control design step.

The identification schemes that are used in the currently available methods, actually do not solve a problem as mentioned by (b), which effectively is a set estimation problem, but rather aim at:

$$\hat{P} = \arg \min_{\tilde{P} \in \mathcal{M}} \|T(P, C) - T(\tilde{P}, C)\| \ .$$

This optimization will not guarantee that the mismatch between plant and nominal model will be sufficiently small to obtain e.g. robust stability. Consequently the robust stability of the obtained controller has to be ascertained before implementing the controller on the plant.

A typical iterative scheme will look as sketched in Fig. 2.

Taking experiments from the plant being controlled by the latest controller is an essential part of the loop. This means that every time a new controller is designed, new experiments are performed with this controller implemented. This is due to the fact that data-information on (P, C) is required to solve the identification problem (b). In all iterative schemes this results in a closed loop identification problem. Note that the separate steps of identification and control design may also incorporate some iteration loops. The diagram also shows a "short" iterative loop of identification and control design without taking new experiments. This loop is employed in a scheme proposed by Rivera [10]. Especially in situations where experiments are expensive and time-consuming, it seems advantageous to fully exploit the possibilities of this "short" loop.

In Zhang, Bitmead and Gevers [16] a scheme is worked out along similar lines as depicted in the figure, with as performance measure a fixed LQ cost criterion.

Schrama [11, 12, 13] adopts the performance measure $\|T(\alpha P, C/\alpha)\|_\infty$ with a scalar weighting α and

$$T(P, C) = \begin{bmatrix} P \\ I \end{bmatrix} [I + CP]^{-1} [C \ \ I] \ .$$

This is the only scheme with a separate step of robustness analysis. In Lee, Anderson et al. [7] a scheme is proposed based on a performance measure:

$$\|\frac{PC}{1 + PC} - T^d\|_\infty$$

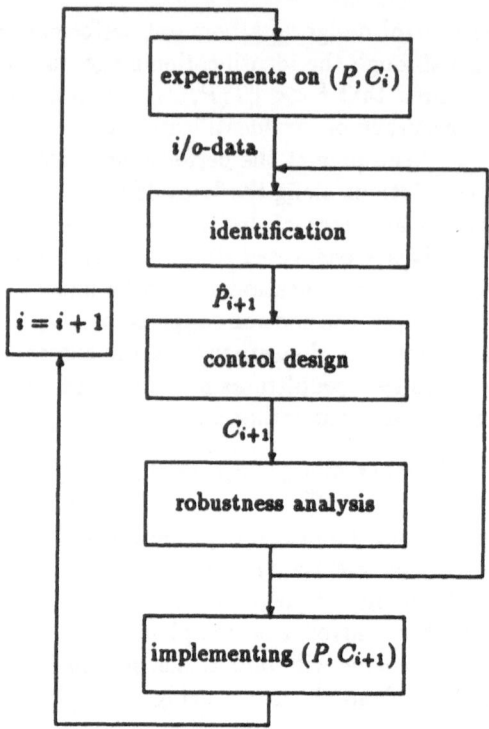

Fig. 2. Iterative scheme for identification and control design

with T^d some prescribed complementary sensitivity function.

Moreover, the latter two schemes have a performance measure that is supposed to change during the iterations; i.e. as the iterations evolve the performance measure is changed to a "higher" level. In the second scheme this is accomplished by gradually increasing the scalar weight α, which generally will refer to a higher bandwidth of the controlled system; in the third scheme this is accomplished by gradually pushing the prescribed complementary sensitivity function T^d to having a higher bandwidth.

Concluding this section, we will provide a list of aspects that can characterize suggested solutions to the problem discussed in this paper. This list is definitely not complete, but could be a nice tool for evaluating the pros and cons of currently available and of new schemes.

- Is there a fixed performance measure, or are the performance requirements gradually strengthened as the iterations evolve?
- What is the performance measure?
- Do the local identification and control design step reflect the same performance criterion?

- Are there guarantees for the stability/performance of (P, C) prior to implementing the controller to the plant?
- Does the performance of $(\hat{P}, C)$ accurately predicts the performance of (P, C)?
- Are bias- and/or variance-type errors accounted for in the identification step?
- Is the identification problem solved in terms of time/frequency - domain data? How is the experiment design solved?
- Is there a guaranteed/monitored improvement of the performance of (P, C) as the iterations proceed? Convergence? Local minima?
- Is there a decision rule that points to the need of new experiments; i.e. running the "inner" loop versus running the "outer" loop?

Final Remarks

In this paper, we have considered a number of aspects that are related to the problem of designing a high performance controller for a plant with unknown dynamics through separate identification and control design. The underlying ideas in this line of research are quite close the area of what is called adaptive control. Whereas in adaptive control the procedures of identification and control design are completely intertwined, the approach discussed in this paper attempts to provide a separate analysis of both steps, and to accomplish a joint performance criterion for both parts.

Iterative procedures in which the performance level of the controlled plant is gradually increased are actually quite appealing. In engineering practice a product is commonly accomplished by a number of successive improvements rather than by a one-step design that starts from scratch. Why should control-engineering be different? The implication of this is that one should not focus on an ideal but possibly unachievable desired control performance. Instead one better aims at an improvement of the performance that has already been achieved.

The usual approach is to model a plant from open-loop data, and to design a controller that achieves a moderate but acceptable performance. How to improve this moderate performance? Often it is taken that "a higher accuracy" is needed in the modeling procedure. This implicitly presumes that the initial controller has to be switched-off, and that a more accurate description is needed of the open-loop behavior of the plant in question. But why should we switch off the controller? There are objections not only for economic or safety reasons, but there is something intrinsically wrong with such a procedure. After all, the moderate performance is "closer" to an improved performance, than the open-loop performance of the plant.

References

1. D.K. de Vries and P.M.J. Van den Hof (1992). "Quantification of model un-

certainty from data: input design, interpolation, and connection with robust control design specifications," In *Proc. 1992 American Contr. Conf.*, June 26-28, 1992, Chicago, IL, pp. 3170-3175.

2. D.K. de Vries and P.M.J. Van den Hof (1992). *Quantification of Uncertainty in Transfer Function Estimation: A Mixed Deterministic-Probabilistic Approach.* Report, Mech. Eng. System and Control Group, Delft Univ. Technology, July 1992.

3. G.C. Goodwin, M. Gevers and B. Ninnes (1992). "Quantifying the error in estimated transfer functions with application to model order selection," *IEEE Trans. Automat. Contr.*, AC-37, pp. 913-928.

4. G. Gu and P.P. Khargonekar (1992). "A class of algorithms for identification in H_∞," *Automatica, vol. 28*, pp. 299-312.

5. R.G. Hakvoort, R.J.P. Schrama and P.M.J. Van den Hof (1992). "Approximate identification in view of LQG feedback design," *Proc. 1992 American Control Conf.*, June 26-28, 1992, Chicago, IL, pp. 2824-2828.

6. A.J. Helmicki, C.A. Jacobson and C.N. Nett (1991). "Control oriented system identification: a worst-case/deterministic approach in H_∞," *IEEE Trans. Automat. Contr.*, AC-36, pp. 1163-1176.

7. A.P. Loh, G.O. Corrêa and I. Postlethwaite (1987). "Estimation of uncertainty bounds for robustness analysis," In *IEE Proc. Pt. D. Control Theory and Applic.*, vol. 134, pp. 9-16.

8. W.S. Lee, B.D.O. Anderson, R.L. Kosut and I.M.Y. Mareels (1992). "On adaptive robust control and control-relevant system identification," In *Proc. 1992 American Control Conf.*, June 26-28, 1992, Chicago, IL, pp. 2834-2841.

9. P.M. Mäkilä (1991). "On identification of stable systems and optimal approximation," *Automatica, vol. 27*, pp. 663-676.

10. D.E. Rivera (1991). "Control-relevant parameter estimation: a systematic procedure for prefilter design," In *Proc. 1991 American Control Conf.*, Boston, MA, pp. 237-241.

11. R.J.P. Schrama (1992). *Approximate Identification and Control Design with Application to a Mechanical System.* Dr. Dissertation, Delft University of Technology, May 1992.

12. R.J.P. Schrama and P.M.J. Van den Hof (1992). "An iterative scheme for identification and control design based on coprime factorizations," *Proc. 1992 American Control Conf.*, June 24-26, 1992, Chicago, IL, pp. 2842-2846.

13. R.J.P. Schrama (1992). "Accurate models for control design: the necessity of an iterative scheme," *IEEE Trans. Automat. Contr.*, AC-37, pp. 991-994.

14. R.J.P. Schrama and P.M.J. Van den Hof (1992). "Iterative identification and control design: a worked out example," *This issue.*

15. G. Stein and J.C. Doyle (1991). "Beyond singular values and loopshapes," *J. Guidance, Control and Dynamics*, Vol. 14, pp. 5-16.

16. Z. Zang, R.R. Bitmead and M. Gevers (1991). "H_2 iterative model refinement and control robustness enhancement," In: *Proc. 30th IEEE Conf. Decision and Control*, Brighton, UK, pp. 279-284.

17. Y. Zhu (1989). "Estimation of transfer functions: asymptotic theory and a bound of model uncertainty," *Int. J. Control, vol. 49*, pp. 2241-2250.

An Informal Review of Model Validation

Roy Smith

Electrical & Computer Engineering Dept.
University of California, Santa Barbara, CA 93106.

Opening Discussion

The following is an informal overview of the model validation problem in the context of robust control and uncertain systems. The philosophical issues are often not clear from a reading of the more technical literature and I hope to provide some understanding of these aspects.

The motivating problem is the design of a control system for a physical plant. Almost all design methods require a model representing the plant behavior, and furthermore, the model must be compatible with the design approach. The more recent approaches, particularly robust control, require models which cannot be provided by the standard identification techniques. This mismatch has hindered the application of robust control and has lead to an increased interest in the problem of modeling uncertain systems.

Most system models contain some characterization of uncertainty. In the traditional least squares based identification methods, this takes the form of random noise signals with assumed stochastic properties. While this type of uncertainty description is compatible with the linear quadratic design approaches, it is ill-suited for robust control methods. It is also worth noting that the standard identification procedures result in models which account for all data used in the procedure.

Robust control models include, in addition to unknown additive noise, bounded perturbations which can be used to describe unmodeled dynamics. This is a powerful formulation as such dynamics can be destabilizing under feedback. Linear models in which all uncertainty is described by additive noise do not have the ability to predict such destabilizing effects. The model perturbations are not entirely arbitrary and are specified as unknown elements of a specified class. Typical perturbation specifications will be discussed subsequently.

The greater richness of this modeling framework carries with it greater difficulties in developing models. A designer must now estimate a bound on the unmodeled dynamics in addition to estimating a nominal model. The standard identification techniques currently do not lead to such bounds on the perturbations, although some work is heading in this direction. The work of Goodwin, Ninness and Salgado [1, 2, 3] or Hjalmarsson and Ljung [4] begins to investigate this area.

System models, whether of the robust control type or not, include a set of model specifications. The model specifies a particular mathematical framework; for example, a frequency domain linear time invariant nominal model. The framework also specifies some of the properties the unknown signals and disturbances: signals of bounded energy or power in the H_∞ case; bounded magnitude in the l_1 case; and stochastic properties in the linear quadratic case. In robust control models the perturbations are also members of a prescribed set: typically linear time invariant perturbations in the H_∞ case; and linear time varying or nonlinear time varying in the l_1 case. Various groups are working on modifying and comparing the consequences of these specifications. For example, studying the effects of removing the time invariance restriction in the H_∞ case. The point that I wish to make is that such specifications are a natural part of the chosen modeling framework.

In using such a model for design purposes one assumes that these specifications also apply to the physical system. Naturally they do not, however the hope is that enough of the behavior of the physical system is captured by the model to enable the design to perform satisfactorily. This assumption is usually also applied to the theoretical study of identification. The properties of an identification algorithm are studied under certain assumptions about the system. This is a natural thing to do — if an algorithm will not work on an idealized system then we have little confidence of it working on a physical system.

The model validation approach is significantly different. No assumptions are made about the nature of the physical system. Rather, measurements are taken, and the assumption that the model describes the system is directly tested. In light of the above discussion, the model will also include specifications on the unknown signal sets and the linearity and time invariance (or lack thereof) of the perturbations, in addition to the numerically given linear time invariant nominal model. Simply stated, the model validation problem is: "Could the model have produced the observed datum?" In a robust control context this is equivalent to asking whether or not there exists an unknown signal vector and an unknown perturbation, within the assumed sets, satisfying the specified bounds, and accounting for the input-output datum.

There are invariably discrepancies between the model and the data and the objective is to determine whether or not these discrepancies invalidate the model. There is no "correct" model of a system — the most that a designer hopes for is that the model describes the system behaviors of importance.

This raises another issue. Model validation can determine whether or not a single experimental datum could have been produced by the model rather than the system. It cannot validate a model of a system for the simple reason that not all behaviors and operating conditions can be observed in a finite number of experiments. Strictly speaking, we can only invalidate a model. Repeated experiments where every datum can be produced by the model does however lead to confidence about the suitability of the model.

Model validation for least squares/stochastic models has been applied for some time. Ljung [4] gives an overview of model validation in the standard identification framework. The fact that all data used in the identification procedure will be accounted for by the model must be considered. One approach, typically used in the chemical process industry, is to take separate data sets solely for the purposes of validation. This is known as cross validation. The temptation is to include this additional data in the generation of a new model. In certain cases taking additional data may be an expensive proposition. Without a cross validation data set, the resulting model can only be compared to a priori assumptions about the system. For example, is the noise level higher than realistically expected? Note that this form of validation requires assumptions about the system — in particular, what is realistic? This is useful, although ideally we want to test the applicability of the a priori assumptions.

Note that the complication that a resulting model will describe all of the data used in its generation does not apply to robust control models. Unfortunately the reason is that we currently have no general means of obtaining such models. At some level an ad hoc approach is applied and any rigor is lost. One reason for this is that the set of models fitting any given datum is overdetermined. In general it is possible to attribute the discrepancies to either the unknown additive signals or the perturbations. A combination of both is probably the most appropriate.

The area of identification in H_∞, originally developed by Helmicki, Jacobson and Nett [6, 7], with additional work by Mäkilä and Partington [8, 9] and Gu and Khargonekar [10], focuses on developing algorithms to provide a robust control model. In attempting to identify a model in this framework, assumptions are usually required to provide a well posed problem. A priori H_∞ perturbation bounds are often assumed and data is taken and analyzed in order to provide a compatible model, meeting the bounds with a small amount of additive noise. The full generality of the robust control framework cannot yet be handled and the perturbation bound does not depend on the experimental data. A solution to the second of these problems will probably require a formalism which deals with multiple experiments, under different operating conditions, in order to appropriately attribute uncertainty to perturbations or additive signals.

As the above discussion suggests, we do not have good systematic methods for developing suitable robust control models. Therefore the modeling and identification procedures used give little measure of the applicability of the model. Model validation is useful here as it allows us to test a given model against data from the physical system.

I hope that the preceding remarks have provided some context for considering model validation for robust control models, which will be the emphasis of the subsequent discussion. A minimal amount of technical detail is given; only enough to illustrate the practical issues. On an historical note, the problem was posed in this form by Smith and Doyle [11, 12]. Krause [13] studied a similarly motivated problem: the implications of test data on determining stability margins. Others have been involved in the subsequent work. Allow me to apologize in advance if I leave out any references.

Robust Control Models

Only a brief outline is given here. The formalism follows that of Doyle [14]. The generic robust control model is given by the input-output relationship,

$$y = [P_{21}\Delta(I - P_{11}\Delta)^{-1}P_{12} + P_{22}]\begin{bmatrix} u \\ w \end{bmatrix},$$

where Δ is an unknown, bounded perturbation. The signal w is unknown and assumed to belong to some specified signal set. In considering model validation, both the input u and output y are known. Measurement noise on either is modeled as a scaled component of w. Note that for some perturbation Δ, the term $(I - P_{11}\Delta)$ may not be invertible. This feature allows the framework to model perturbations that are potentially destabilizing under feedback. This particular model formulation is referred to as a linear fractional transformation (LFT) and will be abbreviated to,

$$y = \mathbf{F}_u(P, \Delta)\begin{bmatrix} u \\ w \end{bmatrix}.$$

This is a useful framework as interconnections of LFTs are simply larger LFTs.

The assumed nature of the uncertainty in the system is captured by the assumptions on w and Δ. As P can be scaled, Δ is taken to be unity norm bounded. Additional structure may also be imposed; multiple perturbations at various locations within a complex system can be modeled as a block diagonal Δ. Refer to Packard [15] for detail on the more general representations. Robust control models are set descriptions in that they include both perturbations and input signal sets specified only by norm bounds.

There are two sets of specifications applied to P, Δ and w. The first of these leads to the popular H_∞/μ analysis and design framework. P and Δ are linear time invariant with $\|\Delta\|_\infty \leq 1$. The unknown signal, w, is of unity bounded energy or power. Analysis and synthesis in this framework is relatively well developed. Refer to [16] for further details.

A more recent framework deals with unknown signals and performance specifications in terms of signal magnitude rather than energy. This leads to the l_1 design framework (refer to [17, 18]). The plant P is linear time invariant, typical represented in the discrete domain, and Δ is linear time varying with $\|\Delta\|_\mathcal{A} \leq 1$. Here $\|.\|_\mathcal{A}$ denotes the induced l_∞ to l_∞ norm. The unknown signals w are of unity bounded magnitude.

In the cases mentioned above, the choice of specifications on Δ has a significant effect on the theoretical and computational stability analysis results. Assuming time invariance in the H_∞/μ case has the effect of decomposing the robust stability analysis problem into a series of independent constant matrix problems. In the l_1 framework, the time varying nature of the perturbation enables the robust stability analysis problem to be treated as a single constant matrix problem.

The subsequent discussion will illustrate that model validation can be posed as a constrained stability analysis. Unfortunately, except in a few special cases, the attractive simplifications mentioned above are lost.

Model Validation Problem — A Formal Statement

It is assumed that P is given and an experimental datum (y, u) is under consideration. The robust control model validation problem can be formulated as follows.

Model Validation: *Given a robustly stable model P, and an experimental datum (u, y), does there exist (w, Δ), $\|w\| \leq 1$, $\|\Delta\| \leq 1$, such that*

$$y = F_u(P, \Delta) \begin{bmatrix} w \\ u \end{bmatrix}.$$

The choice of norms depends on the specifications of the model. This simply asks the question "is there an element of the model set and an element of the unknown input signal set such that the observed datum is produced exactly"?

Note that using the LFT framework makes this formulation general. It can equally well apply to closed loop and MIMO systems.

Characteristics of the Problem

The above problem does not usually have a unique solution in terms of w and Δ. The objective is to determine whether or not a (w, Δ) pair, accounting for the datum, exists. The approaches to this problem are based on optimization: find a (w, Δ) pair meeting the constraints imposed by the model and the datum. The properties of this optimization are strongly influenced by the choice of robust control model specifications.

The robust stability problem — what is the smallest Δ that will destabilize $F_u(P_{11}, \Delta)$? — can similarly be posed as an optimization problem. This approach has been extensively studied by Fan and Tits [19, 20] in the H_∞/μ case. The equality constraint imposed by the input-output datum can be included as an additional constraint in the optimization. Some reformulation of the problem is required in order to do this and this is discussed in [21]. The inclusion of the equality constraint can have significant computational consequences.

The problem was originally solved by Smith and Doyle [11] in an H_∞ framework. This work was done using frequency domain data and a practical application is described by Smith [12, 22]. The input-output constraint can be expressed independently at each frequency. However the requirement that $\|w\|_2 \leq 1$ depends on the signal w at all frequencies. In general this means that the optimization problem must consider all frequencies simultaneously. One means of circumventing this is to search for the minimum $\|w\|_2$, with the constraint that $\|\Delta\|_\infty \leq 1$. This works because the minimum $\|w\|_2$ can be found by minimizing at each frequency independently.

The application of the theory to the l_1 robust control framework is discussed by Smith [23]. Time domain data is used which makes this formulation the more experimentally appealing. In the robust stability analysis for the l_1 case, the search for a worst-case time-varying perturbation, allows the problem to be reformulated as a single matrix problem. The time-varying nature of the perturbation and the fact that the disturbances are bounded independently in time allows the worst case effects to, in effect, be summed together to form a single matrix analysis problem. The inclusion of the model validation input-output constraint destroys the time domain independence of the worst case effects.

In both of the above cases, the robust stability problem is a constant matrix problem of dimension equal to total the input and output dimensions; that is, including the dimensions of Δ and w. If all time or frequency points must be considered simultaneously then these dimensions are multiplied by the number of points in the input-output datum. This may result in a very large matrix problem and efficient computation procedures for such problems are an active research area.

Poolla et al. [24] have also studied model validation in the time domain, using a discrete time H_∞ framework. The use of the time domain is more appealing in its closer connection with experimental data. However their problem solution requires structural constraints on the model, including a SISO system and a restriction on how the perturbation can enter the system. It does result in a large dimensional matrix problem, although it is one which is significantly easier to solve than the constrained robust stability problems discussed above. Zhou and Kimura [25] have also studied this problem using a similar approach. They also include unknown real valued parameters in their formulation.

The algorithmic issues associated with model validation have not yet been clarified to the point where the techniques are readily available.

Identification and Fault Detection

There is no systematic theory for the development of robust control models from experimental data. However, model validation provides a means of assessing models with respect to the data. This is a useful tool for identification as it allows unsuitable models to culled from a group of candidate models. Comparison of the perturbations required for accountability, with a priori uncertainty estimates also gives an ad hoc means of assessing model applicability.

It is possible to extend the model validation ideas to the identification of particular parameters within an uncertain system. This approach was outlined by Smith and Doyle [26]. In essence, the unknown parameters are simply considered as additional variables in the optimization problem. Naturally any computational difficulties in the model validation problem occur here also.

Krause, Khargonekar and Stein [27, 28] and Kosut, Lau and Boyd [29] have also studied the problem of parameter identification in the presence of unmodeled dynamics. A similar formulation is also used by Zhou and Kimura [25]. In their work, the poles of the system are prescribed and the search for parameters in the presence of bounded noise and bounded perturbations is a convex optimization problem.

Model validation is also directly applicable to fault detection. Given a design model and a controller in operation, the model validation theory gives a means of continuously assessing whether or not the physical system is still described by the design model. The techniques discussed here produce the perturbation and noise that come closest to satisfying the specifications of the model. Gradual deterioration in a system may manifest itself as increasing perturbations and/or noise required for accountability of the data. A sudden failure may be identified by a sudden jump in the size of the required perturbations and/or noise.

Closing Comments

Model validation has several appealing philosophical aspects: nothing is assumed about physical reality; and one only gets information by making mistakes. However it is wrong to think of an invalidated model as being of no use. The closed loop iterative identification approaches of Schrama and Van den Hof [30, 31] in the H_∞ case, and Zang, Bitmead and Gevers [32, 33], in the linear quadratic case, have shown that a good open loop model is not necessary for a good closed loop design.

Using an open loop model/system matching criteria may not be appropriate for the purposes of closed loop design. Fortunately, the use of the LFT framework allows us to consider model validation for closed loop experiments. Further investigation of this area is only beginning.

References

1. G. Goodwin and M. Salgado, "Quantification of uncertainty in estimation using an embedding principle," in *Proc. Amer. Control Conf.*, 1989.
2. G. Goodwin, B. Ninness, and M. Salgado, "Quantification of uncertainty in estimation," in *Proc. Amer. Control Conf.*, pp. 2400–2405, 1990.

3. B. M. Ninness and G. C. Goodwin, "Robust frequency response estimation accounting for noise and undermodeling," in *Proc. Amer. Control Conf.*, pp. 2847–2851, 1992.

4. H. Hjalmarsson and L. Ljung, "Estimating model variance in the case of undermodeling," *IEEE Trans. Auto. Control*, vol. 37, pp. 1004–1008, July 1992.

5. L. Ljung, *System Identification, Theory for the User.* Information and System Sciences Series, New Jersey: Prentice-Hall, 1987.

6. A. Helmicki, C. Jacobson, and C. Nett, "H_∞ identification of stable lsi systems: A scheme with direct application to controller design," *Proc. Amer. Control Conf.*, pp. 1428–1434, 1989.

7. A. J. Helmicki, C. A. Jacobson, and C. N. Nett, "Control oriented system identification: A worst-case/deterministic approach in H_∞," *IEEE Trans. Auto. Control*, pp. 1163–1176, 1991.

8. P. Mäkilä and J. Partingtion, "Robust approximation and identification in H^∞," *Proc. Amer. Control Conf.*, pp. 70–76, 1991.

9. P. Mäkilä, "Laguerre methods and H^∞ identification of continuous-time systems," *Int. J. of Control*, vol. 53, pp. 689–707, 1991.

10. G. Gu and P. P. Khargonekar, "Linear and nonlinear algorithms for identification in H^∞ with error bounds," in *Proc. Amer. Control Conf.*, pp. 64–69, 1991.

11. R. S. Smith and J. Doyle, "Model invalidation — a connection between robust control and identification," in *Proc. Amer. Control Conf.*, pp. 1435–1440, 1989.

12. R. S. Smith, *Model Validation for Uncertain Systems.* PhD thesis, California Institute of Technology, 1990.

13. J. M. Krause, "Stability margins with real parameter uncertainty: Test data implications," in *Proc. Amer. Control Conf.*, pp. 1441–1445, 1989.

14. J. Doyle, "Structured uncertainty in control system design," in *Proc. IEEE Control Decision Conf.*, pp. 260–265, 1985.

15. A. K. Packard, *What's new with μ.* PhD thesis, University of California, Berkeley, 1988.

16. The MathWorks, Inc., Natick, MA, *μ-Analysis and Synthesis Toolbox (μ-Tools)*, 1991.

17. M. A. Dahleh and J. B. Pearson, Jr., "l^1-optimal feedback controllers for MIMO discrete-time systems," *IEEE Trans. Auto. Control*, vol. AC-32, pp. 314–322, Apr. 1987.

18. M. Khammash and J. B. Pearson, Jr., "Performance robustness of discrete-time systems with structured uncertainty," *IEEE Trans. Auto. Control*, vol. AC-36, pp. 398–412, Apr. 1991.

19. M. K. H. Fan and A. L. Tits, "Characterization and efficient computation of the structured singular value," *IEEE Trans. Auto. Control*, vol. AC-31, pp. 734–743, 1986.

20. M. K. H. Fan and A. L. Tits, "m-form numerical range and the computation of the structured singular value," *IEEE Trans. Auto. Control*, vol. AC-33, pp. 284–289, 1988.

21. R. S. Smith and J. C. Doyle, "Model validation: A connection between robust control and identification," *IEEE Trans. Auto. Control*, vol. 37, pp. 942–952, July 1992.

22. R. S. Smith, "Model validation for robust control: an experimental process control application." to be presented: *IFAC World Congress*, July 1993.

23. R. S. Smith, "Model validation and parameter identification for systems in H_∞ and l_1," in *Proc. Amer. Control Conf.*, pp. 2852–2856, 1992.

24. K. Poolla, P. Khargonekar, A. Tikku, J. Krause, and K. Nagpal, "A time-domain approach to model validation," in *Proc. Amer. Control Conf.*, pp. 313–317, 1992.

25. T. Zhou and H. Kimura, "Input-output extrapolation-minimization theorem and its application to model validation and robust identification," in *The Modeling of Uncertainty in Control: Proceedings of the 1992 Santa Barbara Workshop* (R. Smith and M. Dahleh, eds.), Springer-Verlag, 1993. in press.

26. R. S. Smith and J. C. Doyle, "Towards a methodology for robust parameter identification," in *Proc. Amer. Control Conf.*, pp. 2394–2399, 1990.

27. J. M. Krause, P. P. Khargonekar, and G. Stein, "Robust parameter adjustment with nonparametric weighted-ball-in-H^∞ uncertainty," *IEEE Trans. Auto. Control*, vol. AC-35, pp. 225–229, 1990.

28. J. M. Krause and P. P. Khargonekar, "Parameter identification in the presence of non-parametric dynamic uncertainty," *Automatica*, vol. 26, pp. 113–124, 1990.

29. R. Kosut, M. Lau, and S. Boyd, "Parameter set identification of systems with uncertain nonparametric dynamics and disturbances," in *Proc. IEEE Control Decision Conf.*, vol. 6, pp. 3162–3167, 1990.

30. R. J. P. Schrama, "Accurate identification for control: the necessity of an iterative scheme," *IEEE Trans. Auto. Control*, vol. 37, pp. 991–994, July 1992.

31. R. J. Schrama and P. M. V. den Hof, "An iterative scheme for identification and control design based on coprime factorizations," in *Proc. Amer. Control Conf.*, pp. 2842–2846, 1992.

32. Z. Zang, R. R. Bitmead, and M. Gevers, "H_2 iterative model refinement and control robustness enhancement," in *Proc. IEEE Control Decision Conf.*, pp. 279–284, 1991.

33. Z. Zang, R. R. Bitmead, and M. Gevers, "Disturbance rejection: on-line refinement of controllers by closed loop modelling," in *Proc. Amer. Control Conf.*, pp. 2929–2833, 1992.

From Data to Control

Munther A. Dahleh[1] and John C. Doyle[2]

[1] Department of Electrical Engineering and Computer Science,
  Massachusetts Institute of Technology, Cambridge, MA 02139

[2] Control and Dynamical Systems, California Institute of Technology,
  Pasadena, CA 91125

The basic control problem for a given process can be stated as follows: *Given some prior information about the process and a set of finite data, design a feedback controller that meets given performance specifications.* Traditionally, this problem has been tackled by the introduction of an intermediate step, namely finding a *model* which describes the process in some precise sense, and then designing a robust controller using the model as the nominal plant.

Rather than assuming that the prior information is true, it is more natural to think of it as a parametrization of model structures from which we desire to explain the data, i.e., a description of our *prejudice*. In this sense, prior information can itself be invalidated by the Data. This distinction is crucial since such information is generally derived from simplified models of the process, and hence is not verifiable. Once a set of finite data is acquired, a set of models that are consistent with the data and the model structure parametrization (prior information) is defined. This set contains all models that are not falsified by the data. Roughly speaking, system identification picks a most powerful unfalsified model where most powerful is defined depending on the objective in mind, in this case it is finding a controller that delivers a given performance level. We also note that the process of finding such a model, and a controller is iterative in nature as more sets of data are acquired.

It is evident that any iterative scheme will generally be based on reducing the set of unfalsified plants until a controller based on the remaining elements can deliver the performance when connected with the actual process, or a decision is made to enlarge the parametrized set of models and/or change the performance requirement. We propose a general scheme that is based on efficiently eliminating models from the set of unfalsified models. Of course, the acquisition of more data systematically reduces this set, although the efficiency of this depends on the data set itself. On the other hand, an unfalsified model is invalidated if there exists a controller that delivers the required performance for this model and the same controller does not meet the performance with the actual process. Given our prejudice, this model is unacceptable. Finally, an unfalsified model to which no controller can be designed to meet the performance specifications is discarded. In this way, if all models are eliminated, we conclude that the performance cannot be met, of course, given our prejudice. Below, an iterative scheme based on this idea is proposed. This scheme is well defined only if we assume that the required performance of any controller connected with the real process can be tested by using a finite number of experiments.

1. Pick a model structure parametrization.
2. Collect a set of Data, and define the set of unfalsified plants.
3. Find a "large" subset of models to which the design procedure produces a controller that delivers the required performance for all models in this set. If no such set exists, go back to (1) and adjust the model structure and/or the performance objective.
4. Test the controller on the real system. If the controller meets the performance, stop. If not, then the above subset is invalidated.
5. Use the data acquired from testing the performance, as well as other sets of data in order to invalidate additional plants.
6. Go to (3).

This scheme defines both an inner and outer loop. Within the inner loop, the performance requirement and the model structure parametrization are fixed and the acquisition of data as well as the design of controllers for subsets of the set of unfalsified models continue to reduce this set until a controller is found, or a decision that the performance cannot be met is made. We then iterate the outer loop. By eliminating large subsets in step (3), the inner loop converges to a decision much faster. It is evident that this loop terminates if the unfalsified set contains only a finite number of models. This may be relaxed to a compactness assumption on the parametrization of model structures.

The process of elimination requires the availability of methods for designing robust controllers for subsets of the set of unfalsified models. It is assumed that for a given subset, which can be a single plant, a decision can be made as to whether or not a controller that meets the performance exists. If the parametrization of models and the performance objective are such that no exact methods exist, we will accept tests based on the existing design methods, as conservative as they may be. The lesser the conservatism of the methods in robust control, the lesser the bias of the above iterative scheme will be.

The step of testing a given controller on the real system generates more sets of data that can be used to invalidate more models, within the inner loop of the above scheme. We may have the ability to conduct more experiments, in which case, they have to be devised in such a way that they have sufficient information to invalidate more unfalsified models. It is not known how to choose such experiments.

The difficulty of the proposed scheme lies primarily in steps (3) and (5). Selecting a large subset of models that are consistent with the given data and to which a controller that meets the performance specifications exists can be a difficult problem. On one hand, it is desirable to make this set as large as possible, however, robust control techniques are likely to give guarantees only for smaller sets. One way of doing this is to parametrize ordered subsets (in terms of size) inside the set of unfalsified models and to increase the size until the performance cannot be met. Another approach would be to pick an element in the set by defining an appropriate notion of a most powerful unfalsified model and then to compute estimates of the radius of uncertainty around that model (within which we optimistically believe that the actual process lies) and then to solve the robust performance problem with respect to that set. Both of these methods are not yet developed, and there is no known notion of a most powerful model from which nonconservative estimates of the radius of uncertainty can be derived.

It is interesting to note that consistency results discussing the convergence of the model to the actual process, when the process lies in the parametrization and the data record is long enough can be formulated for this iterative procedure. Such analysis is crucial in order to raise the level of confidence in any proposed scheme. Only special results in a nonstochastic setting have been developed so far.

In conclusion, it is clear that the major bottleneck is the computational complexity of the various aspects of this procedure. Sample and computational complexity of such algorithms are crucial issues to consider and circumvent. Finally, unless robust control can handle a large class of plant uncertainty in a precise way, such schemes will always be conservative.

Is Robust Control Reliable?

Jeffrey C. Kantor[1] *and Billie F. Spencer, Jr.*[2]

[1] Dept. of Chemical Engineering, University of Notre Dame, Notre Dame, IN 46556

[2] Dept. of Civil Engineering and Geological Sciences, University of Notre Dame, Notre Dame, IN 46556

Robust control has addressed the issue of model uncertainty in control systems mainly through worst-case analysis. While this has led to significant new insight and tools, the emphasis on the worst-case has avoided a probabilistic understanding of robust control. In this essay we suggest that tools for structural reliability analysis can be adapted for this purpose.

Tools for robust control analysis and synthesis have undergone a decade of rapid development. It is now clearly recognized that model uncertainty, adverse dynamical properties (e.g., nonminimum phase), and input/output constraints comprise a triumverate of performance limiting factors. This development has been largely based on H_2/H_∞ and l_1 theories for which the control problem consists of performance specifications, a nominal plant model, and some sort of uncertainty description. The canonical synthesis problem is to find a control which satisfies the performance constraints for all possible plant descriptions, including the worst case.

This abstract problem, however, avoids the practical difficulty of describing the model uncertainty in the first place. On the one hand, there is the task of establishing a tractable analytical framework, and for this reason model uncertainties are often treated as norm-bounded LTI systems, or more recently, LTV systems. As example is coercing an uncertainty in the activation energy of a particular chemical reaction into a bound on the variation of a particular transfer function for a chemical reactor. Extensions of the tools to real-parameter uncertainty generally yield difficult non-convex optimization problems. These issues are thoroughly discussed in other contributions to this volume.

The question we raise here is whether worst-case control analysis is genuinely appropriate for engineering systems. In engineering structures, for example, it is difficult to accurately predict structural damping from a priori information. Factors like mass and stiffness can change in the normal life of the structure. Given multiple uncertain factors, the worst case can be quite bad. Thus worst case control design can require a substantial compromise in performance.

There are two things that stand out about the worst case. First, it is hard to define. Is the worst case an event that would never occur, or one that would occur once a week, once a year, or once in a century? Is the worst case the result of unknown but constant parameters, of sudden unpredictable change, or is it the result of a slow drift that routine control maintenance can remedy? Hard bounds on uncertain parameters are difficult to establish without introducing some sort of probabilistic measure. Is the hard bound absolute, or simply a range for which encompasses a substantial fraction of outcomes? In most applications, the latter case is the proper interpretation.

The second thing is that, even if the cause of the uncertainty is understood and reasonable bounds can be assigned, the likelihood is quite remote that several independent factors will take extreme values simultaneously. For example, suppose that there are three independent uncertain factors for which the 99% one-sided confidence interval is known, and also suppose that the worst case for control analysis is at a vertex of 3 dimensional cube enclosing this region in parameter space. Worst-case control design would require performance sacrifices to accommodate a situation that would occur on the order one time in 10^6. The typical response to this criticism is that it is necessary to use some judgment in identifying the worst case. But this is really a plea to reduce systematic procedures for robust design to the empirical tuning of the various weighting factors. While in skilled hands this can certainly yield good designs, it avoids the central question of developing a quantitative probabilistic description of robust control design.

We have attempted to employ methods for estimating structural reliability to the analysis of structural control systems (Kantor [1], Spencer et al. [2]). Given a model with uncertain real parameters, the methodological tasks were to develop stability criteria, assign probability distributions to the uncertain parameters, then utilize a procedure for the effective numerical estimation of the probability of instability. This was interpreted as a measure of system reliability.

Significant computational issues occur in these sort of reliability computations that have direct importance for control analysis. For example, the stability boundaries are not convex when expressed in parameter space, so the computation of a worst-case can be quite difficult. The unlikelihood of failure means that Monte Carlo style calculations can be quite expensive. In appropriate coordinates, the most likely failure is a point on the stability boundary nearest the nominal case. The probability for failure can be computed to good accuracy by approximating the instability criterion in the neighborhood of that point; then analytically calculating the probability content for the region of instability. First and second order reliability methods (FORM/SORM) provide accurate and efficient computational procedures.

For this application, it turns out that the probabilistic reliability computation can be quite efficient even though the stability criteria are quite complex. As opposed to Monte Carlo simulation, the computational effort required for FORM/SORM is independent of the probability of instability. Moreover, the methods are most accurate for small probabilities which is the case of most interest.

Thus our suggestion is that probabilistic methods for reliability analysis might be usefully employed for control systems. When the sources of model error are understood, we expect that this approach will yield more realistic tradeoffs between control performance, stability, and uncertainty.

References

1. Kantor, J. C., and B. F. Spencer, Jr.. *"On Real Parameter Stability Margins and Their Computation."* Report 90-1, Department of Chemical Engineering, University of Notre Dame, 1990. To appear in *International Journal of Control.*
2. Spencer, B. F., M. K. Sain, J. C. Kantor, and C. Montemagno, "Probabilistic Stability Measures for Controlled Structures Subject to Real Parameter Uncertainties." submitted to *Smart Materials and Structures,* 1992.

Modeling Uncertainty in Control Systems: A Process Control Perspective

Daniel E. Rivera

Department of Chemical, Bio and Materials Engineering and Control Systems
Engineering Laboratory, Computer-Integrated Manufacturing Systems
Research Center, Arizona State University,Tempe, Arizona 85287-6006

The upswing in robust control research witnessed during the past decade
has also included members of the academic process control community. Robust process control, however, stands at a crossroad, its ability to impact
the industrial engineering community imperiled by a number of issues, one
of which is uncertainty modeling. The topic of this workshop therefore has
a profound effect on the future impact of modern robust control theory on
process control practice.

This situation is not new to chemical process control. The rapid rise of
optimal control theory during the 60's was followed by a period of reflection
and skepticism in the early 70's; the critique articles by Foss [1] and Lee and
Weekman [2] are written evidence of the mood of the times. It is interesting
to note that many of the criticisms made by these authors remain true today,
almost two decades later. In current industrial practice, optimal control theory is manifested in an extremely limited form through the concept of model
predictive control [3]. It can be safely said that the general theory of LQG
regulation and estimation remain a mystery to the large majority of chemical
engineers engaged in control tasks in the process industries. If current trends
continue, the situation will certainly be the same for robust control theory.

A promising direction for robust process control lies in the field of *intelligent control*, which is qualitatively understood as the ability of a control
system to learn about its environment and process the information to reduce
uncertainty, plan, and execute control action in a safe and reliable manner
[4]. It is becoming increasingly clear that any such system will involve a combination of analytic, heuristic, and symbolic elements; methodologies that
incorporate all three elements in an integrated, coherent fashion are most
likely to succeed in this endeavor. One can speculate that robust control theory and uncertainty modeling will only find life in the process industries if
they are successfully integrated into an intelligent control system.

In recent years, there has been increasing activity towards translating
bias and variance measures in identification into uncertainty descriptions for
robust control design. Ljung [4] has performed seminal work which clearly indicates the role of design variables such as the input signal power spectrum,
the choice of model structure, and the choice of prefilter on the power spec-

trum of the prediction error in the parameter estimation problem. Recent publications such as the work of Wahlberg [6], Goodwin et al. [7], Helmicki et al. [8] and Bayard [9] show a wide diversity of approaches to the computation of error bounds, which include both stochastic and deterministic methodologies.

From a research standpoint, there is a need to examine more carefully this increasing number of uncertainty estimation techniques and seeking ways to fully integrate them into the overall control design procedure. Particularly, it is of interest to further understand the role of design variables (e.g., choice of lag windows, truncation parameters, role of prefilter and input signal power) to determine which of these proposed methods are most useful and under what conditions are they least conservative.

A simple example illustrates the need to further understand how judicious choices of design variables are necessary for the practical use of uncertainty estimation in control design. We consider a set of refinery data generated from a distillation column belonging to a Shell Oil refinery. The data (in terms of deviation variables) is shown in Fig. 1. This data shows the effect of manipulating reflux flow on the overhead temperature of the column. 197 observations were collected after a pseudo-random binary sequence was applied to the reflux flow controller. We consider the estimation of a 30-lag Finite Impulse Response (FIR) model using both prefiltered and nonprefiltered data. In the prefiltered case, a low-pass prefilter as developed in [10] is used. Spectral analysis (Tukey lag window with a truncation parameter of 20) is used to estimate the bias contribution, while for the variance term a stochastic approach based on the work of [11] is used; this is contrasted with a deterministic bound as described by [4].

The results are shown in Fig. 2. The effect of the low pass filter is to reduce the uncertainty in the low frequencies, while resulting in a corresponding increase in the intermediate and high frequencies. Clearly, prefiltering represents an influential design variable that can have much impact regarding the usefulness of a model for control. A disturbing aspect of the computed norm bounds for both the prefiltered and nonprefiltered cases, however, is the significant contribution of the variance term to the overall uncertainty in spite of the fact that the time series displays a high signal-to-noise ratio. The deterministic variance estimate is especially conservative, and based on this bound one would conclude that robust control of this plant is not possible. The stochastic variance estimate is lower, but still significant. Guidelines for better choices of lag window and truncation parameter in spectral analysis are clearly needed if uncertainty estimation is going to be of value to process control engineers.

Beyond the previous example (which addresses the usage of existing techniques), it is clear that numerous issues remain untouched by current research activity which must be addressed if uncertainty estimation is to have widespread applicability for robust process control problems. We motivate some of these issues by examining a real-life process system, such as the

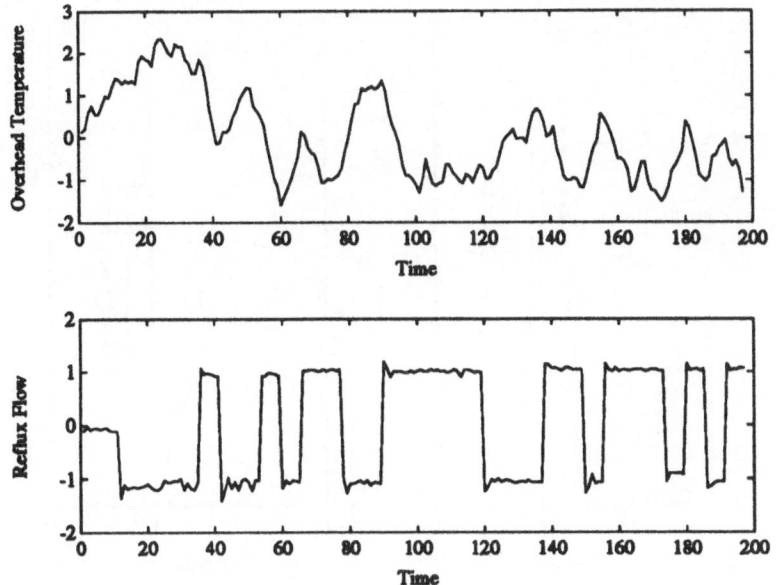

Fig. 1. Refinery column identification data. Top: overhead temperature; bottom: reflux flow

26-tray methanol-isopropanol column which forms part of ASU's Unit Operations Laboratory (Fig. 3). The methanol-isopropanol column represents a versatile benchmark problem that addresses many of the issues of importance to control in the process industries. The main objective of the column is to generate high-purity methanol in the overhead stream of the column, while minimizing the amount of methanol that leaves in the bottoms stream. Reflux flow and reboiler steam flow are available as manipulated variables. Methanol composition in the overhead stream is determined via an on-line differential refractometer; thermocouples located on every other tray yield temperature readings which represent useful secondary measurements for control. The column is instrumented such that a variety of control structures beyond simple feedback (i.e., feedforward, cascade, inferential) can be examined. Among the distinguishing features of this system are

Slow dynamics. The step response between reflux flow and methanol concentration in the distillate is overdamped with a 30 minute time constant. The slow dynamics place a significant constraint on any uncertainty estimation procedure. Any methodology that calls for repeated experimentation is clearly impractical. On the other hand, expressions that are derived for finite sets (instead of asymptotic conditions) are highly desirable.

Nonlinearity. The plant is a multivariable, nonlinear system, with the severity of nonlinearity increasing as the demands on purity increase. Further-

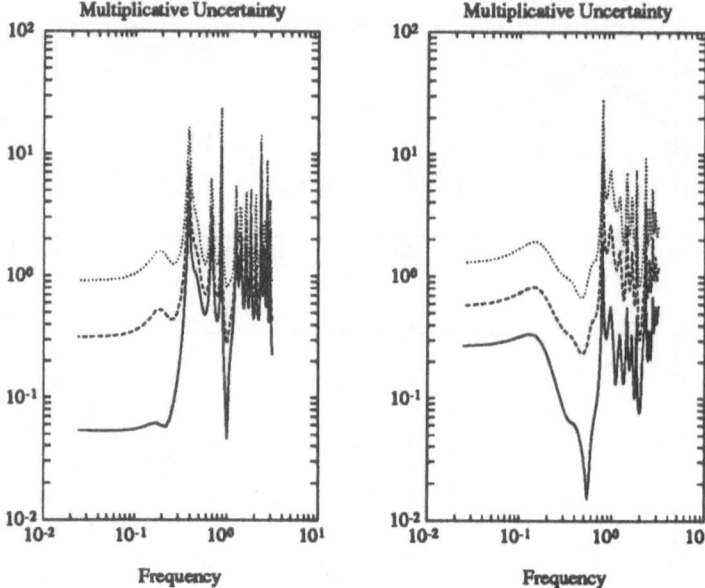

Fig. 2. Multiplicative uncertainty norm bound $\bar{l}_m$. Left diagram: prefiltered results; right diagram: no prefiltering. Solid line: bias bound; dashed line: bias + variance bound (Jenkins-Watts); dotted line: bias + variance bound (Ljung)

more, at high purities the plant is ill-conditioned, which creates a high degree of sensitivity to actuator uncertainty in the closed-loop system. Even if linear models are estimated from plant data, the residuals will reflect the plant's nonlinear character, which must accounted for in some fashion by the uncertainty estimate.

Nonideal disturbances. Disturbances entering the system are rarely adequately described by the classical assumption of independent, white noise signals passing through linear filters. Instead, a variety of stochastic and deterministic disturbances, potentially correlated with each other, affect this system. Among these disturbances are sensor noise (stochastic), ambient temperature changes (combined deterministic-stochastic), and feed flowrate changes (piecewise deterministic). Many significant disturbances are measurable, however, which provides some advantages.

In addition to the issues mentioned previously, a number of plant-independent issues also influence the usefulness of uncertainty estimation in the process industries. Among these are:

Operating restrictions. Plant testing is usually carried out while the plant is in normal operation. Hence any external signal that is introduced to the plant must not create significant deviations between the controlled variables and their setpoints. Uncertainty estimation from data generated in the closed-loop is a problem of interest.

User skill requirements. The set of intended users of identification technology is composed mostly of B.S.-level chemical engineers with limited training in advanced control topics. Techniques that take advantage of process understanding (the engineer's strength) without requiring a graduate-level education are desirable.

Control requirements. The nature and type of control systems implemented in process plants strongly influences the requirements for uncertainty estimation. Process control engineers are usually satisfied with overdamped closed-loop responses with time constants similar to the open-loop speed-of-response of the plant. PID and model predictive controllers are the control algorithms favored by process control engineers.

In conclusion, the impact of robust control theory on the process control community is currently hampered by the limiting assumptions of existing techniques. It is hoped that future research activity in uncertainty modeling will address the issues presented in this essay which are of vast importance to process control.

References

1. Foss, A.S. "Critique of Chemical Process Control Theory," *AIChE Journal*, 19, 209, 1973.
2. Lee, W. and V.W. Weekman. "Advanced Control Practice in the Chemical Industry: A View from Industry," *AIChE Journal*, 22, 27, 1976.
3. García, C.E., D.M. Prett, and M. Morari. "Model Predictive Control: Theory and Practice - a Survey," *Automatica*, 25, 335, 1989.
4. "NSF/EPRI Workshop on Intelligent Control Systems," Final Report, Edited by R. Shoureshi and D. Wormley, October, 1990.
5. Ljung, L. System Identification: Theory for the User, Prentice-Hall, New Jersey, 1987.
6. Wahlberg, B. "System Identification Using Laguerre Models," *IEEE Trans. Autom. Cntrl.*, 36, 551, 1991.
7. Goodwin, G.C., B. Ninness, and M.E. Salgado, "Quantification of Uncertainty in Estimation," *Proc. 1990 American Control Conference*, San Diego.
8. Helmicki, A.J., C.A. Jacobson, and C.N. Nett, "Control-Oriented System Identification: A Worst-Case/Deterministic Approach in H_∞," *IEEE Trans. Autom. Cntrl.*, 36, 1163, 1991.
9. Bayard, D.S., "Statistical plant set estimation using Schroeder-phased multisinusoidal input design," *1992 American Control Conference*, Chicago, pgs. 2988-2995, also *J. Applied Mathematics and Computation*, forthcoming.
10. Rivera, D.E., J.F. Pollard, and C.E. García, "Control-Relevant Prefiltering: A Systematic Design Approach and Case Study," *IEEE Trans. Autom. Cntrl.*, Special Issue on System Identification for Robust Control Design, 37, 964, 1992.

11. Jenkins, G.M. and D.G. Watts. Spectral Analysis and its Applications, Holden-Day, San Francisco, 1969.

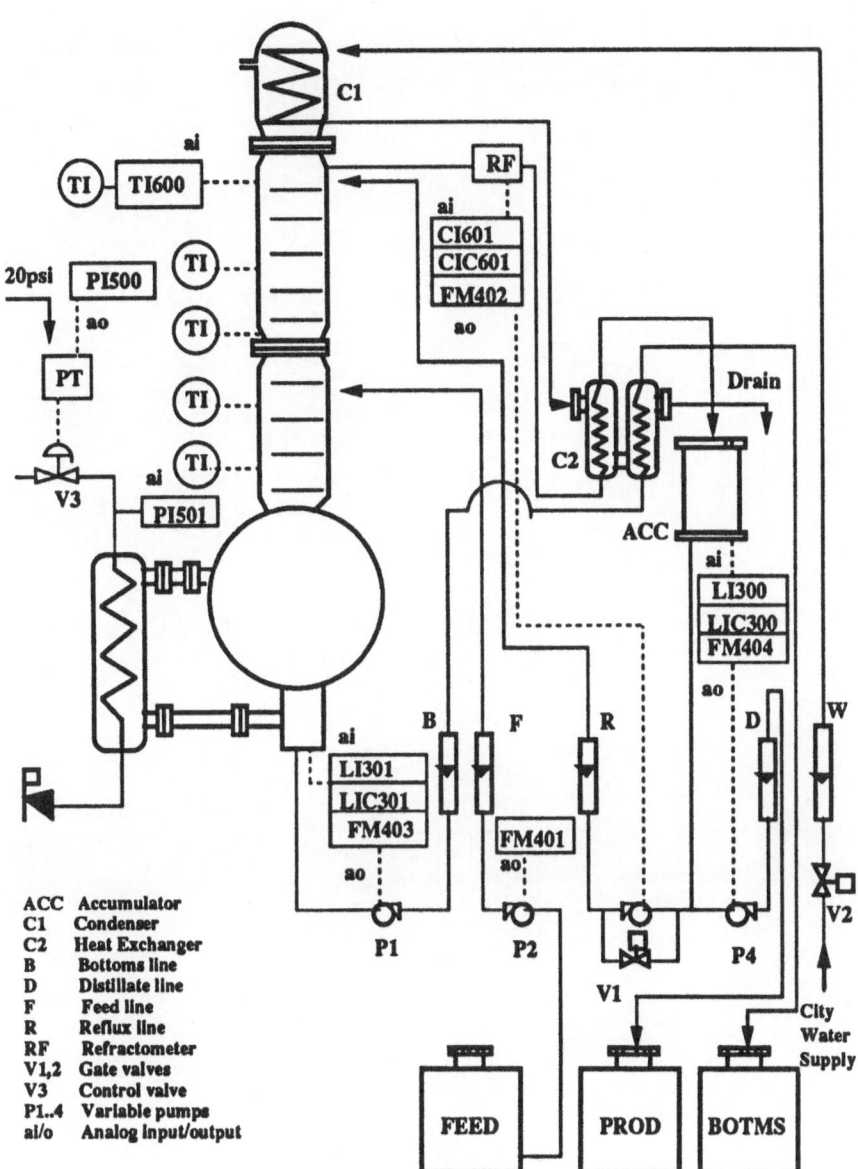

Fig. 3. Binary distillation column schematic

Part II

Technical Papers

A Note on H_∞ System Identification with Probabilistic A Priori Information*

Clas A. Jacobson and Gilead Tadmor

Department of Electrical and Computer Engineering,
Northeastern University, Boston, MA 02115

1 Introduction

This paper is concerned with control oriented system identification and continues a line of work begun in [4, 4, 11]. The purpose of control oriented system identification is to produce models that are compatible with robust control synthesis which for the purposes of this paper means simply that both a nominal model as well as a quantification of model uncertainty must be produced by an algorithm. The quantification of model uncertainty must of course be done in a metric that is useful for controller synthesis and in particular for robust stabilization. The general problem formulation thus consists of four elements: (1) a delineation of the class of systems under consideration as models, (2) a delineation of a priori information that is given concerning the system to be identified, (3) a delineation of experimental information that is assumed available and (4) a delineation of the error measure that is used for expressing model fidelity and for quantifying model uncertainty. The particular problem formulation in this paper adopts the setting in [6] with the exception of the specification of a priori information. The purpose of this paper is to examine the case where the given a priori information is specified in a probabilistic manner.

The problem formulation in [6] presents the specific where the underlying model set consists of linear shift invariant exponentially stable systems, the experimental information consists of corrupted point frequency estimates and the quantification of error is in the H_∞ norm. The a priori information that is assumed available in [6] regarding the unknown system to be identified consists of a lower bound on the decay rate and an upper bound on a worst case amplification of the system. This information is brought to bear as it guarantees smoothness of the transfer function, which is obviously a necessary requirement for interpolation based identification. The purpose of the

---
* This work was supported in part by ARO, AFOSR and NSF. This work is dated September 1992.

current paper is to examine a similar problem formulation when the a priori
information on smoothness is given in a probabilistic manner.

The probabilistic description of a priori information for H_∞ system identification can be understood by detailing the determination of such prior
information and the proper usage of such information for robust control design. A scenario concerning measurement of prior information is given that
renders the determination of a probability distribution a statistical question
concerning an ensemble of possible systems. This distribution of systems will
consequently induce a conditional distribution of those systems that agree
with a set of measurements. The issue of quantifying the error above is then
clear given this conditional probability: the identification error is given over
the set of possible systems up to a prespecified tolerance. In short, precise
meaning is given to statements as "the H_∞ norm of the error is at most α
with probability β" where α and β are the prespecified tolerances conditioned
by the plant measurements. This type of specification of identification error
is still naturally useful for robust control synthesis, hence, this probabilistic specification of a priori information represents a contribution to control
oriented system identification.

The probabilistic problem formulation given in this paper for the H_∞
system identification problem has been motivated in part by the abstract
probabilistic setting given in the works of information based complexity (IBC)
[15, 16]. The interested reader is referred to these works for an alternate
development of the probabilistic setting than that given in the current paper,
in particular, for a treatment of a different class of statistical information.
One significant difference between the IBC treatment and this paper in the
development of the probabilistic setting is the determination of the σ-algebra
on which the probability measure giving a priori information is defined. The
σ-algebra in IBC is the Borel σ-algebra, as determined by the metric topology
whereas in the current paper the σ-algebra is considerably coarser and is
determined by simple measurements that may be made on a representative
sampling of the systems under consideration.

This paper is organized as follows. Section 2 presents the problem formulation in detail and specifically presents a discussion of probabilistic a priori
information. Section 3 presents a solution to the identification problem considered in this paper. A construction in which a probabilistic problem may
be considered as an allied worst case problem is given and this construction
allows worst case analysis and algorithm design to be utilized. Conclusions
are presented in Sect. 5.

2 Problem Formulation

The purpose of this section is to detail the components of the control oriented
system identification problem considered in this paper. This includes the
delineation of the class of systems and experimental information in Sect. 2.1

and a discussion of a priori information and error measures in Sects. 2.2 and 2.3.

2.1 Preliminaries

Notation $\mathbf{R}, \mathbf{C}, \mathbf{Z}$ denote the real, complex numbers, and integers respectively. $\mathbf{D}, \mathbf{T}$ denote the open unit disk (centered at the origin) in the complex plane and the boundary or unit circle respectively. $\mathbf{D}_\rho$ denotes the disk centered at the origin of radius ρ. H_∞ denotes the space of bounded analytic functions on the unit disk and $H_{\infty,\rho}$ denotes the space of bounded analytic functions on the disk of radius ρ. ℓ_∞ denotes the space of bounded sequences on $\mathbf{Z}_+$. T_n denotes the truncation operator $T_n : \ell_\infty \to C^n$. $BX(r)$ denotes the ball of radius r in the normed space X (centered at the origin).

The class of systems considered in this note are those elements of H_∞ that admit analytic continuations to a disk $\mathbf{D}_\rho$, $\rho > 1$. This corresponds precisely to the class of exponentially stable linear shift invariant systems [3]. This set will be denoted by

$$H_+(\mathbf{D}) := \bigcup_{\rho>1} H_{\infty,\rho} \ .$$

The assumed experimental information consists of a finite set of corrupted point frequency estimates of the system,

$$E(h, \eta) = \left\{ h(e^{i\omega_k}) + \eta_k \right\}_{k=1}^n$$

where $h \in H_+, \eta \in T_n B\ell_\infty(\epsilon)$. For simplicity in what follows $E(h, \eta)$ will be denoted by d (for datum).

2.2 H_∞ Error Quantification and A Priori Information: A Review

The concept of worst case identification error detailed in [4, 6, 11] is useful to review as it captures clearly the basic elements of the identification problem, moreover, even though this type of error criterion will not be directly used in the problem formulation of this paper the worst case problem setting will be important as will be seen. The reason for this is that it will be shown that a probabilistic setting can be interpreted as a worst case problem through an explicit construction.

The worst case identification problem error is quantified as follows. Let $F \subseteq H_+$ be the set of problem elements satisfying the given a priori information conditions. Let $A : C^n \to H_+$ be an algorithm operating on the specified data[2] and producing a model. The error of this algorithm is defined as

$$e_{\text{local}}^{\text{wor}}(A, d) = \sup \left\{ \|f - A(d)\|_\infty \ : \ f \in F, \eta \in T_n B\ell_\infty(\epsilon), E(f, \eta) = d \right\}. \tag{1}$$

---

[2] Actually, as treated in [5, 6] the algorithm could be a map from the data and the a priori information, corresponding respectively to untuned and tuned algorithms respectively, but here the dependence on the a priori information if present, is being suppressed. The imprecision is not important for this discussion.

The local designation is given as this error is measured for one piece of observed data, of natural interest is also the global error

$$e_{global}^{wor}(A) = \sup e_{local}^{wor}(A, d)$$

where the supremum is taken over all observed data elements.

The error of identification can also be presented as follows. Let $E^{-1}(d) \subseteq H_+$ denote the set

$$E^{-1}(d) = \{f \in F \ : \ E(f, \eta) = d \text{ for some } \eta \in T_n Bl_\infty(\epsilon)\} \ .$$

This inverse image of the data is the set of all systems satisfying the a priori information ($f \subseteq F$) that agree with the observed data ($E(f, \eta) = d$). This set is not a singleton due to the partial and corrupt nature of the a priori and experimental information. The use of the set $E^{-1}(d)$ is as follows. The error criterion (1) is naturally termed a worst case error as it measures the worst excursion, quantified by the H_∞ norm, over the set of possible systems that could have generated the observed data. The natural questions that arise are the determination of lower bounds for this error as well as optimal algorithms, namely, algorithms that achieve the lower bound. The focus in this review will be on one concept from [15] which is useful in motivating the construction of lower bounds and algorithms below, namely, the radius of information. The radius of information is a measure of the "size" of the set $E^{-1}(d)$, particularly the smallest covering ball in H_∞ defined as follows.

Definition 2.1 *The deterministic local radius of information of a data record $d \in C^n$ is*

$$rad_{local}^{wor}(d) = \inf_{f \in H_+} \sup_{x \in E^{-1}(d)} \|f - x\|_\infty \ .$$

The deterministic global radius of information is

$$rad_{global}^{wor} = \sup_{\substack{f \in F \\ \eta \in T_n Bl_\infty(\epsilon) \\ d = E(f, \eta)}} rad_{local}^{wor}(d)$$

where $F \subseteq H_+$ is the global set of problem elements considered.

The importance of this concept is that the radius of information is a tight lower bound for the identification error (1) [15, p. 43]. The radius of information is algorithm independent and focuses on the set of uncertainty E^{-1} rather on the algorithm description. This focus is key to understanding the role that partiality and corruptedness play in system identification.

The difficulty of identification expressed by the radius of information is critically dependent on the type of a priori and experimental information. The dependence should be strong enough so that the identification error converges to zero as the experimental information increases and the corruption decreases, precisely, as the cardinality of the data record $n \to \infty$ and the corruption of the experimental data $\epsilon \to 0$ the radius should converge to zero. A fundamental issue is therefore the determination of kind of information that guarantees this convergence property. This may be elaborated on as follows.

The boundedness of the radius of information clearly requires that the modulus of continuity of elements of the set $E^{-1}(d)$ be uniformly bounded. This means that the quantity

$$\omega(f, \Delta) = \sup_{|x-y| \leq \Delta} |f(x) - f(y)| \tag{2}$$

is bounded uniformly for $f \in E^{-1}(d)$ where Δ is the distance on sample points on the unit circle. Although it is not explicitly obvious the utility of the a priori information assumed in [6, 4, 11, 12] is in providing a bound on $\omega(f, \Delta)$. This paper proceeds from a more direct assumption on the a priori information that is used to bound $\omega(f, \Delta)$ than previous works. The specification that is utilized in this paper is a uniform bound, say M, on the derivative of problem elements which by the mean value theorem implies

$$\omega(f, \Delta) \leq M\Delta$$

so that convergence of the radius of information as discussed above will result. It is interesting to note that what is being utilized is the Lipschitz continuity of the class of systems under consideration so a direct use could also be a quantification of the Lipschitz constants of the problem elements as a priori information.

A slightly different viewpoint is that the a priori information defines a set of problem elements and the experimental information fixes the structure of an ϵ-net [15, 17] on this set of problem elements. If the number of samples increases, $n \to \infty$, without corruption then the set of problem elements admits an arbitrarily fine covering by a finite number of balls of radius ϵ, or, is compact. The compactness conditions though, while equivalent, partially hide the fact that a bound on the variation of the functions between sample points is the necessary a priori information.

2.3 Probabilistic H_∞ A Priori Information and Error Quantification

This section presents the type of a priori information that is utilized in the system identification algorithms given in this paper and an error quantification that is consistent with this type of a priori information. Roughly speaking, the type of a priori information that is assumed in this paper specifies the statistical distribution of a class of systems and identification (and in particular, experimentation) is performed on one member out of this class . The H_∞ norm of the identification error is to be quantified up to a given probability of assurance. The technical details to complete this program rest on the specification of the required probability distribution on the class of models under consideration and the interplay of this measure with the experimental operator that yields the experimental information. The problem formulation adopted in the current paper is motivated by the considerations of measuring a prior distribution (discussed in Sect. 4 below) and this consideration requires that the a priori information in this paper is a probability measure defined on a coarser σ-algebra (see [9] for a discussion of measure theory appropriate for this discussion) than the Borel subsets of H_∞.

The a priori information that is assumed available in this paper consists of a probability measure on $\mathbf{R}_+$ denoted by μ. This measure is used to assign probabilities to events in the σ-algebra $\mathcal{A}$ generated by sets of the form $\{h \in H_+ : \alpha \leq \|wh'\|_\infty \leq \beta\}$ where w is a fixed weight function and h' denotes the derivative of h. The probability of an event such as this is denoted by $P(\cdot)$ and is determined by

$$P\{h \in H_+ : \alpha \leq \|wh'\|_\infty \leq \beta\} = \mu([\alpha, \beta]) .$$

The purpose of this section is to discuss the definition of a consistent error criterion for identification utilizing this type of a priori information.

The concept of quantifying the error in a manner consistent with this type of a priori information is quite simple. The idea is to define the uncertainty set $E^{-1}(d)$ as in Sect. 2.2 above but with $F = H_+$, and, up to a given probability that allows subsets of $E^{-1}(d)$ to be deleted, measure the radius of the uncertainty set. An apparent technical difficulty in undertaking this straightforward task is that the probability $P(\cdot)$ is defined on a different σ-algebra than the Borel subsets of H_+, hence, $E^{-1}(d)$ is not in general a measurable set with respect to the σ-algebra on which $P(\cdot)$ is defined. This situation is not a mathematical pathology but arises from the natural distinction that an event (more precisely, measurable subset of the probability giving the a priori information) corresponds to a physical or a distinguishable experiment. The consequence is that in certain circumstances it is not possible or meaningful to define a measure on the same σ-algebra, namely the Borel subsets, that defines the metric topology, in this case H_∞, but rather a considerably coarser σ-algebra.

The above discussion gives a heuristic overview. The details of the error criterion are now given. Let $E^{-1}(d)$ denote the inverse image of the observed data, as above (with H_+ substituting for F)

$$E^{-1}(d) = \{h \in H_+ : E(h, \eta) = d \text{ for some } \eta \in T_n B\ell_\infty(\epsilon)\} . \qquad (3)$$

Let $E^{*-1}(d)$ denote the smallest $\mathcal{A}$-measurable extension of $E^{-1}(d)$ as

$$E^{*-1}(d) = \bigcap_{\substack{S \in \mathcal{A} \\ E^{-1}(d) \subseteq S}} S . \qquad (4)$$

In keeping with this notation let Ω^* denote the smallest $\mathcal{A}$-measurable extension of a set Ω. The computation of this set will be dealt with in Sect. 3, for the moment the point is the well defined existence of this extension. Now the outcome of identification is a model of a particular system and also the experimentation is to be done on a particular system. However, the statistical a priori information is given on a class or ensemble of systems. The consequence is that it is of interest to measure the conditional probability of subsets $\Omega \subseteq E^{*-1}(d)$. The conditional probability is denoted by P_d (conditioned by d for datum) and is defined by

$$P_d(\Omega) = P(\Omega | E^{*-1}(d)) \tag{5}$$

$$= \frac{P(\Omega \cap E^{*-1}(d))}{P(E^{*-1}(d))} \tag{6}$$

where $\Omega \in \mathcal{A}$ and assuming that $P(E^{*-1}(d)) > 0$.

A formal problem statement can now be given for the remainder of the paper.

Definition 2.2 *Let $0 \leq \delta \leq 1$ be specified. The probabilistic error of an algorithm $A : \mathbf{C}^n \to H_+$ is then defined as*

$$e_{local}^{prob}(A, \delta, d) = \inf \left\{ \alpha \ : \ P_d \left\{ h \in E^{-1}(d) \ : \ \|A(d) - h\|_\infty > \alpha \right\}^* < \delta \right\} \ .$$

The first problem to be considered now is to quantify, explicitly as a function of the probabilistic tolerance δ and the data cardinality n, the error $e_{local}^{prob}(A, \delta, d)$ for a given algorithm A. The second problem is to determine the lower bound of $e_{local}^{prob}(A, \delta, d)$ over all algorithms A and, using this lower bound, produce an optimal algorithm where optimal refers to the fact that no algorithm producing a lower error exists. These goals will be dealt with in the following sections. Section 3.1 presents an analysis of the problem formulation and constructs a worst case problem with error equivalent to the probabilistic error criterion just given. Section 3.2 presents an (almost) optimal error algorithm for the probabilistic problem formulation just presented by dealing with the allied worst case problem constructed from the probabilistic one.

3 Problem Solution

This section presents a solution of the probabilistic identification problem introduced in Sect. 2. The probabilistic error criterion is analyzed in Sect. 3.1 from the perspective of the probabilistic radius of information. A construction is presented that determines an allied worst case problem description similar to that in [6] with a worst case error equal to the probabilistic error. This construction allows an efficient near optimal algorithm to be given in Sect. 3.2.

3.1 Analysis of Probabilistic Identification Error

This section shows that the probabilistic error may be converted, under the problem formulation considered in Sect. 2, to a worst case setting with a "hard" bound on the derivative of the problem elements. This conversion allows known techniques for the worst case problem to be applied and, perhaps more importantly, allows the structure of the probabilistic problem to be more clearly seen.

The point of view that is most useful in uncovering the structure of the probabilistic setting in this paper is that of the radius of information. In analogy to the worst case radius presented in Sect. 2.2 the probabilistic radius of information can be given as

$$\text{rad}^{\text{prob}}_{\text{local}}(\delta, d) = \inf_{\substack{f \in H_+ \\ \Omega \subseteq E^{*-1}(d) \ : \ P_d(\Omega) \leq \delta}} \sup_{x \in E^{-1}(d) \setminus \Omega} \|f - x\|_\infty \quad (7)$$

where the set $E^{-1}(d)$ and $E^{*-1}(d)$ are as defined in (3) and (4).

In words this concept quantifies the "size" of the uncertainty set $E^{-1}(d)$ in an analogous fashion to the worst case setting only here a set Ω is deleted from E^{-1} in order to reduce the H_∞ norm in a maximal fashion. The restriction on the set Ω is that the conditional probability must be bounded by the preassigned probability or tolerance δ.

The main point of this section is the determination of the best set Ω given the problem formulation presented in Sect. 2. It turns out that the concept is very simple — delete the elements with large modulus of continuity up to the specified tolerance δ and consistent with the given probability distribution. This is done as follows.

Define $\alpha_{\min}$ to be the smallest bound on the weighted derivative consistent with the observed data d,

$$\alpha_{\min}(d) = \inf \left\{ \|wh'\|_\infty \ : \ h \in E^{-1}(d) \right\} \quad (8)$$

where $E^{-1}(d)$ is defined as in (3). It can be easily observed that $E^{*-1}(d) = \{h \in H_+ \ : \ \|wh'\|_\infty \geq \alpha_{\min}\}$. To compute the conditional probability of subsets of $E^{*-1}(d)$ assume that $P(E^{*-1}(d)) = \mu([\alpha_{\min}, \infty)) > 0$ (else d will be an inconsistent measurement!) and let $\Omega \subseteq E^{*-1}(d)$ be $\mathcal{A}$-measurable. Then

$$P_d(\Omega) = \frac{P(\Omega)}{\mu([\alpha_{\min}, \infty))}$$

or, the conditional probability for this problem formulation is the normalized probability of the event Ω. Now a moments thought will convince one of the fact that if $\Omega = \{h \in H_+ \ : \ \gamma \leq \|wh'\|_\infty \leq \beta\}$ then the norm closure of Ω depends only on β, hence the "best" set to delete in the probabilistic radius of information is given as follows.

Fact 1 *Let α_{min} be given as in (8) and assume that $\mu([\alpha_{min}, \infty)) > 0$. Define*

$$\alpha_{max}(\delta, d) = \inf\{\alpha \; : \; \mu([\alpha, \infty))/\mu([\alpha_{min}, \infty)) \leq \delta\}$$
$$= \inf\{\alpha \; : \; P_d\{h \in H_+ \; : \; \|wh'\|_\infty > \alpha\} \leq \delta\}$$

and

$$\Omega(\delta, d) = \{h \in H_+ \; : \; \alpha_{max} \leq \|wh'\|_\infty\} \; .$$

Under these conditions the probabilistic radius of information (7) is given by deleting the set $\Omega(\delta, d)$ from $E^{-1}(d)$.

In other words the probabilistic radius of information is actually a worst case radius of information as given in Sect. 2.2 because the set Ω may be completely specified and computed. Indeed, as remarked above the H_∞ norm closure of the set Ω depends only on the upper bound of the derivative the lower bound is irrelevant and the determination of the radius of information depends therefore only on the "hard" bound $\alpha_{\max}$.

Remark The assumption that $P(E^{*-1}(d)) > 0$ is a consistency condition. Given the probability distribution of the class of problem elements the condition that the probability of the system agreeing with the data is zero expresses a data invalidation condition. This condition has been studied in other similar situations [2, 13] and the complete details of the appearance here of the data validation problem will be dealt with in future work.

Remark It is interesting to realize that the earlier works, particularly [6] are completely consistent with the above construction when $\delta = 0$. Specifically, in [6] a priori conditions are assumed that give a hard bound on the norm of the derivative $\|h'\|_\infty$ (so called (M, ρ) constants, see [6] for details). Now the assumption in [6] and related worst case papers is that the data is actually generated by the plant and is consistent. Hence, by interpreting a worst case bound on the derivative as a probability distribution with support $[0, \|h'\|_\infty]$ the probabilistic radius with 100 % certainty is exactly the situation studied in [6] (actually any probability distribution with support on $[0, K]$ will give the same conclusion with K the "hard" bound on the derivative, $\|h'\|_\infty \leq K$). The probabilistic setting addressed in this paper is thus a natural generalization or relaxation of the worst case setting.

3.2 Algorithm Description

The purpose of this section is to utilize the structural result presented in the preceding section to develop optimal algorithms for the problem formulation presented in Sect. 2. The reduction of the probabilistic setting to a worst case setting allows the well studied case of worst case algorithms to be brought to bear on this subject.

The simplest type of construction is that of an interpolatory algorithm
[15, p. 51]. An interpolatory algorithm is a mapping from the data to an
element from the set of possible systems that agrees with the given data,
namely, $E^{-1}(d)$ and hence is said to be interpolatory. The purpose of this
section is to construct an interpolatory algorithm for the system identification
problem posed in Sect. 2.

A precise definition in the context of the present discussion goes as follows.

Definition 3.1 *Let $\alpha_{max}(\delta, d)$ be as defined in Fact 1. An interpolatory algorithm A_{int} is a mapping from the information $d = E(h, \eta)$ to the set H_+*

$$A_{int} : C^n \to H_+$$

such that

1. $\left\| w(z) A'_{int}(d) \right\|_\infty \leq \alpha_{max}(\delta, d)$
2. $A_{int}(d) \in E^{-1}(d)$

It is a straightforward fact [15, p. 51] that interpolatory algorithms are
within a factor of two of optimal as measured by

$$\text{rad}_{local}^{prob}(\delta, d) \leq e_{local}^{prob}(A_{int}, \delta, d) \leq 2\text{rad}_{local}^{prob}(\delta, d) \ .$$

The problem of constructing an interpolatory algorithm may be carried out
by a convex program [7] which may be in turn solved by an ellipsoid algorithm
[1, 10]. Consider the following optimization problem in the $\{\beta_l\}$ coefficients
of the polynomial (FIR) identified model $m(z) = \sum_{l=0}^{N} \beta_l z^l$.

$$\min_{\{\beta_l\}} \max_k \left| \sum_{l=0}^{N} \beta_l e^{i\omega_k l} - d_k \right| \tag{9}$$

subject to

$$\left\| w(z) \frac{d}{dz} \sum_{l=0}^{N} \beta_l z^l \right\|_\infty \leq \alpha_{max} \ . \tag{10}$$

This is clearly a convex program. Let the solution of the convex program
be denoted by $\phi(\delta, d) = m(z) = \sum_{k=0}^{N} \alpha_k z^k$. The following fact gives the
performance of this algorithm for the case where the weight function is unity.

Theorem 2. *The algorithm $\phi(\delta, d)$ is an interpolatory algorithm with the corruption on the data given by $\tilde{\epsilon}$ below and hence within a factor of two of
optimal, moreover, an error bound for this algorithm is given as*

$$e_{local}^{prob}(\phi, \delta, d) \leq \tilde{\epsilon} \left(\frac{\bar{\Delta}\alpha_{max}}{2} + 1 \right)$$

where

$$\tilde{\epsilon} = \epsilon + 2\sqrt{\frac{3\pi^3\alpha_{max}^3}{\beta N}}$$

where

$$\beta = \begin{cases} 1 & \frac{1}{\alpha_{max}} \leq \pi \\ 1 - (1 - \alpha_{max}\pi)^3 & \frac{1}{\alpha_{max}} > \pi \end{cases}$$

$$\bar{\Delta} = \max_k |\omega_{k+1} - \omega_k| .$$

Proof. The performance of the algorithm as stated above is determined by the truncation error of using an order N polynomial as a basis. The error bound for the algorithm is derived as follows. Note that the $\|\cdot\|_2, \|\cdot\|_\infty$ norms on the unit circle are related as

$$\|f\|_2^2 \geq \frac{\|f\|_\infty^2 \beta}{3\pi\alpha_{max}} .$$

Now the claim is that a polynomial m of order N may be constructed so that

$$\|f - m\|_\infty \leq 2\sqrt{\frac{3\pi^3\alpha_{max}^3}{\beta N}} .$$

To see this note that by taking m to have coefficients equal to the first N Fourier coefficients of f

$$\|f - m\|_2^2 = \left\|\sum_{k=N+1}^{\infty} c_k z^k\right\|_2^2$$

$$= \sum_{k=N+1}^{\infty} |c_k|^2$$

$$\leq \sum_{k=N+1}^{\infty} \frac{\|f'\|_1^2}{k^2}$$

$$\leq \frac{(2\pi\|f'\|_\infty)^2}{N}$$

$$\leq \frac{(2\pi\alpha_{max})^2}{N} .$$

where the bound on the Fourier coefficients is standard [8, p. 24] and the result follows. The fact that the algorithm is interpolatory with $\tilde{\epsilon}$ the corruption on the data is clear from the preceding discussion on the truncation error of using a polynomial as a basis for the identified model.   □

This error bound then shows that the algorithm is actually asymptotically optimal in terms of the global radius, not to within a factor of two, but will as $\epsilon \to 0, n \to \infty$ achieve the worst case lower bound of the global radius of information.

4 Determination of A Priori Distributions

This section presents an idealized experiment that yields the probability distributions utilized as a priori information in Sect. 2. The discussion is meant as an indication of the plausibility of the intended methodologies and is deliberately oversimplified: in particular, noise is not considered here. The idealized experiment consists of testing an ensemble of similar systems to gather information on the derivatives of the family. This family could arise for example as the collection of test samples from a manufacturing process. This family could also arise from a slowly time varying system where the time variation is slow enough so that test inputs could be applied at one time scale and the variation of the systems in time is at another longer time scale. In either case a collection or ensemble of similar systems is available for testing purposes.

The experiment on the ensemble consists of measuring the frequency response over a grid of points $\{e^{i\omega_k}\}_{k=0}^{N-1}$ that are evenly spaced, say, e^{ω_k} are the Nth roots of unity. This experiment could be performed by applying discrete unit pulses to each system and, by taking the discrete Fourier transform, obtain the required point frequency samples.

Having obtained the estimates of the point frequency samples form the approximation of the Lipschitz constants on the grid, that is, for each system f_i in the ensemble form

$$|f_i(\omega_k) - f_i(\omega_{k+1})| \leq L_{k,i}|\omega_k - \omega_{k+1}|$$

obtaining a family of constants $\{L_{k,i}\}$.

Observe that this information is what is being utilized for the a priori information given by the derivative in the body of the paper above. The derivative information is utilized in the paper for the bound on the variation which arises from the application of the mean value theorem and the idealized experiment above measures the variation (on a discrete grid) for the ensemble of systems available for test. The remaining items to be covered are the weight function and the probability distribution itself.

The weight function can be seen in this experiment as normalizing the variation over frequency so that a probability distribution given on the norm of the functions in H_+ actually represents the variation on Lipschitz constants given locally. For simplicity this weight will be dropped and a cruder estimate will be derived as follows.

For each fixed $k = 0, 1, \ldots, N - 1$ a family of constants $\{L_{k,i}\}$ indexed by i is available. A histogram may be made of these constants to visualize the distribution, conceptually, this histogram is to be fitted by an appropriate probability density. An algorithm based on (penalized) maximum likelihood and convergence result as the number of samples increases can be found in the book of Thompson and Tapia [14].

5 Conclusions

This paper has presented results concerned with system identification in H_∞. The contribution of this paper has been to initiate a probabilistic description of a priori information to the H_∞ system identification problem. Results have been presented to show that under certain natural specifications of a priori information the probabilistic problem setting can be converted to an allied worst case problem formulation. An algorithm has been presented to solve the resulting worst case problem.

The groundwork that has been laid in this short note motivates a number of natural investigations to be completed that now rest on a solid logical foundation. These problems include the natural case of probabilistic descriptions of corrupting noise on the data. Another problem is the connection of the probabilistic ideas contained in this paper with the model falsification problems.

References

1. S.P. Boyd and C.H. Barratt. *Linear Controller Design*. Prentice-Hall, New York, 1991.
2. J. Chen, C.N. Nett, and M.K.H. Fan. Worst-case system identification in H_∞: Validation of a priori information, essentially optimal algorithms, and error bounds. *IEEE Transactions on Automatic Control*. submitted.
3. V. H. L. Cheng and C. A. Desoer. Discrete time convolution control systems. *Int. J. Control*, 35:367–407, 1982.
4. G.Gu and P.P. Khargonekar. Linear and nonlinear algorithms for identification in H_∞ with error bounds. *IEEE Trans. Automatic Control*, 37:953–963, 1992.
5. A. J. Helmicki, C. A. Jacobson, and C. N. Nett. Identification in H_∞: Linear algorithms. In *Proc. of the 1990 Amer. Contr. Conf.*, pages 2418–2423, San Diego, 1990.
6. A. J. Helmicki, C.A. Jacobson, and C.N. Nett. Control oriented system identification: A worst-case/deterministic approach in H_∞. *IEEE Transactions on Automatic Control*, 36(10), October 1991.
7. R.B. Holmes. *A Course on Optimization and Best Approximation*. Springer-Verlag, New York, 1972.
8. Y. Katznelson. *An Introduction to Harmonic Analysis*. Dover Press, 1976.
9. S. Lang. *Real Analysis*. Addison–Wesley, 1983.
10. A.S. Nemirovsky and D.B. Yudin. *Problem Complexity and Method Efficiency in Optimization*. Wiley-Interscience, New York, 1983.
11. J.R. Partington. Robust identification in H_∞. *J. Math. Anal. Appl.*, pages 428–441, 1992.
12. J.R. Partington. Worst-case identification in Banach spaces. *Systems & Control Letters*, pages 423–428, 1992.
13. R.S. Smith and J.C. Doyle. Model validation: A connection between robust control and identification. *IEEE Transactions on Automatic Control*, 37:942–952, 1992.

14. J.R. Thompson and R.A. Tapia. *Nonparametric Function Estimation, Modeling, and Simulation*. SIAM, Philadelpha, 1990.

15. J. F. Traub, G. W. Wasilkowski, and H. Woźniakowski. *Information-Based Complexity*. Academic Press, 1988.

16. H. Woźniakowski. Probabilistic setting of information-based complexity. *J. Complexity*, 2:255–269, 1986.

17. G Zames. On the metric complexity of causal linear systems: ϵ-Entropy and ϵ-dimension for continuous time. *IEEE Trans. on Auto. Contr.*, AC-24:222–230, 1970.

A Worst Case Identification Method Using Time Series Data

Jie Chen, Carl N. Nett, and Michael K.H. Fan

Schools of Aerospace and Electrical Engineering,
Georgia Institute of Technology, Atlanta, GA 30332-0250

Abstract:

In this paper we formulate and solve a control-oriented system identification problem for single-input, single-output, linear, shift-invariant, distributed parameter plants. In this problem the available a priori information is minimal, consisting only of worst-case/deterministic, time dependent, upper and lower bounds on the plant impulse response and the additive output noise. The available a posteriori information consists of a corrupt finite output time series obtained in response to a known, non-zero but otherwise arbitrary, applied input. A novel system identification method is presented for this problem. This method yields an "uncertain model" of the plant, which is comprised of a nominal plant model, a bounded additive output noise, and a bounded additive model uncertainty measured in both the l_1 and H_∞ system norms. Both the identification method and the resulting uncertain model are computationally simple and possess certain desirable properties.

1 Introduction

The primary goal of this paper is to develop an identification method for identifying, from available a priori and a posteriori information, an "uncertain plant model" which can be used for robust control design. The essential characteristic of this uncertain plant model is that it must "explain" exactly all possible a posteriori information obtainable by experimentation on the underlying physical plant. This includes not only a posteriori information observed prior to the identification of the uncertain plant model, but also a posteriori information which may be observed subsequent to the identification of the uncertain plant model. An uncertain plant model will typically involve a nominal design model, a noise[1] specification, and a model error specification, and nearly all existing robust control design methods dictate that these specifications be stated as explicit, worst-case/deterministic bounds on the levels

---

[1] Here the term "noise" is used in a general sense to refer to noise, disturbances, effects of non-zero initial conditions, and so on.

of noise and model error [4, 5]. For this reason a worst-case/deterministic approach to system identification is adopted in this paper. Note that such an approach to system identification has been previously adopted by many authors; see, e.g., [6, 8, 16] and the references therein. The specific identification problem under consideration in this paper pertains to single-input, single-output, linear, shift-invariant, distributed parameter plants. The available a priori information consist of worst-case/deterministic, time dependent, upper and lower bounds on the plant impulse response and the additive output noise. The available a posteriori information consists of a corrupt finite output time series obtained in response to a known, non-zero but otherwise arbitrary, applied input.

In general it will not be possible to identify an uncertain plant model possessing the essential characteristic described above unless the available a priori information is known to be "correct" for the underlying physical plant. As such, as the first step toward achieving the primary goal of this paper it is necessary to develop results which would provide practicing engineers with insights which help reduce the "engineering leap of faith" inherent in specifying the a priori information in a given application. Toward this end, in this paper available a priori and a posteriori information are defined to be consistent if there exists a plant model and noise sample, each of which satisfies the available a priori information, and which additionally yield the available a posteriori information when the plant model is driven by the known input and the resulting plant model output is additively corrupted by the noise sample. A computable necessary and sufficient condition is then derived under which available a priori and a posteriori information are consistent. This condition requires only that a single linear programming problem be solved.

When the available a priori and a posteriori information are consistent, the computation alluded above yields as a by-product a plant model and a noise sample which establish consistency of the available a priori and a posteriori information. This plant model is not unique. Rather, a continuum of distinct plant models exist which may be used to establish consistency and which form a set of "indistinguishable" plant models [2], referring to the fact that the available a priori and a posteriori information can not rigorously support any distinction between the members of this set. For the specific identification problem considered here, it is shown that this set is a simple polytope defined by a set of linear inequalities. An uncertain model description may be obtained by characterizing this set in terms of a nominal plant model, a model error specification, and the noise a priori information. The model error specification considered in this paper consists of an additive model error structure and the ℓ_1 and H_∞ system norms used to measure the error. Our motivation for working with these two system norms stems from the fact they are the system norms of choice in the currently popular robust control design methods [4, 5].

In characterizing the set of indistinguishable plant models in terms of a nominal plant model together with a model error specification, one possibility is to select the center and radius of the set of indistinguishable plant models as the nominal plant model and the model error specification, respectively. This choice is optimal in the sense that, considering the definition of the radius of a set, it minimizes the level of the model error specification needed to "cover" the set of indistinguishable plant models. Unfortunately, however, computation of the center of a polytope was shown in [2] to be extremely demanding of computational resources. A more tractable approach is to select any member of the set of indistinguishable plant models as the nominal plant model, and then set the model error specification equal to the diameter of the set of indistinguishable plant models. Since the diameter of a set is at most twice its radius [15], this approach can be regarded as being optimal, in the same sense described above, to within a factor of two, and this will be the case independent of the chosen system norm [2], given that the set of indistinguishable plant models is wholly independent of the system norm selected to characterize the error.

Though for the specific identification problem considered in this paper a member of the set of indistinguishable plant models can be readily computed by solving a linear programming problem, computation of the diameter of the set of indistinguishable plant models is also known to be extremely difficult [1, 2, 9]. As such, we provide upper bounds, in terms of both the ℓ_1 and H_∞ system norms, on the diameter of the set of indistinguishable plant models, which can then be used as the model error specification corresponding to the nominal plant model. Additionally, we also derive lower bounds on the diameter of the set of indistinguishable plant models. The upper and lower bounds are then used to establish necessary and sufficient conditions for convergence of the identification method, whereas convergence means that the model error specification tends to zero in the limit as the experiment duration and noise level tend jointly to infinity and zero, respectively. When the applied input is non-zero, the necessary and sufficient conditions for convergence are identical for the two system norms and it requires that the available plant a priori information asymptotically define the plant impulse response at a sufficiently fast rate (cf. Sec. 2). In other words, the specified time dependent upper and lower bounds on the plant impulse response must tend toward each other, at a sufficiently fast rate, as time tends to infinity.

Finally, the following notation is adopted in this paper. Let $\mathbf{Z}$ denote the set of integers, $\mathbf{Z}_+ := \{k \in \mathbf{Z} : k \geq 0\}$, and $\mathbf{Z}_{+,n} := \{k \in \mathbf{Z}_+ : 0 \leq k \leq n-1\}$. Let $\mathbf{R}$ denote the set of real numbers, and $\mathbf{R}^n$ denote the space of n dimensional real vectors. Let $\mathbf{S}_n(a, b) := \{x \in \mathbf{R}^n : a_i \leq x_i \leq b_i, i \in \mathbf{Z}_{+,n}\}$. Let $\ell_{1,\rho}$ denote the normed space $\{f : \mathbf{Z}_+ \to \mathbf{R} \mid \|f\|_{1,\rho} := \sum_{k=0}^{\infty} |f(k)|\rho^k < \infty\}$, $\ell_1 := \ell_{1,1}$, and let $\ell_{\infty,\rho}$ denote the normed space $\{f : \mathbf{Z}_+ \to \mathbf{R} \mid \|f\|_{\infty,\rho} := \sup_{k \in \mathbf{Z}_+} |f(k)|\rho^k < \infty\}$, $\ell_\infty := \ell_{\infty,1}$. Additionally, define the operator $T_n : \ell_\infty \to \mathbf{R}^n$ such that $(T_n f)(k) = f(k)$ for $k \in \mathbf{Z}_{+,n}$.

2 Preliminaries

The class of systems under consideration in this paper corresponds to causal, single-input, single-output, linear, shift invariant, distributed parameter systems. This class of systems can be identified with a normed linear space $\mathbf{X}$ which contains one-sided sequences and is endowed with a system induced norm $\|\cdot\|$. A system $h \in \mathbf{X}$ is represented by its impulse response $\{h(k)\}_{k=0}^{\infty}$. Alternatively, it is represented by its corresponding transform $\hat{h}(z)$, where[2]

$$\hat{h}(z) := \sum_{k=0}^{\infty} h(k)z^k \ . \tag{1}$$

The system to be identified is assumed to belong to $\mathbf{X}$. Additionally, its impulse response $\{h(k)\}_{k=0}^{\infty}$ is assumed to satisfy $M_k^- \le h(k) \le M_k^+$, $\forall k \in \mathbf{Z}_+$, where $\{M_k^-\}_{k=0}^{\infty}$ and $\{M_k^+\}_{k=0}^{\infty}$ are two pre-specified sequences which represent the available plant a priori information. More compactly, we may write $h \in \mathbf{S}_\infty(M^-, M^+) \subset \mathbf{X}$, and thus identify the plant a priori information with the hyperrectangle $\mathbf{S}_\infty(M^-, M^+)$.

The experimental procedure considered consists of applying an arbitrary non-zero input $u \in \ell_\infty$ to the system $h \in \mathbf{X}$ to generate an output $h * u \in \ell_\infty$. An additively corrupted version of $h * u$ is observed over a finite duration n, yielding a data record $y \in \mathbf{R}^n$. Let the corrupting noise be denoted by $v \in \ell_\infty$. Then, the experimental procedure can be precisely written as

$$E_n(h, v) := T_n\left((h * u) + v\right) \ .$$

The a priori information on the output noise $v \in \ell_\infty$ consists of two pre-specified sequences $\{\epsilon_k^-\}_{k=0}^{\infty} \in \ell_\infty$ and $\{\epsilon_k^+\}_{k=0}^{\infty} \in \ell_\infty$. The noise sequence $\{v(k)\}_{k=0}^{\infty}$ is assumed to satisfy $\epsilon_k^- \le v(k) \le \epsilon_k^+$, $\forall k \in \mathbf{Z}_+$. Consequently the noise a priori information can be identified with the hyperrectangle $\mathbf{S}_\infty(\epsilon^-, \epsilon^+) \subset \ell_\infty$. The experiment operator $E_n(\cdot, \cdot): \mathbf{X} \times \ell_\infty \to \mathbf{R}^n$ is linear in the ordered pair (h, v) and yields a data record $y \in \mathbf{R}^n$. Let U be the lower triangle Toeplitz matrix formed by the input u,

$$U := \begin{bmatrix} u(0) & 0 & \cdots & 0 \\ u(1) & u(0) & \cdots & 0 \\ \vdots & \vdots & \ddots & \vdots \\ u(n-1) & u(n-2) & \cdots & u(0) \end{bmatrix} \ .$$

Then, the data record $y \in \mathbf{R}^n$ can be conveniently expressed as $y = UT_n h + T_n v$, where $h \in \mathbf{S}_\infty(M^-, M^+)$, and $v \in \mathbf{S}_\infty(\epsilon^-, \epsilon^+)$. The set of all possible such data records is given by

$$\mathbf{Y} := \left\{ y \in \mathbf{R}^n : y = UT_n h + T_n v \text{ for some } h \in \mathbf{S}_\infty(M^-, M^+), v \in \mathbf{S}_\infty(\epsilon^-, \epsilon^+) \right\} \ .$$

---

[2] This transform corresponds to the standard z-transform evaluated at $1/z$.

Note that by nature the experiment operator $E_n(\cdot, \cdot)$ is a many-to-one mapping, and thus a given data record $y \in \mathbf{R}^n$ in general does not uniquely specify an element in $\mathbf{S}_\infty(M^-, M^+)$. Instead, there will be a set of possible indistinguishable systems that will agree with the data record. The set of indistinguishable systems that both satisfy the plant a priori information and agree with the experimental data $y \in \mathbf{R}^n$ is denoted by $\mathbf{P}(y)$, where

$$\mathbf{P}(y) := \left\{ h \in \mathbf{S}_\infty(M^-, M^+) : \ y = UT_n h + T_n v \ \text{ for some } v \in \mathbf{S}_\infty(\epsilon^-, \epsilon^+) \right\} \ .$$

Let $\mathbf{P}_n(y) := \{T_n h : \ h \in \mathbf{P}(y)\}$. Then, $\mathbf{P}_n(y)$ consists of the truncated series in $\mathbf{P}(y)$. It is easy to see that $\mathbf{P}_n(y)$ is a *convex polytope* in $\mathbf{R}^n$. Implicitly, the sets $\mathbf{P}(y)$ and $\mathbf{P}_n(y)$ depend upon the plant and noise a priori information in addition to the data record.

Given the above a priori and experimental information, our task in this paper is as follows. First, we want to determine whether the a priori and experimental information are *consistent*. This addresses the question whether the set $\mathbf{P}(y)$ is nonempty for a given y. The problem of consistency between the a priori information and data records is an important one and will be formulated and solved in Sect. 4. With this accomplished, our next task aims at developing "optimal" or "nearly optimal" identification algorithms which map the data record into an identified model. Additionally, we provide bounds on the corresponding identification errors. An identification algorithm is a mapping[3] $A_n : \ \mathbf{R}^n \to \mathbf{X}$ which operates on the data record and selects an element $A_n(y) \in \mathbf{X}$ as the identified nominal plant. In general, an identification algorithm is required to napproximate the set of indistinguishable systems $\mathbf{P}(y)$, as this is precisely the set of plant models which are unfalsified by the available a priori and a posteriori information (recall the discussion in the Introduction). The worst case identification error (occasionally shortened to identification error, or simply error) associated with a given algorithm A_n is defined as the approximation error of $A_n(y)$ to the set $\mathbf{P}(y)$.

Definition 2.1 *The local identification error associated with an algorithm A_n at a data record y is defined as*

$$e(A_n; y; \| \cdot \|) := \sup_{h \in \mathbf{P}(y)} \|h - A_n(y)\|$$

and the global identification error associated with A_n is defined as

$$e(A_n; \| \cdot \|) := \sup_{y \in \mathbf{Y}} e(A_n; y; \| \cdot \|).$$

Clearly, the local identification error is defined with respect to a given (fixed) data record, while the global identification error corresponds to the worst

---

[3] It should be noted that both identification algorithms and errors implicitly depend upon the plant and noise a priori information. However, to simplify the notation, this dependence is not made explicit.

case data record. It can be readily verified [12, 15] that the global error can be alternatively expressed as

$$e(A_n; \| \cdot \|) = \sup_{h \in S_\infty(M^-, M^+)} \sup_{v \in S_\infty(\epsilon^-, \epsilon^+)} \| h - A_n(E_n(h, v)) \| .$$

In relation to an identification algorithm, we introduce the following notions concerning the performance aspects of the algorithm. First, the issue of algorithm optimality is clarified.

Definition 2.2 *The optimal local identification error at a data record y is defined as*

$$e^*(y; \| \cdot \|) := \inf_{\{A_n : A_n(y) \in \mathbf{X}\}} e(A_n; y; \| \cdot \|).$$

An algorithm A_n^* is said to be locally optimal at y if $e(A_n^*; y; \| \cdot \|) = e^*(y; \| \cdot \|)$. Similarly, the optimal global identification error is defined as

$$e^*(\| \cdot \|) := \inf_{\{A_n : A_n(y) \in \mathbf{X}\}} e(A_n; \| \cdot \|),$$

and A_n^* is said to be globally optimal if $e(A_n^*; \| \cdot \|) = e^*(\| \cdot \|)$. Finally, an algorithm is said to be strongly optimal if it is locally optimal at each $y \in \mathbf{Y}$.

Note that these notions of identification errors and optimal algorithms are adapted from existing literature in information based complexity and system identification theories [6, 8, 12, 15]. It is useful to emphasize that a strongly optimal algorithm minimizes the local identification error for each data record and hence also the global identification error as well.

Finally, we define precisely a notion of algorithm convergence.

Definition 2.3 *Let*

$$\delta M_k := M_k^+ - M_k^-, \tag{2}$$

and

$$\delta \epsilon_k := \epsilon_k^+ - \epsilon_k^-. \tag{3}$$

Denote $\delta M = \max_k \delta M_k$ and $\delta \epsilon := \max_k \delta \epsilon_k$. Then, an algorithm A_n is said to be convergent if

$$\lim_{\substack{\delta M \to 0 \\ \delta \epsilon \to 0}} e(A_n; \| \cdot \|) = 0,$$

and

$$\lim_{\substack{n \to \infty \\ \delta \epsilon \to 0}} e(A_n; \| \cdot \|) = 0 .$$

This definition is motivated by [8] and constitutes a slight generalization to the corresponding concept in [6], whose interpretation also follows analogously. Roughly speaking, the convergence requirement means that the worst case error for a convergent algorithm should tend to zero as the available partial plant and corrupted experimental information tend to be complete and uncorrupted.

3 Characterizations of Optimal Algorithms

The above notions of optimal algorithms and optimal identification errors
can be alternatively characterized in the framework of information based
complexity (IBC) theory [15]. Recall that the radius of a set A [15] in a
normed linear space $(\mathbf{X}, \|\cdot\|)$ is defined by

$$r(A, \|\cdot\|) := \inf_{x \in \mathbf{X}} \sup_{a \in A} \|x - a\|$$

and the diameter of A is defined by

$$d(A, \|\cdot\|) := \sup_{x, a \in A} \|x - a\| \ .$$

A center of A is an element c such that $r(A, \|\cdot\|) = \sup_{a \in A} \|c - a\|$. It is a
well-known fact (see, e.g., [15]) that for any set A in a normed linear space,

$$r(A, \|\cdot\|) \leq d(A, \|\cdot\|) \leq 2\, r(A, \|\cdot\|) \ . \tag{1}$$

Furthermore, for any $x \in A$,

$$r(A, \|\cdot\|) \leq \sup_{a \in A} \|x - a\| \leq d(A, \|\cdot\|) \ . \tag{2}$$

In relation to the set of indistinguishable systems $\mathbf{P}(y)$, we readily notice the
following fact.

Fact 1 *For any data record y,*

$$e^*(y; \|\cdot\|) = r(\mathbf{P}(y); \|\cdot\|) \tag{3}$$

$$\frac{1}{2}\, d(\mathbf{P}(y); \|\cdot\|) \leq e^*(y; \|\cdot\|) \leq d(\mathbf{P}(y); \|\cdot\|) \ . \tag{4}$$

These results follow directly by definition and are standard in IBC theory [15].
In [15], $r(\mathbf{P}(y); \|\cdot\|)$ and $d(\mathbf{P}(y); \|\cdot\|)$ are termed local radius and diameter
of information, respectively. Analogously, the quantities $\sup_{y \in \mathbf{Y}} r(\mathbf{P}(y); \|\cdot\|)$
and $\sup_{y \in \mathbf{Y}} d(\mathbf{P}(y); \|\cdot\|)$ are called global radius and diameter of informa-
tion, respectively. Fact 3.1 indicates that the optimal identification errors
can be determined by computing the corresponding radii of information. An
algorithm that selects the center of $\mathbf{P}(y)$ will be locally optimal at y. An algorithm that selects the center of $\mathbf{P}(y)$ for all possible data is called a *central
algorithm* [8, 12, 15]. Hence, by definition, an algorithm is strongly optimal
if and only if it is central. Since the center of a set is difficult to find [15], the
task of constructing a central algorithm is generally difficult.

In addition to central algorithms, also of interest is a class of *essentially
strongly optimal* algorithms which we call *interpolatory algorithms* [15]. An
interpolatory algorithm is one that maps the data into the set of indistinguishable systems. In general, an interpolatory algorithm is easy to compute
and moreover, it possesses some appealing properties. Noticing Fact 3.1 and
(3.1–2), we easily conclude the following

Fact 2 *Let A_n be an algorithm such that $A_n(y) \in P(y)$. Then,*

$$e^*(y; \|\cdot\|) \le e(A_n; y; \|\cdot\|) \le 2e^*(y; \|\cdot\|) . \tag{5}$$

The implication of this result is that *any* interpolatory algorithm will be strongly optimal to within a factor of 2, in the sense that the local identification error associated with the algorithm is tight to within a factor of 2 of the optimal local error for all possible data records. Furthermore, owing to its algebraic nature, an interpolatory algorithm is strongly optimal to within a factor of 2 in all norms.

4 Consistency of Information

The issue of consistency between the a priori information and data record is concerned with the following question: given a model set characterized by the plant a priori information and given also a data record, will there be an element in the model set that will agree with the data for the given input and the allowable noise level? More specifically, given the model set defined by $S_\infty(M^-, M^+)$, the noise set $S_\infty(\epsilon^-, \epsilon^+)$, and given also a data record y, will there be an element $h \in S_\infty(M^-, M^+)$ so that $y = T_n((h * u) + v)$? Clearly, this addresses precisely the issue whether the set $P(y)$ is empty. Formally, the consistency of data and a priori information is defined as follows.

Definition 4.1 *A data record y is said to be consistent with the plant and noise a priori information if $P(y) \ne \emptyset$.*

From this definition, it is important to note that the data and a priori information should be consistent in order for the identification problem to be meaningful. It can be readily recognized that $P(y) \ne \emptyset$ if and only if $P_n(y) \ne \emptyset$. This observation leads to the following computable necessary and sufficient condition for consistency of the data and a priori information.

Theorem 3. *A data record y is consistent with the a priori information if and only if the linear inequalities*

$$M_k^- \le h(k) \le M_k^+, \quad k = 0, 1, \cdots, n-1, \tag{1}$$

$$\epsilon_k^- \le y(k) - \sum_{j=0}^{k} u(k-j)h(j) \le \epsilon_k^+, \quad k = 0, 1, \cdots, n-1 \tag{2}$$

admit a solution $\{h(k)\}_{k=0}^{n-1}$.

As a consequence of Theorem 4.1, determining the consistency of the data and a priori information amounts to solving a set of linear inequalities, which in turn may be solved efficiently using the linear programming method (see, e.g., [13]). Various necessary or sufficient conditions for the data and a priori information to be consistent can be derived based on Theorem 4.1, and several of such conditions were given and interpreted in [2]. Additionally, a computable expression was also derived in [2] for determining the minimal noise level[4] for which available a priori and a posteriori information are consistent.

In closing, we note that the solution of the above consistency problem, together with the noise specification and the diameter of $P(y)$, yields immediately an uncertain model. Hence, the procedure of solving the data consistency problem and that of computing the diameter of $P(y)$ may together be considered as a "synthesis" procedure for uncertain plant models from the available a priori and a posteriori information. This is the very goal which the present identification method aims at achieving. Clearly, such an uncertain model explains not only the observed a posteriori information up to the time instant n, but also the future a posteriori information, owing to the fact that $P(y)$ constitutes the totality of plants that explain both the observed as well as the future a posteriori information. In these aspects, our identification method differs from the treatment of [14], where the emphasis is on the analysis of given uncertain plant models relative to whether or not they explain available a posteriori information, and little consideration is given to the role played by a priori information in allowing the model to explain future a posteriori information.

5 Identification Algorithms and Errors

In this section we give an essentially strongly optimal algorithm and derive corresponding error bounds for the identification problem posed in Sect. 2. This algorithm is motivated from Fact 3.2 and requires *only* solving the data consistency problem. More specifically, it follows from Fact 3.2 that *any* element of $P(y)$ serves as an identified model that yields a local identification error to within a factor of 2 of the optimal local error. Hence, the following result is clear.

Theorem 4. *Let* $\{h(k)\}_{k=0}^{n-1}$ *be a solution of the linear inequalities (4.1-2). Construct* A_n *so that*

$$[A_n(y)]_k = \begin{cases} h(k) & 0 \leq k \leq n-1 \\ \frac{M_k^+ + M_k^-}{2} & k \geq n \end{cases} \tag{1}$$

Then, for any data record y,

$$e^*(y; \|\cdot\|) \leq e(A_n; y; \|\cdot\|) \leq 2\, e^*(y; \|\cdot\|) \ . \tag{2}$$

The algorithm given in Theorem 5.1 is easy to compute since the data consistency problem must be solved a priori and it can be done efficiently

---

[4] Here the noise level is defined as the maximal difference, over time, between the noise upper and lower bounds.

by solving a *single* linear program. Once the data consistency problem is solved, an identified nominal model is obtained immediately. In addition, this algorithm is clearly interpolatory and hence it possesses all desirable properties associated with interpolatory algorithms. Specifically, the algorithm is strongly optimal to within a factor of 2 in all norms. This is particularly useful because the model identified via this algorithm can be applied equally well to ℓ_1, H_∞, or H_2 design purposes.

Having given the above interpolatory algorithm, in the remainder of this section we derive explicit error bounds associated with this algorithm. Note that although the algorithm (5.1) is essentially strongly optimal for any norms, the identification errors do depend upon norms in general. We shall consider first the ℓ_1 norm case. Our first result shows that the identification error can be obtained by solving a maximization problem over the set $\mathbf{P}_n(y)$.

Lemma 5. *Let $A_n(y)$ be given by (5.1). Then,*

$$e(A_n; y; \|\cdot\|_1) = \frac{1}{2} \sum_{k=n}^{\infty} \delta M_k + \max_{T_n h \in \mathbf{P}_n(y)} \|T_n h - T_n A_n(y)\|_1 . \tag{3}$$

As shown in [2], the exact computation of $e(A_n; y; \|\cdot\|_1)$ according to (5.3) is in general computationally infeasible. This is because computing $\max_{T_n h \in \mathbf{P}_n(y)} \|T_n h - T_n A_n(y)\|_1$ amounts to solving a norm maximization problem over a polytope. It is known that the computational complexity of such an problem grows exponentially with the dimension of the polytope [1]. For this reason, we seek both lower and upper bounds for the global diameter $\sup_{y \in Y} d(\mathbf{P}(y); \|\cdot\|_1)$. From Lemma 5.1 and the relation

$$\frac{1}{2} d(\mathbf{P}_n(y); \|\cdot\|_1) \leq \max_{T_n h \in \mathbf{P}_n(y)} \|T_n h - T_n A_n(y)\|_1 \leq d(\mathbf{P}_n(y); \|\cdot\|_1),$$

it follows that such bounds also provide upper and lower bounds for the global identification error. The following result is needed in our derivation and it follows as an easy consequence of Theorem 18.1 in [7].

Lemma 6. *Let $u(0) \neq 0$. Then U^{-1} is also a lower triangular Toeplitz matrix.*

It should be noted that the assumption $u(0) \neq 0$ can be imposed without loss of generality. This is because a non-zero input is used and the data is gathered from the first instant of the experiment at which the input is nonzero. As a result of Lemma 5.2, we may write

$$U^{-1} = \begin{bmatrix} a_0 & 0 & \cdots & 0 \\ a_1 & a_0 & \cdots & 0 \\ \vdots & \vdots & \ddots & \vdots \\ a_{n-1} & a_{n-2} & \cdots & a_0 \end{bmatrix} .$$

The sequence $\{a_k\}_{k=0}^{n-1}$ is completely determined from the matrix U^{-1}. The lower and upper bounds for $e(A_n; \|\cdot\|_1)$ are given below and they together establish the convergence property of the algorithm (5.1) in the ℓ_1 case.

Theorem 7. *Let $A_n(y)$ be given by (5.1), and let δM_k and $\delta \epsilon_k$ be defined by (2.2-3). Then,*

$$\frac{1}{2}\sum_{k=n}^{\infty}\delta M_k + \frac{1}{2}\max_{0\leq k\leq n-1}\min\left\{\sum_{j=0}^{k}|a_{k-j}|\delta\epsilon_j, \delta M_k\right\} \leq e(A_n; \|\cdot\|_1)$$

$$\leq \frac{1}{2}\sum_{k=n}^{\infty}\delta M_k + \sum_{k=0}^{n-1}\min\left\{\sum_{j=0}^{k}|a_{k-j}|\delta\epsilon_j, \delta M_k\right\}. \tag{4}$$

Furthermore, a necessary and sufficient condition for A_n to be convergent in ℓ_1 is that $\{\delta M_k\}_{k=0}^{\infty} \in \ell_1$.

One appealing feature with the above convergence condition is that it is actually *guaranteed* by the assumed plant a priori information in the ℓ_1 case. Indeed, it is easy to realize that a necessary and sufficient condition for $\{M_k^-\}_{k=0}^{\infty} \in (\mathbf{X}, \|\cdot\|_1)$ and $\{M_k^+\}_{k=0}^{\infty} \in (\mathbf{X}, \|\cdot\|_1)$ is that $\{M_k^*\}_{k=0}^{\infty} \in (\mathbf{X}, \|\cdot\|_1)$, where $M_k^* := \max\{|M_k^-|, |M_k^+|\}$. This implies that $\{\delta M_k\}_{k=0}^{\infty} \in \ell_1$. Hence in the ℓ_1 identification, any interpolatory algorithm will be convergent independently of experimental inputs provided that they are non-zero.

According to the well-known fact $\|\cdot\|_{H_\infty} \leq \|\cdot\|_1$ [3], it is clear that for any algorithm A_n,

$$e(A_n; y; \|\cdot\|_{H_\infty}) \leq e(A_n; y; \|\cdot\|_1) .$$

Hence, the upper bound in (5.4) also bounds the global identification error $e(A_n; \|\cdot\|_{H_\infty})$, and the convergence condition in Theorem 5.2 is sufficient in H_∞ case. It turns out that this same condition is necessary in H_∞ case as well. The necessity in this case can also be established by deriving a lower bound.

Theorem 8. *Let $A_n(y)$ be given by (5.1), and let δM_k and $\delta \epsilon_k$ be defined by (2.2-3). Then,*

$$\frac{1}{2}\sum_{k=n}^{\infty}\delta M_k \leq e(A_n; \|\cdot\|_{H_\infty}) \leq \frac{1}{2}\sum_{k=n}^{\infty}\delta M_k + \sum_{k=0}^{n-1}\min\left\{\sum_{j=0}^{k}|a_{k-j}|\delta\epsilon_j, \delta M_k\right\} . \tag{5}$$

A necessary and sufficient condition for A_n to be convergent in H_∞ is that $\{\delta M_k\}_{k=0}^{\infty} \in \ell_1$.

Similar to the ℓ_1 identification problem, the necessary and sufficient condition given above is also guaranteed in the H_∞ case if the plant a priori information satisfies some additional properties. A discussion on these properties are given in [2], and several cases of interest were presented. Again, in those cases, any interpolatory algorithm will be convergent in H_∞ for any nonzero inputs.

6 Conclusion

We have presented a theoretical basis for identifying, from available a priori and a posteriori information, an "uncertain plant model" which can be used for robust control design. Our main results are i) a technique for determining the consistency of the available a priori and a posteriori information, and ii) an identification algorithm of the interpolatory class. This identification algorithm is worst-case strongly optimal to within a factor of two.

Our main motivation for deriving an interpolatory algorithm and error bounds stems partly from the computational complexity in computing the optimal algorithm and error. In [2], it was shown that for the identification problem considered in this paper, computation of the central algorithm and optimal error requires explicit use of the vertices of $\mathbf{P}_n(y)$, and this problem is in general computationally intractable, so is use of exact diameters of information as error bounds. Note, however, that for certain special experimental inputs such as impulse and step signals, the central algorithm and optimal error can nevertheless be computed readily and explicitly [2].

References

1. H. L. Bodlaender, P. Gritzmann, V. Klee, and J. Van Leeuwen. *Combinatorica*, vol. 10, no. 2, pp. 203-225, 1990.
2. J. Chen, C.N. Nett, and M.K.H. Fan, *Proc. 1992 ACC*, pp. 279-285, June 1992.
3. C. A. Desoer and M. Vidyasagar. *Feedback Systems: Input-Output Properties.* Academic Press, New York, 1975.
4. P. Dorato, ed., *Robust Control.* IEEE Press, 1987.
5. P. Dorato and R. Yedavalli, eds., *Recent Advances in Robust Control.* IEEE Press, 1990.
6. A. Helmicki, C. Jacobson, and C. Nett. *IEEE TAC*, vol. 36, no. 10, pp. 1163-1176, Oct. 1991.
7. I.S. Iohvidov. *Hankel and Toeplitz Matrices and Forms.* Birkhäuser, 1982.
8. C. A. Jacobson and C. N. Nett. *Proc. 1991 ACC*, pp. 3152-3157, June 1991.
9. T. H. Matheiss and D. S. Rubin. *Math. Oper. Res.*, vol. 5, no. 2, pp. 167-185, 1980.
10. M. Milanese and A. Vicino. *Automatica*, vol. 27, no. 6, pp. 997-1009, 1991.
11. M. Milanese, R. Tempo, and A. Vicino, *J. Complexity*, vol. 2, pp. 78-94, 1986.
12. M. Milanese and R. Tempo, and prediction, *IEEE TAC*, vol. 30, pp. 730-738, 1985.
13. K. G. Murty. *Linear Programming.* John Wiley & Sons, New York, 1983.
14. R. Smith and J. C. Doyle, *Proc. 1989 ACC*, pp. 1435-1440, June 1989.
15. J. F. Traub, G. W. Wasilkowski, and H. Wozniakowski. *Information-Based Complexity.* Academic Press, 1988.
16. D.C.N. Tse, M.A. Dahleh and J.N. Tsitsiklis, *Proc. 30th IEEE CDC*, pp. 623-628, Dec. 1991.
17. V.A. Yemelichev, M.M. Kovalev, and M.K. Kravtsov. *Polytopes, Graphs and Optimization.* Cambridge University Press, 1984.

Identification in $\mathcal{H}^\infty$ Using Time-Domain Measurement Data*

Guoxiang Gu

Department of Electrical and Computer Engineering, Louisiana State University, Baton Rouge, LA 70803–5901, USA

1 Introduction

This paper is concerned with a particular control oriented identification problem formulated by Helmicki, Jacobson and Nett [11]: given a finite number of noisy experimental frequency response data, find an algorithm which not only identifies the nominal plant model, but also quantifies the worst-case identification error in $\mathcal{H}^\infty$ norm. Further, the algorithm is required to have the property that the worst case identification error converge to zero as the noise level goes to zero and the number of experimental data points goes to infinity. This particular identification problem is termed as identification in $\mathcal{H}^\infty$ which is mainly motivated by the need of modern robust control theory which has its origin in [26]. The research work along this direction constitutes an important part of the robust identification. Earlier work on identification in $\mathcal{H}^\infty$ has concentrated on the two-stage algorithm [7, 8, 11, 18, 19]. Today, identification algorithms are more diversified which include interpolatory algorithm [2, 3, 10] and algorithms developed in [12, 17, 19, 25].

While robust identification enjoys its popularity [5, 16, 17, 21, 23, 24, 27], identification in $\mathcal{H}^\infty$ does not gain much recognition in the conventional identification community. A common criticism to identification in $\mathcal{H}^\infty$ is that it is expensive to obtain point-wise frequency response data of the plant and it is difficult to quantify the noise level of the frequency response estimate based on input/output measurements. This paper addresses the above issue by considering identification of feedback systems. It is assumed that the true unknown plant model is linear time invariant possibly unstable infinite-dimensional and the closed-loop system is stable. Our contribution is the re-formulation of identification in $\mathcal{H}^\infty$ for feedback systems as pioneered by Mäkilä and Partington [17]. Similarly to [17], our objective is to identify the normalized coprime factors of the true unknown plant and to quantify the worst-case identification error in terms of the directed gap ball [4]. A

* Supported in part by National Science Foundation under grant number 9110636.

simple time-domain experimental procedure will be developed to obtain frequency response data of the coprime factors of the true unknown plant without point-wise sine-dwell experiment. It will be shown that under certain mild assumptions on the stabilizing compensator and the stability of the closed-loop system, the time-domain measurements can be used directly to identify the normalized coprime factors of the true unknown plant, and that various existing algorithms for identification in $\mathcal{H}^\infty$ are convergent (see Sect. 2 for its precise definition) in the directed gap ball. Further, an improved explicit error bound is derived for the identification of feedback systems using a specific two-stage nonlinear algorithm. For simplity reason, only discrete-time models will be studied but generalization to continuous-time case is straightforward. It is hoped that the results reported in this paper would ease the criticism to identification in $\mathcal{H}^\infty$ from the conventional identification community.

2 Preliminaries

In this section, a simple time-domain experiment will be considered that can replace the point-wise sine-dwell experiment for identification of feedback systems. The gap metric will then be discussed which will be used to quantify the identification error. Finally, some key notions for identification in $\mathcal{H}^\infty$ will be defined to facilitate the development in Sect. 3.

2.1 Time-domain Experiment

Let $\hat{h}(z)$ be the transfer function of a given linear shift-invariant stable system with impulse response $\{h_k\}_{t=0}^\infty$. The time-domain experiment procedure proposed in this part of the paper consists of one single periodic input rather than lengthy point-wise sine-dwell experiment as given next.

Proposition 1. *Suppose the linear shift-invariant system h is ℓ^∞-BIBO stable. If the exciting input signal is $\{u(t)\}_{t=0}^{N-1}$ which is periodic with period N. Then the true steady-state output is $\{y(t)\}_{t=0}^{N-1}$ which is also periodic with the same period N. Let $\{\hat{u}_k\}_{k=0}^{N-1}$ and $\{\hat{y}_k\}_{k=0}^{N-1}$ be the N-point discrete Fourier transforms (DFT) of $\{u(t)\}_{t=0}^{N-1}$ and $\{y(t)\}_{t=0}^{N-1}$ respectively. Then there holds*

$$\hat{y}_k = \hat{h}(z_k)\hat{u}_k, \quad z_k = e^{-j2k\pi/N} \qquad (1)$$

where $0 \leq k \leq N-1$.

Proof. With periodic input $\{u(t)\}_{t=0}^{N-1}$, the true steady state output response of the system satisfies the convolution equation

$$y(t) = \sum_{\tau=0}^\infty h_\tau v(t-\tau) \ .$$

Set $\tau = \ell N + m$ where $\ell = 0, 1, ...,$ and $m = 0, 1, ..., N - 1$. The above convolution can be written as

$$y(t) = \sum_{\ell=0}^{\infty} \sum_{m=0}^{N-1} h_{\ell N+m} u(t - \ell N - m) = \sum_{m=0}^{N-1} \left(\sum_{\ell=0}^{\infty} h_{\ell N+m} \right) u(t - \ell N - m) \ .$$

Denoting $h_N(m) = \sum_{\ell=0}^{\infty} h_{\ell N+m}$ for $0 \le m \le N - 1$, $h_N(m)$ exists and is finite for each m by the ℓ^∞-BIBO stability of the system. In light of [13, 6], $\{h_N(m)\}_{m=0}^{N-1}$ is the N-point inverse DFT of the uniformly spaced frequency response sample sequence $\{\hat{h}(e^{j\omega_k})\}_{k=0}^{N-1}$ with $\omega_k = 2k\pi/N$. Using the periodicity of $u(t)$, we have circular convolution of size N

$$y(t) = \sum_{m=0}^{N-1} h_N(m) u(t - m) \ . \tag{2}$$

Hence, applying DFT on both sides of the above equation, (1) holds true. □

The above result indicates that in perfect measurement case, the frequency response of the system can be estimated using

$$\hat{h}(e^{j\omega_k}) = \frac{\hat{y}_k}{\hat{u}_k}, \ 0 \le k \le N - 1$$

provided that $\hat{u}_k \ne 0$ for each k. This suggests the use of simple elegant time-domain experiment as in Proposition 1 instead of lengthy point-wise sine-dwell experiment. However, this time-domain experiment has an inherent drawback in presence of noisy measurements. Let $\{\eta(t)\}_{t=0}^{N-1}$ be the periodic part of the noise present at the output and $\{\hat{\eta}\}_{k=0}^{N-1}$ be the N-point DFT of $\{\eta(t)\}_{t=0}^{N-1}$. Then, the noisy measurement at the outout and its DFT are

$$y_E(t) = y(t) + \eta(t), \quad \text{and} \quad \hat{y}_k^E = \hat{y}_k + \hat{\eta}_k \tag{3}$$

respectively where $y(t)$ and $\hat{y}_k$ are same as in (2) and (1) with $0 \le k, t \le N-1$. The noisy frequency response estimate can then be modified as

$$\hat{h}_E(e^{j\omega_k}) = \frac{\hat{y}_k}{\hat{u}_k} + \frac{\hat{\eta}_k}{\hat{u}_k}, \ 0 \le k \le N - 1 \ . \tag{4}$$

Hence, small error in frequency response estimate at each k requires that the magnitude of $\hat{u}_k$ be large for all k. However, it is virtually impossible to find a uniformly bounded periodic sequence $\{u(t)\}_{t=0}^{N-1}$ for which its N-point DFT $\hat{u}_k$ has large magnitude for all k especially in the case N is large. The requirement on uniform boundedness of the input sequence is due to the saturation of practical systems. Such issue is also addressed in Ljung's book [14]. Although it seems that for open-loop identification, accurate frequency response estimate in (4) is possible only by using point-wise sine-dwell experiment. It will be shown in Sect. 3 that the above proposed time-domain experimental procedure does work for identification of feedback systems.

Remark. It should be pointed out that an upper bound on $\{\hat{\eta}_k\}_{k=0}^{\infty}$ can often be estimated by analyzing the spectral properties of the sensor noise. Hence, one may assume that a sequence $\{\hat{w}_k\}_{k=0}^{N-1}$ exists such that

$$|\hat{\eta}_k \hat{w}_k| \leq \epsilon, \ 0 \leq k \leq N - 1 \tag{5}$$

for some $\epsilon > 0$. Further, if the system to be identified has multiple inputs and multiple outputs, the above time domain experimental procedure can still be used but (4) and (5) should be modified accordingly.

2.2 Gap Metric

For identification of feedback systems, a frequently used measure in quantification of identification error is the gap metric [4, 17]. Consider feedback system in Fig. 1 where $\hat{P}(z)$ represents a linear time-invariant possibly unstable infinite-dimensional system. Suppose $\hat{P}(z)$ admits a normalized right coprime factorization

$$\hat{P}(z) = \hat{N}(z)\hat{D}(z)^{-1}, \ N(z^{-1})^T N(z) + D(z^{-1})^T D(z) = I \ .$$

Let $\hat{P}_1(z) = \hat{N}_1(z)\hat{D}_1(z)^{-1}$ be a perturbed plant model with $\hat{N}_1(z)$ and $\hat{D}_1(z)$ a pair of normalized right coprime factors for $\hat{P}_1(z)$. Then the gap between $\hat{P}$ and $\hat{P}_1$ is defined to be

$$\delta(\hat{P}, \hat{P}_1) = \max\left\{\delta(\hat{P}, \hat{P}_1), \delta(\hat{P}_1, \hat{P})\right\}$$

where δ is the directed gap as defined by

$$\delta(\hat{P}, \hat{P}_1) = \inf_{\hat{Q} \in \mathcal{H}^{\infty}} \left\| \begin{pmatrix} \hat{D} \\ \hat{N} \end{pmatrix} - \begin{pmatrix} \hat{D}_1 \\ \hat{N}_1 \end{pmatrix} \hat{Q} \right\|_{\infty} . \tag{6}$$

We will not elaborate the gap metric here in details but would like to point out that the uncertainty in normalized coprime factors is characterized by the directed gap ball. The next result is due to Georgiou and Smith [4].

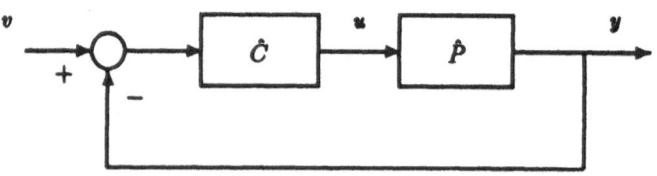

Fig. 1.

Lemma 2. *Let the directed gap ball and gap ball be defined by*

$$\mathcal{B}(\hat{P}, b) := \left\{ \hat{P}_1 : \vec{\delta}(\hat{P}, \hat{P}_1) < b \right\}$$

$$B(\hat{P}, b) := \left\{ \hat{P}_1 : \delta(\hat{P}, \hat{P}_1) < b \right\}$$

respectively. Let $\hat{P}$ have normalized coprime factorization $\hat{P}(z) = \hat{N}(z)\hat{D}(z)^{-1}$. Then for all $0 < b \le 1$ and $\hat{\Delta}_N, \hat{\Delta}_D \in \mathcal{H}^\infty$,

$$\mathcal{B}(\hat{P}, b) = \left\{ \hat{P}_1 : \hat{P}_1 = (\hat{N} + \hat{\Delta}_N)(\hat{D} + \hat{\Delta}_D)^{-1}, \; \left\| \begin{pmatrix} \hat{\Delta}_D \\ \hat{\Delta}_N \end{pmatrix} \right\|_\infty < b \right\}.$$

Further, if $b \le \inf\limits_{|z|>1} \sigma_{\min} \left(\dfrac{\hat{D}(z)}{\hat{N}(z)} \right)$, then $\mathcal{B}(\hat{P}, b) = B(\hat{P}, b)$.

It is noted that the directed gap ball never contains more elements than the gap ball does for the same radius b. Hence, the directed gap may give tight error bound for its use in identification of feedback systems. Further, since the directed gap ball is same as the uncertain system with perturbations in normalized coprime factors in the case $0 < b \le 1$, and the robust control of coprime factor uncertain systems has been studied in [4, 9], it is believed that the directed gap ball is more appropriate for quantification of the identification error here.

2.3 Notions for Identification in $\mathcal{H}^\infty$

Identification in $\mathcal{H}^\infty$ is concerned with identification of the nominal model and quantification of the worst case identification error in $\mathcal{H}^\infty$ norm. We will consider only scalar systems but extension to multivariable case is trivial.

Simply speaking, $\mathcal{H}^\infty$ is the collection of all transfer functions of the stable and causal systems which have bounded frequency response. To be specific, let the true unknown plant $\hat{h}(z)$ be stable and belong to a set $\mathcal{S}$ which is a relatively compact subset of $\mathcal{H}^\infty$. The set $\mathcal{S}$ is specified by the a priori information on the true unknown plant. A commonly used set is

$$\mathcal{S} = \mathcal{H}(M, \rho) = \left\{ \hat{h}(z) : \hat{h}(z) \in \mathcal{H}^\infty, \|\hat{h}(z\rho^{-1})\|_\infty \le M \right\} \tag{7}$$

where $\rho > 1$ and $M > 0$. The set $\mathcal{H}(M, \rho)$ is often called the set of exponentially stable systems because for any $\hat{h} \in \mathcal{H}(M, \rho)$, its impulse response satisfies

$$|h_t| \le M\rho^{-t}, \; 0 \le t \le \infty.$$

However, it should be noted that a system has impulse response decaying exponentially as above may not belong to $\mathcal{H}(M, \rho)$. A simple example is when $h_t = M\rho^{-t}$ which decays exponentially but its $\mathcal{Z}$-transform does not belong to $\mathcal{H}(M, \rho)$.

Identification in $\mathcal{H}^\infty$ also assumes that the noisy frequency response estimate of the true unknown system is available at a finite number of frequency samples and the level of the corrupting noise is known a priori. That is, we have

$$E_k^N = \hat{h}(e^{j\omega_k}) + \hat{\eta}_k, \quad |\hat{\eta}_k| \le \epsilon \tag{8}$$

for $0 \le k \le N - 1$. Let $\hat{h}_{id}$ be the identified model obtained with Algorithm A_N based on experimental data (8) and possibly on a priori information. The first notion for identification in $\mathcal{H}^\infty$ is the worst case identification error defined by

$$e_N(A_N) := \sup \left\{ \|\hat{h} - \hat{h}_{id}\|_\infty : \hat{h} \in S, |\hat{\eta}_k| \le \epsilon \right\} . \tag{9}$$

The performance of the identification algorithm is measured by $e_N(A_N)$. The second notion is the convergence.

Definition 3. An identification algorithm A_N is said to be convergent if the worst case identification error $e_N(A_N)$ satisfies

$$\lim_{N\to\infty} \lim_{\epsilon\to 0} e_N(A_N) = \lim_{\epsilon\to 0} \lim_{N\to\infty} e_N(A_N) = 0 .$$

Further, if in addition the identification algorithm A_N does not depend on a priori information on S and is convergent, then A_N is said to be (untuned and) robustly convergent.

3 Main Results

In this section, we consider identification of feedback systems as studied in [17]. The class of systems under consideration consists of all linear shift-invariant systems which are exponentially stablizable. We will assume throughout this section that the closed-loop system in Fig. 1 is exponentially stable in the sense of (7). Using the time domain experimental procedure in Proposition 1 in Sect. 2, we excite the system with periodic input $\{v(t)\}_{t=0}^{N-1}$. The experimental data consists of collected signals $\{y_E(t)\}_{t=0}^{N-1}$ and $\{u_E(t)\}_{t=0}^{N-1}$ (see Fig. 1). Let the true transfer function from the exciting signal $v(t)$ to the output of the controller $u(t)$ be $\hat{G}(z)$ and the true transfer function from the exciting signal $v(t)$ to the controlled output $y(t)$ be $\hat{T}(z)$. Then Proposition 1 implies that

$$\hat{u}_k^E = \hat{G}(z_k)\hat{v}_k + \hat{\eta}_k^u \quad \text{and} \quad \hat{y}_k^E = \hat{T}(z_k)\hat{v}_k + \hat{\eta}_k^y \tag{10}$$

for $0 \le k \le N - 1$ where $\{\hat{v}_k\}_{k=0}^{N-1}$, $\{\hat{u}_k\}_{k=0}^{N-1}$ and $\{\hat{y}_k\}_{k=0}^{N-1}$ are N-point DFTs of $\{v(t)\}_{t=0}^{N-1}$, $\{u(t)\}_{t=0}^{N-1}$ and $\{y(t)\}_{t=0}^{N-1}$ respectively.

As discussed in Sect. 2, our objective is to identify the normalized coprime factors of $\hat{P}(z)$ which is linear shift invariant possibly unstable and to quantify the modeling error in terms of the directed gap ball. To use the experimental data in (10) for our identification problem, The following assumptions will be used.

Assumption 1: There exists a weighting function $\hat{w} \in \mathcal{H}(M_w, \rho_w)$ for some $M_w > 0$ and $\rho_w > 1$, with $\hat{w}^{-1} \in \mathcal{H}^\infty$ such that

$$|\hat{w}(e^{j\omega_k})| \sqrt{|\eta_y^k|^2 + |\eta_u^k|^2} \le \epsilon, \quad k = 1, 2, ..., N$$

for some $\epsilon > 0$.

Assumption 2: The stabilizing controller satisfies that $\hat{C}^{-1} \in \mathcal{H}^\infty$. Further, the closed-loop system is exponentially stable satisfying

$$\hat{T}(z) \in \mathcal{H}(M_t, \rho_t) \quad \text{and} \quad \hat{G} \in \mathcal{H}(M_g, \rho_g)$$

for some $M_g, M_t > 0$ and $\rho_g, \rho_t > 1$ where

$$\hat{T} = \hat{P}\hat{C}(I + \hat{P}\hat{C})^{-1}, \quad \hat{G} = \hat{C}(I + \hat{P}\hat{C})^{-1} .$$

Assumption 3: The sequence $\{\hat{v}_k\}$ in experimental data (10) are equally spaced frequency response samples of a known real rational function $\hat{v}(z) \in \mathcal{H}^\infty$ and $\hat{v}^{-1} \in \mathcal{H}(M_v, \rho_v)$ for some $M_v > 0$ and $\rho_v > 1$.

Assumption 1 is necessary if one intends to quantify the identification error in $\mathcal{H}^\infty$ norm. Further, the spectrum of the sensor noise can be analyzed generically which may lead to the weighting function $\hat{w}$ and the weighted noise level ϵ. Assumption 2 is basically exponential stabilizability of the the plant $\hat{P}$. It is noted that the minimum phase assumption on controller $\hat{C}$ is in contrast to the stability assumption on controller $\hat{C}$ as in [17]. Assumption 3 is always feasible provided one can perform the sine-dwell experiments on the physical plant. Assumptions 1-3 implies that in perfect measurement case, there exist $M_u, M_y > 0$ and $\rho > 1$ such that the $\mathcal{Z}$-transforms of weighted outputs satisfy

$$\hat{u}_w = \hat{u}\hat{w} = \hat{G} \in \mathcal{H}(M_y, \rho) \quad \text{and} \quad \hat{y}_w = \hat{y}\hat{w} = \hat{T} \in \mathcal{H}(M_y, \rho)$$

if $\hat{v}$ is chosen such that $\hat{v}\hat{w} \equiv 1$. In fact, with $\hat{v}\hat{w} \equiv 1$, Assumption 3 follows from Assumption 1. Further, Assumption 2 implies that the transfer functions $\hat{G}$ and $\hat{T}$ are (right) coprime. This is due to the fact that with $\hat{X} = I$ and $\hat{Y} = \hat{C}^{-1}$,

$$\hat{X}\hat{T} + \hat{Y}\hat{G} = I .$$

For noisy samples, there holds the relation

$$\begin{pmatrix} \hat{u}_w^E(z_k) \\ \hat{y}_w^E(z_k) \end{pmatrix} = \begin{pmatrix} \hat{G}(z_k) \\ \hat{T}(z_k) \end{pmatrix} + \begin{pmatrix} \eta_y^k \\ \eta_u^k \end{pmatrix} \hat{w}(z_k), \quad k = 0, 1, 2, ..., N-1 . \quad (11)$$

Hence, the time-domain measurement data with periodic input is equivalent to the above experimental frequency response data with noise uniformly bounded by $\epsilon > 0$.

Remark. The implication of Assumptions 1-3 and $\hat{v}\hat{w} \equiv 1$ is that

$$\hat{H}(z) = \begin{pmatrix} \hat{G}(z) \\ \hat{T}(z) \end{pmatrix} \in \mathcal{H}(M, \rho), \quad M = \sqrt{M_u^2 + M_y^2}$$

and the noisy frequency response estimate of $\hat{H}(z_k)$ is available as in (11) for $0 \leq k \leq N - 1$. Let $\hat{H}_{id} \in \mathcal{H}^\infty$ be the identified model of $\hat{H}$ and

$$\begin{pmatrix} \hat{D}_{id} \\ \hat{N}_{id} \end{pmatrix} = \hat{H}_{id}\hat{R}^{-1}$$

with $\hat{R}$ as the spectral factor of $H_{id}(z^{-1})^T H_{id}(z)$. Then, $\hat{P}_{id} = \hat{N}_{id}\hat{D}_{id}^{-1}$ is the normalized coprime factorization of the identified model. Hence, the directed gap ball for identification error is

$$\mathbf{B}(\hat{P}_{id}, \hat{H}) = \inf_{\hat{Q} \in \mathcal{H}^\infty} \|\hat{H}_{id}\hat{R}^{-1} - \hat{H}\hat{Q}\|_\infty \leq \|\hat{R}^{-1}\|_\infty \|\hat{H}_{id} - \hat{H}\|_\infty \ .$$

Let $e_N(A_N)$ be the worst case identification error for the true unknown system $\hat{H}$ with algorithm A_N. The worst case directed gap ball is then (by abuse of notation)

$$\mathbf{B}(\hat{P}_{id}, \mathcal{H}(M, \rho), \epsilon) = \sup_{\hat{H} \in \mathcal{H}(M, \rho), |\hat{\eta}_k| \leq \epsilon} \mathbf{B}(\hat{P}_{id}, \hat{H}) \leq \|\hat{R}^{-1}\|_\infty e_N(A_N) \quad (12)$$

where $|\hat{\eta}_k| = |\hat{w}_k|\sqrt{|\hat{\eta}_k^u|^2 + |\hat{\eta}_k^y|^2}$. We are thus led to identify $\hat{H}(z)$, the coprime factors of $\hat{P}$ based on experimental data (11) and to quantify the worst case identification error in $\mathcal{H}^\infty$ norm which is exactly the problem of identification in $\mathcal{H}^\infty$.

The next result tells us that any convergent algorithm A_N for identification in $\mathcal{H}^\infty$ converges for identification of feedback systems.

Theorem 4. *Suppose that Assumptions 1-3 hold true. Let $\hat{H}_{id}$ be the identified model based on experimental data (11) and $e_N(A_N)$ be the worst case identification error as defined in (9) with convergent algorithm A_N. Then, the worst case identification error quantified with the directed gap metric converges to zero asymptotically as the period N of the input signal $\{v(t)\}_{t=0}^{N-1}$ goes to infinity and the noise level goes to zero where $\{v(t)\}_{t=0}^{N-1}$ is the N-point inverse DFT of $\{\hat{w}(z_k)^{-1}\}_{k=0}^{N-1}$. Further, if the controller $\hat{C} \equiv 1$ and $e_N(A_N) < 1/2$, the worst-case identification error for directed gap metric as in (12) is bounded by*

$$\mathbf{B}(\hat{P}_{id}, \mathcal{H}(M, \rho), \epsilon) \leq \frac{2e_N(A_N)}{1 - 2e_N(A_N)} \ .$$

Because of the page limit, the proof is omitted. But the readers should not have difficulty to deduce the above results.

It is noted that the performance of the algorithm is related to the stabilizing controller. In fact, the quantities $\|\hat{R}^{-1}\|_\infty$ and $e_N(A_N)$ are determined by the performance of the feedback system. It is possible that the exponential stability of the closed-loop system is a quite strong assumption. However, Theorem 4 does apply to a more broad class of feedback systems. We will use a specific two-stage nonlinear algorithm to analyze the worst case identification error for feedback systems next.

Suppose that the stabilized feedback system belongs to $S \neq \mathcal{H}(M, \rho)$. We will assume that S is admissible set as in [7]. That is, let $\mathcal{P}_n$ be the collection of all polynomials of z^{-1} with degree no larger than n. Then, the set S is admissible if

$$\lim_{N > n \to \infty} d_n = \lim_{N > n \to \infty} \sup_{\hat{H} \in S} \left\{ \inf_{\hat{p}_n \in \mathcal{P}_n} \|\hat{p}_n - \hat{H}\|_\infty \right\} = 0 .$$

It is shown in [1] that the admissibility is equivalent to the relative compactness which is related to the ϵ-net [26] and n-width in approximation theory [20]. The next result follows from [8].

Corollary 5. *Let the window function used in the two-stage nonlinear algorithm be trapezoidal as given by*

$$w_{n,k} = \begin{cases} 1 & 0 \leq k \leq 2m \\ (n + m - k)/(n - m) & 2m \leq k \leq n + m \\ 1 + k/(n - m) & m - n \leq k \leq 0 \end{cases}$$

with $0 \leq m < n \leq N - m$. Let the anticausal part of the identified model at the first stage be $\hat{H}_{pi}^a$. Then the worst case identification error for $\hat{H}$ (consists of coprime factors of the true unknown plant) with the two-stage nonlinear algorithm is bounded by

$$e_N \leq (\epsilon + d_{2m})\sqrt{\frac{n + m}{n - m}} + d_{2m} + \|\hat{H}_{id}^a\|_H$$

where $\| \cdot \|_H$ denotes Hankel norm.

Since d_{2m} denotes the optimal approximation error using $2m$th order FIR models, the performance of this specific two-stage nonlinear algorithm ties to the smoothness of the frequency response of the true feedback system. Further, if $m + n = N$ and $n = 3m$ are chosen, we have

$$e_N \leq 2\sqrt{2}(\epsilon + d_{N/2}) + 2d_{N/2}$$

in light of the analysis in [7]. For the case $S = \mathcal{H}(M, \rho)$, then n-width theory in [20] can be used which yields $d_{N/2} = M\rho^{-(1+N/2)}$. The above error bound is quite tight and it can be used in (12) to estimate the worst case directed gap ball for identification of feedback systems.

Most of the results on identification in $\mathcal{H}^\infty$ are not applicable to on-line identification directly and on-line identification is studied mostly in adaptive estimation of parametric ARMA model. However, the results in this paper brings the possibility to on-line recursive identification in $\mathcal{H}^\infty$. In particular, the periodicity of the input signal $v(t)$ resembles the sufficient richness of the input signal as assumed in adaptive identification. Moreover, the identification of the coprime factors presented in this paper also yields a nominal ARMA model. Although the similarity between identification in $\mathcal{H}^\infty$ is not so obvious, it provides the opportunity for the future research to unify these two different areas which are the subareas of robust identification.

4 Conclusion

The problem of identification in $\mathcal{H}^\infty$ is considered for feedback systems. A novel time-domain experimental procedure is developed to obtain the frequency response data of the coprime factors of the true unknown (possibly unstable) plant. The identified model is represented by its normalized coprime factors and the worst case identification error between the true unknown plant and the identified model is quantified by the gap metric (with the radius of the directed gap ball). Under certain mild assumptions on the closed-loop system, the convergence of the various existing algorithms from identification in $\mathcal{H}^\infty$ is established for identification of feedback systems. Further, an explicit identification error bound for the two-stage nonlinear algorithm is given.

References

1. H. Akcay, G. Gu and P. P. Khargonekar, "Identification in $\mathcal{H}^\infty$ with nonuniformly spaced frequency response measurements," *Proceedings of American Control Conference*, 246-250, June 1992.

2. J. Chen, C.N. Nett, and M.K.H. Fan, "Worst-Case System identification in H_∞: Validation of A priori Information, Essentially Optimal Algorithms, and Error Bounds," *Proceedings of American Control Conference*, 251-257, June 1992.

3. J. Chen, G. Gu and C. Nett, "Worst case identification of continuous-time systems via interpolation," submitted to *Automatica*.

4. T. T. Georgiou and M. C. Smith, "Optimal robustness in the gap metric," *IEEE Trans. Automat. Contr.*, vol. 35, 673-685, 1990.

5. G. C. Goodwin, B. Ninnes, and M. E. Salgado, "Quantification of uncertainty in estimation," *Proc. of the 1990 Amer. Contr. Conf.*, pp. 2400-2405, San Diego, CA.

6. G. Gu, P. P. Khargonekar and E. B. Lee, "Approximation of infinite dimensional systems," *IEEE Trans. Automatic Control*, vol. 34, pp. 610-618, June 1989.

7. G. Gu and P. P. Khargonekar, "A class of algorithms for identification in $\mathcal{H}^\infty$," *Automatica*, vol. 28, pp. 199-312, March 1992.

8. G. Gu, P. P. Khargonekar and Y. Li, "Robust convergence of two-stage non-linear algorithms for identification in $\mathcal{H}^\infty$, *Syst. and Control Lett.*, vol. 18, 253-263, April 1992.

9. D. McFarlane and K. Glover, *Robust Controller Design Using Normalized Coprime Factor Plant Descriptions*, (Lecture Notes in Control and Information Sciences, Vol. 138). New York: Springer Verlag, 1989.

10. G. Gu, D. Xiong, and K. Zhou, "Identification in $\mathcal{H}^\infty$ Using Pick's Interpolation," submitted to *Syst. Contr. Lett.*.

11. A. J. Helmicki, C. A. Jacobson and C. N. Nett, "Control oriented system identification: a worst-case/deterministic approach in $\mathcal{H}^\infty$," *IEEE Trans. Automat. Contr.*, vol. 36, 1163-1176, Oct. 1991.

12. A. J. Helmicki, C. A. Jacobson, and C. N. Nett, "Identification in $\mathcal{H}^\infty$: linear algorithms," *Proc. of the 1990 American Control Conf.*, pp. 2418-2423, San Diego, CA.

13. P. Henrici, "Fact Fourier methods in computational complex analysis," *SIAM Rev.*, vol. 21, pp. 481-527, 1979.

14. L. Ljung, *System Identification: Theory for the User*, Prentice-Hall, Englewood Cliffs, 1987.

15. D. McFarlane and K. Glover, "A loop shaping design procedure using $\mathcal{H}^\infty$ synthesis," *IEEE Trans. Automat. Contr.*, vol. 37, pp. 759-769, June 1992.

16. J. M. Krause and P. P. Khargonekar, "Parameter identification in the presence of non-parametric dynamic uncertainty," *Automatica*, vol. 26, 113-124, 1990.

17. P. M. Mäkilä and J. R. Partington, "Robust identification of stabilizable systems," *Proceedings of 30th Conference on Decision and Control*, 629-633, 1991.

18. J. R. Partington, "Robust identification in $\mathcal{H}^\infty$," *J. Math. Anal. and Appl.*, vol. 166, 428-441, 1992.

19. J. R. Partington, "Algorithms for identification in $\mathcal{H}^\infty$ with unequally spaced function measurements", preprint.

20. A. Pinkus, *n-Widths in Approximation Theory*, Chapt. VIII, Springer-Verlag, Berlin, 1985.

21. K. Poolla, P. Khargonekar, A. Tikku, J. Krause, and K. Nagpal, "A time approach to model validation," to appear in the *Proc. 1992 American Control Conference*.

22. R. S. Smith and J. C. Doyle, "Towards a methodology for robust parameter identification," *Proc. America. Contr. Conf.*, pp. 2395-2399, 1990.

23. Special Issue on System Identification for Robust Control Design, *IEEE Trans. Automat. Contr.*, vol. 37, July 1992.

24. D. N. C. Tse, M. A. Dahleh, and J. N. Tsitsiklis, "Optimal Asymptotic Identification under Bounded Disturbances", Preprint, LIDS, MIT, Cambridge, MA. An abridged version is in *Proc. 1991 American Control Conference*, 1786-1787, Boston, MA.

25. D. Xiong, *Worst Case Identification in $\mathcal{H}^\infty$*, M.S. Thesis, Department of Electrical Engineering, Louisiana State University, July 1992.

26. G. Zames, "On the metric complexity of causal linear systems: ϵ-entropy and ϵ-dimension for continuous time" *IEEE Trans. on Auto. Contr.*, vol. AC-24, 222-230, 1979.

27. T. Zhou and H. Kimura, "Identification for robust control in time-domain," submitted for publication, Dept. of Mechanical Engineering, Osaka University, Osaka, JAPAN, 1992.

Identification of Feedback Systems from Time Series

Pertti Mäkilä[1] *and Jonathan Partington*[2]

[1] Åbo Akademi University, Department of Engineering, SF-20500 Åbo, Finland

[2] School of Mathematics, University of Leeds, Leeds LS2 9JT, UK

1 Introduction

Recently problems of identification of feedback systems for robust control design have attracted a lot of attention in the literature (see e.g. [25, 9, 14, 16, 45, 13, 8, 10, 30, 37, 43, 35] and the references therein). As there seems to be presently no general agreement as to what the correct theoretical framework(s) for defining and addressing the various issues in this field should be, it is necessary to start here by stating our view of some of the concepts and their relationships.

Worst-case identification is a broad concept which is concerned with the worst-case behaviour in identification over classes/sets of systems and experimental error (noise etc.). This is reminiscent of the philosophy in modern robust control [4] and in various parts of approximation theory and optimal algorithms theory [38]. Our joint work here [30, 31, 22] has dealt especially with issues of quantitative approximate identification of feedback systems under coprime factor uncertainty. We have also studied [20, 30, 31] issues of worst-case asymptotic analysis of identification to reveal intrinsic limitations to identification algorithm performance [39, 18, 27, 12, 19]. This complements the rich and most useful theory of asymptotic analysis in stochastic system identification [15].

Worst-case identification has been criticized [33] from the point of view of high time-complexity of the identification experiments to reach a certain guaranteed worst-case a priori computed identification error level. However the high time-complexity is not usually an issue in many application areas, such as in industrial process control, where low-order models are sufficient for control design. It is obviously an issue if one uses an inappropriate model structure or a bad input design! But it is precisely these issues that can be addressed in a systematic way with worst-case analysis of identification problems [20, 19, 31]. Obviously, in applications one can take advantage of a posteriori computed identification error levels using methods of set membership or parameter bounding identification (see e.g. [25, 23, 22] and the references therein). A more valid criticism of worst-case identification is that so far only some quite simple error models have been dealt with in a systematic way. It would be important to study the effect of other error structures than the basic bounded error model.

It should be remembered that the primary goal of identification for the purpose of robust control design is that it should provide the modelling information necessary for successful control design. Thus it would be more appropriate perhaps to measure the performance of an identification algorithm based on the inferred closed-loop performance of the control design obtained. This is, in fact, the natural way to study identification for the purpose of robust control design as it gives natural measures for the required accuracy of identification [30, 34, 35, 31, 22].

Then depending on what type of robust control design is appropriate in the application in question, it is possible to define several notions of identification algorithm (set-valued or singleton-valued algorithms) and identification algorithm performance and optimality. In the case that an analytical design method based on a nominal model is used, it is natural to study singleton-valued identification algorithms [22, 35]. If a set-valued robust design method, such as μ-synthesis, is used, it is more appropriate to study set-valued identification methods [37, 25, 23].

Finally as the results of any identification experiment depend on the priors used about the unknown system and the experimental error, see e.g. [44, 43] for a discussion, it is necessary to study problems of priors falsification and model validation [37] in the context of identification for robust control design.

This paper deals with identification of stabilizable feedback systems, and to a smaller degree with robustness of feedback stabilization of linear systems. Here we shall be concerned with the case when stabilization is equivalent to the closed-loop system being a bounded-input bounded-output (BIBO) stable operator, cf. l^1 optimal control [41, 2].

There are several ways to represent system uncertainty to deal with both stable and unstable systems: one is by considering perturbations of the graph of the system, another is by looking at perturbations of coprime factorizations of the system. The quantitative measure for the size of the perturbations depends on the particular space in which the graph, or the coprime factorization, is defined. We mention here the rich theory developed in the l^2 (L^2) Hilbert space set-up (see e.g. [47, 40, 7, 6, 26]). Furthermore, several papers dealing with various aspects of identification of systems in the gap, graph and/or chordal metrics have appeared recently [17, 39, 30]. It appears possible to develop an equally rich theory with many applications in the l^∞ input/output signal space set-up [3, 42, 1, 21].

There is a large literature on the identification of controlled autoregressive (ARX) and controlled autoregressive moving average (ARMAX) models of systems. Mainstream identification theory deals mainly with questions of consistency and convergence, in a stochastic sense, of ARX and ARMAX models obtained by prediction error type identification methods [15]. Set membership identification, or parameter bounding identification, provides a deterministic perspective for the identification of ARX and ARMAX models (see e.g. the surveys: [25, 23]). In the present work we are interested in a special type of ARX model: namely, in ARX models in which the AR part and the X part are coprime (cf. [9, 34, 36, 17, 30]).

The proofs of most of the results presented here can be found in [21, 31].

2 Systems and Distance Functions

A linear discrete-time system is defined as a linear convolution operator $G :$ $l^p \to l^p$ with $1 \le p \le \infty$. As usual the linear system G is called l^p stable if

$$\|G\|_{<p>} \equiv \sup_{x \in l^p, x \ne 0} \frac{\|Gx\|_{l^p}}{\|x\|_{l^p}} < \infty. \tag{1}$$

Here $\|G\|_{<p>}$ is the induced operator norm, or the system gain, over l^p. We shall often simplify the notation somewhat and write simply $\|G\|$.

Let S^p denote the Banach space of linear shift-invariant causal l^p stable systems equipped with the operator norm (1). It is well-known that S^∞ is isometrically isomorphic to l^1.

A convenient way of representing both l^p stable and unstable systems ($p \in [1, \infty]$) is to consider the quotient field $F(S^p)$ of S^p. $F(S^p)$ can be thought of as the set of all pairs (P, Q), $Q \ne 0$, of elements in S^p. Two elements (P, Q), (R, S) in $F(S^p)$ are equal if $PS = RQ$. Equivalently , we can say that $F(S^p)$ is the set of all systems with transfer functions of the form $\hat{P}(z)/\hat{Q}(z)$, where $P, Q \in S^p$, and $\hat{Q} \ne 0$. Here $\hat{P}(z) = \sum_{k \ge 0} p_k z^k$, where $\{p_k\}$ are the coefficients of the (unit) impulse response of P.

The system $G \in F(S^p)$ is said to have a coprime factorization (c.f.) (N, D) over S^p, if $\hat{G} = \hat{N}/\hat{D}$, $\hat{D} \ne 0$, $N, D \in S^p$, and there exist $X, Y \in S^p$ such that $NX + DY = 1$. Let $CF(S^p)$ denote the set of all causal systems in $F(S^p)$ that have a c.f. over S^p. Note that if $G \in S^\infty$ (S^1) then $G \in S^p$ for any $p \in [1, \infty]$. It follows that if $G \in CF(S^\infty)$ $(CF(S^1))$ then $G \in CF(S^p)$ for any $p \in [1, \infty]$. Furthermore, if $G \in CF(S^\infty)$ $(CF(S^1))$ has a c.f. (N, D) over S^∞ (S^1) then (N, D) is also a c.f. of G over S^p for any $p \in [1, \infty]$. We shall often say simply that G has a c.f. when it is clear over which S^p space the c.f. is defined.

The graph (in l_2^p) of a system $G \in CF(S^p)$ is defined as the set of l_2^p-bounded input/output pairs corresponding to G, i.e.

$$gr(G; l^p) = \left\{ \begin{bmatrix} u \\ y \end{bmatrix} \in l_2^p \mid y = Gu \right\}. \tag{2}$$

The graph of G can then be expressed as (obvious generalization of [40])

$$gr(G; l^p) = \left\{ \begin{bmatrix} Dv \\ Nv \end{bmatrix} \in l_2^p \mid v \in l^p \right\}, \tag{3}$$

where (N, D) is any c.f. of G over S^p.

It is possible to introduce the so-called graph topology in the set $CF(S^p)$, making $CF(S^p)$ a topological space [40] exactly as in the thoroughly studied finite energy l^2 (H^2) set-up. Here we are concerned with the S^∞ setting.

We now define a collection of distance functions which can be used to measure convergence in the graph topology.

A c.f. (N, D) of $G \in CF(S^\infty)$ over S^∞ is said to be normalized (cf. [40]) if $\hat{N}(\bar{z})\hat{N}(z) + \hat{D}(\bar{z})\hat{D}(z) = 1$ for any $|z| = 1$. The following result is established in [21].

Theorem 1. *Let $G \in CF(S^\infty)$. Then G has a normalized c.f. over S^∞ which is unique to within multiplication by ± 1.*

Let $G_1, G_2 \in CF(S^\infty)$. Let (N_i, D_i) be a normalized c.f. of G_i, $i = 1, 2$, over S^∞. Denote

$$A_i = \begin{bmatrix} D_i \\ N_i \end{bmatrix}, \ i = 1, 2. \tag{4}$$

Introduce the directed quantity

$$\mathbf{d}(G_1, G_2) = \inf_{Q \in S^\infty, \|Q\|_{<\infty>} \leq 1} \|A_1 - A_2 Q\|_{<\infty>}. \tag{5}$$

Define

$$d(G_1, G_2) = \max\{\mathbf{d}(G_1, G_2), \mathbf{d}(G_2, G_1)\}, \tag{6}$$

where the notation is as in (4). It is easy to see that the d satisfies all the requirements for a metric. Note that the metric defined above is analogous to the the graph metric of the l^2 (S^2) setting [40]. Thus it is natural to call d the graph metric in our setting, too. Introduce also the quantity

$$\rho(G_1, G_2) = \inf_{Q \in S^\infty} \|A_1 - A_2 Q\|_{<\infty>}, \tag{7}$$

where the notation is as in (4). Finally, define

$$\rho(G_1, G_2) = \max\{\rho(G_1, G_2), \rho(G_2, G_1)\}. \tag{8}$$

This quantity, called here the rho function, is analogous to the Georgiou formula for the gap metric in the l^2 case [5]. If we define $\left\| \begin{bmatrix} X \\ Y \end{bmatrix} \right\|_{<\infty>} \equiv \max\{\|X\|_{<\infty>}, \|Y\|_{<\infty>}\}$, where $X, Y \in S^\infty$, the quantity $\rho(G_1, G_2)$ can be determined using the Dahleh-Pearson theory [2, 3].

Define the directed gap between the graphs of G_1 and G_2 in $CF(S^1)$ (note that we are using S^1 not S^∞, and thus an l^1 input/output space setting) as

$$\delta(G_1, G_2) = \sup_{x \in B(G_1)} \inf_{y \in gr(G_2; l^1)} \|x - y\|_{l_2^1} \tag{9}$$

where $B(G_1) = \{x \in gr(G_1; l^1) \mid \|x\|_{l_2^1} = 1\}$. In the sequel we shall denote the norm $\|\cdot\|_{l_2^1}$ in $l_2^1 = l^1 \times l^1$ simply as $\|\cdot\|$.

The (subspace) gap between G_1 and G_2 is defined as

$$\delta(G_1, G_2) = \max\{\delta(G_1, G_2), \delta(G_2, G_1)\}. \tag{10}$$

The gap between two systems is thus the usual gap between closed subspaces of Banach spaces.

It is also possible to define a projection gap γ using the fact that the graph of a system is a complemented subspace even in the S^∞ case; this is done in [21] where another normalized gap function, denoted by κ, is also introduced.

The following result (which is proved in [21]) relates the various distance notions given in this section.

Theorem 2. *Let $G \in CF(S^\infty)$, and let $\{G_i\}$ be a sequence in $CF(S^\infty)$. Then the following are equivalent:*

(i) $G_i \to G$ in the S^∞ graph topology;
(ii) $d(G_i, G) \to 0$;
(iii) $\rho(G_i, G) \to 0$;
(iv) $\delta(G_i, G) \to 0$;
(v) $\gamma(G_i, G) \to 0$;
(vi) $\kappa(G_i, G) \to 0$.

Thus all the distance functions presented here can be used to quantify the identification error in a way compatible with the requirements of robust control design.

3 Worst-case Identification of Stable Systems

Before turning our attention to stabilizable systems in general, let us reconsider the question of robust identification of BIBO stable systems from the point of view of input design. We suppose that we are given the stable model $y = h * u + v$, with $h \in l_1$ unknown, $v \in l_\infty$ comprising the noise, with $u \in l_\infty$ being the input, and with y the measured output. It is desired to choose $u(t)$ for $t \geq 0$ such that given $y(0), \ldots, y(n)$ we may construct an identified model $\tilde{h}_n$ such that the following *robust convergence* condition is satisfied.

$$\lim_{n \to \infty, \epsilon \to 0} \sup_{\|v\|_\infty \leq \epsilon} \|\tilde{h}_n - h\|_1 = 0. \tag{11}$$

This condition automatically implies a uniform bound for h lying in any relatively compact set [30, 28]. It is known [28, 12] that with u chosen to be an impulse or step no such algorithm can exist. On the other hand it is known [18, 39] that some input designs (e.g. Galois sequences) do guarantee the existence of such an algorithm. The following result gives a necessary and sufficient condition (in fact unknown inputs $u(t)$ for $t < 0$ can also be allowed, provided that they are bounded). Some related results are given in [29].

Theorem 3. *Given $u \in l_\infty$ and output measurements $y_0, y_1, \ldots,$ where $y = h * u + v$ and v is noise, as above, then there is a robustly convergent identification algorithm using y if and only if u satisfies:*

$$\exists C > 0 \qquad \text{such that for all } k \in l_1, \qquad \|k * u\|_\infty \geq C\|k\|_1. \qquad (12)$$

In particular it follows from results given in [29] that any complete model set (that is, any sequence of subspaces whose union is dense in l_1) can be used as a basis for a robustly convergent identification algorithm when one exists. The simplest is that in which the models have the form $(h_0, \ldots, h_{p-1}, 0, 0, \ldots)$, but given certain a priori information on the system it may be more convenient to use an alternative model set. Likewise it is possible to devise algorithms tuned to any closed absolutely convex set in which the true system is assumed to lie.

It is desirable that an identified model be of low order, and it is therefore generally necessary to perform a subsequent model reduction step. As in the H_∞ case, treated in [20] one can use the inequality $\|h\|_1 \leq 2n\|\Gamma\|$ (where Γ is the Hankel operator corresponding to a degree-n system with impulse response h) to deduce the following.

Theorem 4. *Let $h, \bar{h}$ be in l_1, and let $\hat{h}$ be the impulse reponse of an nth order optimal Hankel approximation to $\bar{h}$. Then*

$$\|h - \hat{h}\|_1 \leq (8n + 1)d_n(h) + 8n\|h - \bar{h}\|_1, \qquad (13)$$

where $d_n(h)$ is the minimum l_1 distance from h to a degree-n system.

Thus if h is well approximable (in particular if h is exponentially stable), a model reduction step on the identified model can be used successfully.

4 ARX Models

A convenient way to represent the input/output signal dependency of causal linear shift-invariant systems in $CF(S^\infty)$ is to use controlled autoregressive (ARX) models. Thus consider

$$A(q^{-1})y(t) = B(q^{-1})u(t) + v(t), \qquad (14)$$

where q^{-1} is the backward shift operator (i.e. $q^{-1}y(t) = y(t-1)$ etc.), y is the output, u is the input, v is a bounded disturbance, and $A(z)$, $B(z)$, interpreted as complex-valued functions of the complex variable z, are functions analytic in the open unit disk with absolutely convergent Fourier series. Here it is usual to take $A(0) \neq 0$. (Then in digital control applications $B(0) = 0$ due to strict causality requirements.)

Here it will be convenient to write equation (14) as a convolution operator equation

$$Ay = Bu + v, \qquad (15)$$

where the meaning of the symbols is obvious from (14). A and B are l^∞ stable operators.

Now let (X_0, Y_0) denote a coprime factorization (c.f.) of a stabilizing controller K_0 over S^∞. Let $N_0, D_0 \in S^\infty$ be such that the Bezout identity $N_0 X_0 + D_0 Y_0 = 1$ is satisfied. By the Youla parametrization of all plants in $CF(S^\infty)$ stabilizable by K_0, we can express (15) as

$$(D_0 - RX_0)y = (N_0 + RY_0)u + v, \tag{16}$$

where $R \in S^\infty$. This can be written in the more convenient form

$$D_0 y - N_0 u = R(X_0 y + Y_0 u) + v. \tag{17}$$

The unknown system G is given by the Youla parametrization

$$G = (N_0 + RY_0)(D_0 - RX_0)^{-1}. \tag{18}$$

Thus we see that "all" we need to do is to identify R accurately enough in the S^∞ norm (which is the same as to identify accurately the impulse response of R in the l^1 norm), and we are then guaranteed to get a good approximation to the unknown system G in the S^∞ graph topology.

The R-scheme as described above becomes particularly transparent when the input is chosen as follows [9]. Define

$$u(t) = [K_0(r_1 - y) + r_2](t), \tag{19}$$

where r_1 and r_2 are bounded reference inputs. The reference inputs r_1 and r_2 are to act as probing signals to guarantee sufficient information about the c.f. (N, D) of the unknown plant, $N = N_0 + RY_0$, and $D = D_0 - RX_0$. With the above choice of u, we see that

$$D_0 y - N_0 u = R(X_0 r_1 + Y_0 r_2) + v. \tag{20}$$

Let $\chi = \begin{bmatrix} \Delta u \\ \Delta y \end{bmatrix}$ denote the closed-loop effect of the reference inputs to u and y (this is obtained by setting the noise (disturbance) $v = 0$ (due to linearity)). Take $r_1 = N_0 w$ and $r_2 = D_0 w$, where w is a bounded signal. Then

$$D_0 y - N_0 u = Rw + v, \tag{21}$$

and $\chi = \begin{bmatrix} D \\ N \end{bmatrix} w$, i.e. the probing action is concentrated completely into the graph of the unknown system. Thus any element of the graph of G can be generated by a proper choice of w. Note that the common reference signal generator w acts now as a simple bounded input in (21).

Let $\mathcal{A}$ denote an untuned closed-loop identification algorithm mapping the assumed experimental information $Y_N = \{(D_0 y)(t) - (N_0 u)(t)\}_{t=0}^{N-1}$, $W_N = \{w(t)\}_{k=0}^{N-1}$ into a model $\mathcal{A}(Y_N, W_N) \in CF(S^\infty)$ of the unknown system G. Note that it is natural to call here an algorithm $\mathcal{A}$ as untuned when it only uses the a priori information about the unknown system that it is stabilizable by a known controller K_0.

Theorem 5. *Let $C > 0$ given. There exists a bounded reference input w satisfying $\sup_{t \geq 0} |w(t)| \leq C$, and an untuned closed-loop identification algorithm $\mathcal{A}$ such that*

$$\lim_{N \to \infty, \epsilon \downarrow 0} \sup_{\substack{|v(t)| \leq \epsilon \\ |w(t)| \leq \alpha, t < 0}} m(G - \mathcal{A}(Y_N, W_N)) = 0 \tag{22}$$

for any $G \in CF(S^\infty)$ stabilizable by $K_0 \in CF(S^\infty)$, and for any $\alpha \geq 0$. Here $m = \mathrm{d}, d, \rho, \rho, \delta, \gamma,$ or κ.

This result follows directly from the above description of the $R -$ scheme, using the results of Sect. 2, and an the existence of robustly convergent l^1 identification algorithms cf. [18, 39]. Thus if there is small experimental uncertainty, we can identify the unknown system G accurately (here it must be added that some additional prior information should be available about the unknown system G in order to derive useful worst-case identification error bounds after a given finite number N of data points - it suffices to have e.g. information which defines a relatively compact set of R operators, see e.g. [18, 12]. Explicit bounds relating identification errors in R to errors in G can be found in [31].

References

1. Dahleh, M.A.: BIBO stability robustness in the presence of coprime factor perturbations. Preprint (1990).
2. Dahleh, M.A. and Pearson, J.B.: l^1 optimal feedback controllers for MIMO discrete-time systems. IEEE Trans. Automat. Control **AC-32** (1987) 314–322.
3. Dahleh, M.A. and Pearson, J.B.: Optimal rejection of persistent disturbances, robust stability, and mixed sensitivity minimization. IEEE Trans. Automat. Control, **AC-33** (1988) 722–731.
4. Doyle, J.C., Francis, B.A. and Tannenbaum, A.: Feedback Control Theory. MacMillan, New York (1992).
5. Georgiou, T.T.: On the computation of the gap metric. Systems and Control Letters 11 (1988) 253–257.
6. Georgiou, T.T. and Smith, M.C.: Optimal robustness in the gap metric. IEEE Trans. Automat. Control **AC-35** (1990) 673–686.
7. Glover, K. and McFarlane, D.C.: Robust stabilization of normalized coprime factor plant descriptions with H_∞-bounded uncertainty. IEEE Trans. Automat. Control, **AC-34** (1989) 821–830.
8. Goodwin, G.C., Ninness, B. and Salgado, M.E.: Quantification of uncertainty in estimation. Proc. 1990 American Control Conf., San Diego.
9. Hansen, F., Franklin, G. and Kosut, R.: Closed-loop identification via the fractional representation. Proc. 1989 American Control Conf., Pittsburgh.
10. Helmicki, A.J., Jacobson, C.A. and Nett, C.N.: Control oriented system identification: A worst-case/ deterministic approach in H^∞. IEEE Trans. Automat. Control **AC-36** (1991) 1163–1176.
11. Jacobson, C.A. and Nett, C.N.: Worst case system identification in l^1: optimal algorithms and error bounds. Proc. 1991 American Control Conf., Boston.

12. Jacobson, C.A., Nett, C.N. and Partington, J.R.: Worst case system identification in l^1: optimal algorithms and error bounds. Systems and Control Letters (to appear).

13. Kosut, R.L., Lau, M., and Boyd, S.: Identification of systems with parametric and nonparametric uncertainty. Proc. 1990 American Control Conf.

14. Krause, J.M. and Khargonekar, P.P.: Parameter identification in the presence of nonparametric uncertainty. Automatica 26 (1990) 113–124.

15. Ljung, L.: System Identification. Theory for the User. Prentice-Hall, Englewood-Cliffs (1987).

16. Mäkilä, P.M.: Approximation and identification of continuous-time systems. Int. J. Control 52 (1990) 669–687.

17. Mäkilä, P.M.: Identification of stabilizable systems: closed-loop approximation. Int. J. Control 54 (1991) 577–592.

18. Mäkilä, P.M.: Robust identification and Galois sequences. Int. J. Control 54 (1991) 1189–1200.

19. Mäkilä, P.M.: Worst-case input-output identification. Int. J. Control (to appear).

20. Mäkilä, P.M. and Partington, J.R.: Robust approximation and identification in H^∞. Proc. 1991 American Control Conf., Boston, Vol. 1, 70–76.

21. Mäkilä, P.M. and Partington, J.R.: Robust stabilization − BIBO stability, distance notions and robustness optimization. Automatica (to appear).

22. Mäkilä, P.M. and Partington, J.R.: On bounded error identification of feedback systems. Submitted.

23. Milanese, M. and Vicino, A.: Optimal estimation theory for dynamic systems with set membership uncertainty: An overview. Automatica, 27 (1991) 997–1009.

24. Mo, S.H. and Norton, J.P.: Fast and robust algorithm to compute exact polytope parameter bounds. Mathematics and Computers in Simulation 32 (1990) 481–493.

25. Norton, J.P.: Identification and application of bounded-parameter models. Automatica 23 (1987) 497–507.

26. Partington, J.R.: Approximation of unstable infinite-dimensional systems using coprime factors. Systems and Control Letters 16 (1991) 89–96.

27. Partington, J.R.: Robust identification in H^∞. J. Math. Anal. Appl. 166 (1992) 428–441.

28. Partington, J.R.: Worst case identification in Banach spaces. Systems and Control Letters 18 (1992) 423–428.

29. Partington, J.R.: Interpolation in normed spaces from the values of linear functionals. Submitted.

30. Partington, J.R. and Mäkilä, P.M.: Robust identification of strongly stabilizable systems. Proc. IEEE Conference on Decision and Control, Brighton (1991) 629–633. Also IEEE Trans. Automat. Control. (to appear).

31. Partington, J.R. and Mäkilä, P.M.: Worst-case identification of feedback systems from closed-loop time series. Submitted. A shortened version appears in Proc. 1992 American Control Conference, Chicago, Vol. 1, 301–306.

32. Poolla, K., P. Khargonekar, Tikku, A., Krause, J. and Nagpal, K.: A time-domain approach to model validation. Proc. 1992 American Control Conf., Chicago.

33. Poolla, K. and Tikku, A.: On the time complexity of worst-case system identification. Preprint.

34. Schrama, R.J.P.: Control-oriented approximate closed-loop identification via fractional representations. Proc. 1991 American Control Conf., Boston.

35. Schrama, R.J.P.: Approximate Identification and Control Design with application to a mechanical system. Ph.D. Thesis, Delft Univ. of Technology (1992).

36. Schrama, R.J.P., Bongers, P.M.M.: Experimental robustness analysis based on coprime factorizations. Selected Topics in Identification, Modelling and Control, vol. 3. Progress Rep. Mechanical Engineering Systems and Control Group, Delft Univ. of Technology (1991).

37. Smith, R.S. and Doyle, J.C.: Model validation : a connection between robust control design and identification. IEEE Trans. Automat. Control, AC-37 (1992) 942–952.

38. Traub, J.F., Wasilowski, G.W. and Woźniakowski, H.: Information-based complexity. Academic Press, New York (1988).

39. Tse, D.N.C., Dahleh, M.A. and Tsitsiklis, J.N.: Optimal asymptotic identification under bounded disturbances. Proc. 30th IEEE Conf. on Decision and Control, Brighton (1991). Also IEEE Trans. Automat. Control (to appear).

40. Vidyasagar, M.: Control System Synthesis. The MIT Press, Cambridge, Massachusetts (1985).

41. Vidyasagar, M.: Optimal rejection of persistent bounded disturbances. IEEE Trans. Automat. Control, AC-31 (1986) 527–534.

42. Vidyasagar, M. and Anderson, B.D.O.: Approximation and stabilization of distributed systems by lumped systems. Systems and Control Letters 12 (1989) 95–101.

43. Wahlberg, B. and Ljung. L.: Hard frequency-domain model error bounds from least-squares like identification techniques. IEEE Trans. Automat. Control. To appear.

44. Willems, J.C.: From time series to linear system − Part III. Approximate modelling. Automatica 23 (1987) 87–115.

45. Younce, R.C. and Rohrs, C.E.: Identification with non-parametric uncertainty. International Conference on Circuits and Systems, New Orleans (1990).

46. Zames, G.: On the metric complexity of causal linear systems : ϵ-entropy and ϵ-dimension for continuous time. IEEE Trans. Automat. Control AC-24 (1979) 222–230.

47. Zames, G. and El-Sakkary, A.K.: Unstable systems and feedback: The gap metric. Proc. Allerton Conference (1980).

Input-Output Extrapolation-Minimization Theorem and Its Applications to Model Validation and Robust Identification

Tong Zhou and Hidenori Kimura

Dept. of Mechanical Engineering for Computer-Controlled Machinery, Osaka University, 2–1, Yamada-oka, Suita, Osaka 565, Japan.

1 Introduction

In control system design, it is very important to obtain a model describing the dynamics of the plant to be controlled, no matter what synthesis method is used in the controller design, e.g., $\mathcal{LQG}$ control, $\mathcal{LQG}/\mathcal{LTR}$ control, adaptive control, predictive control, $\mathcal{H}^\infty$ control, $\mathcal{L}^1$ control, etc. It is just this reason that system identification has been a very active research area of control theory (Eykhoff et al. [4], Ljung [10]).

A common knowledge in control engineering is that we can not obtain exact model even for simple plants, due to the effect of noise, parameter change, etc. Fortunately, even though we can not obtain the exact model of a plant, we can still design a high quality controller with the help of robust control theory, provided that we can obtain a suitable nominal model of the plant and the bound of the uncertainty in the nominal model (Dorato et al. [3], Kimura [8]).

Compared with the traditional identification, the so called *robust identification* has many new features, e.g., (Goodwin et al. [5], Helmicki et al. [6]) (Krause et al. [9], Smith et al. [12]) (Zhou et al. [14, 15]). The most significant one is that in robust identification procedure, a *deterministic* approach is preferable to a stochastic one which has been the common framework of the traditional identification.

Briefly speaking, robust identification includes two steps. First, design identification experiment based on some prior information about the plant; second, obtain a suitable nominal model and its uncertainty bound using the experimental data and prior information.

On the other hand, model validation is to check whether the model set defined by a given nominal model and a given uncertainty bound includes a model which matches the input-output data obtained from experiment, along with a noise sequence which belongs to a given noise set.

In this paper, we formulate and tackle a problem which is related to both robust identification and model validation. Although we only discuss single-input/single-output discrete-time system here, the results can be immediately extended to multi-input/multi-output case.

The problem is formulated in Sect. 2, and the main results will be given in Sect. 3. In Sect. 4, we shall discuss some applications of the main results, and our paper will be concluded in Sect. 5.

Notations:

$\mathcal{H}^\infty$: The Hardy space of bounded analytic functions in the open unit disk.
$\|F(z)\|_\infty$: The H^∞-norm of a transfer function $F(z)$ belonging to $\mathcal{H}^\infty$.
$\mathcal{BH}^\infty$: $\mathcal{BH}^\infty = \{F(z) \mid F(z) \in \mathcal{H}^\infty,\ \|F(z)\|_\infty \leq 1\}$.
$\bar{\sigma}(X)$: The maximum singular value of a matrix X.
$x_-(z)$: $x_-(z) := \sum_1^n x_j z^{j-n-1}$ for a vector $x = [x_1\ x_2\ \cdots\ x_n]^T$.
$x_+(z)$: $x_+(z) := \sum_1^n x_j z^{n-j}$ for a vector $x = [x_1\ x_2\ \cdots\ x_n]^T$.
$x(z)$: $x(z) := \sum_1^n x_j z^{j-1}$ for a vector $x = [x_1\ x_2\ \cdots\ x_n]^T$.
$T(n)$: $T(n) := \begin{bmatrix} 0_{1\times n} & 0_{1\times 1} \\ I_{n\times n} & 0_{n\times 1} \end{bmatrix}$.

2 Problem Formulation

In robust control theory, three kinds of unstructured uncertainty are often used, i.e., additive uncertainty, multiplicative uncertainty, and coprime factor description uncertainty (Dorato et al. [3], Kimura [8]).

$$G(z) = G_0(z) + \Delta_a(z)W_a(z), \tag{1}$$

$$G(z) = G_0(z)[I + \Delta_m(z)W_m(z)], \tag{2}$$

$$G(z) = [D_0(z) + \Delta_d(z)W_d(z)]^{-1}[N_0(z) + \Delta_n(z)W_n(z)], \tag{3}$$

where $W_a(z)$, $W_m(z)$, $W_d(z)$, $W_n(z)$ reflect the magnitude frequency characteristics of the uncertainty, and $\Delta_a(z)$, $\Delta_m(z)$, $[\Delta_d(z)\ \Delta_n(z)]$ belong to some "ball". That is,

$$\|\Delta_a(z)\| \leq \gamma, \quad \|\Delta_m(z)\| \leq \gamma, \quad \|[\Delta_d(z)\ \Delta_n(z)]\| \leq \gamma,$$

for some suitable norm and constant γ.

For the compatibility with $\mathcal{H}^\infty$ control, the "ball" is measured by H^∞-norm in this paper.

Let $U(z)$, $Y(z)$ represent the input and the corresponding output of a plant, respectively. Then, (1), (2) and (3) can be rewritten as

$$Y(z) - G_0(z)U(z) = \Delta_a(z)W_a(z)U(z), \tag{4}$$

$$Y(z) - G_0(z)U(z) = \Delta_m(z)G_0(z)W_m(z)U(z), \tag{5}$$

$$D_0(z)Y(z) - N_0(z)U(z) = [\Delta_d(z)\ \Delta_n(z)]\begin{bmatrix} -W_d(z)Y(z) \\ W_n(z)U(z) \end{bmatrix}. \tag{6}$$

Model validation is in fact to check whether there is a $\Delta_a(z)$ $(\Delta_m(z)$, $[\Delta_d(z) \quad \Delta_n(z)])$ in the given "ball" that makes $G(z)$ match the given experimental data. On the other hand, robust identification is essentially to find suitable $G_0(z)$ $([D_0(z) \quad N_0(z)])$ and $W_a(z)$ $(W_m(z), [W_d(z) \quad W_n(z)])$ such that the uncertainty "ball" is as small as possible, because in controller design, a tighter model set is desirable (Dorato et al. [3], Kimura [8]).

In model validation, the "radius" of the "ball", γ, can be normalized to 1, without loss of generality.

On the other hand, robust identification and model validation are generally performed using partial information about the plant, because we can not obtain perfect information of the plant through experiment. For example, the input and the output of the plant can only be measured in a finite time interval, etc.

Now, we formulate the following problem,

Problem Given input-output data (u_0, y_0), (u_1, y_1), $\cdots\cdots$, (u_n, y_n), find the minimal H^∞-norm of the transfer functions which match the prescribed input-output data.

3 Main Results

To solve the problem formulated above, we first note the following fact.

Fact 1 *Let* $G(z) = \sum_0^\infty g_i z^i$ *be the transfer function of a plant. Suppose* $u_0 \neq 0$. *Then* $G(z)$ *matches the given input-output data* (u_0, y_0), (u_1, y_1), $\cdots\cdots$, (u_n, y_n), *if and only if* $g = U^{-1}y$, *where* $g := [g_0 \quad g_1 \quad \cdots \quad g_n]^T$, $y := [y_0 \quad y_1 \quad \cdots \quad y_n]^T$, *and* $U = \sum_0^n u_j T^j(n)$.

The next theorem is well-known in classical extrapolation theory (Adamja et al. [1], Kimura [7]).

Theorem 2 (Schur-Takagi-AAK). *For a given vector* $a = [a_0\, a_1\, \cdots\, a_n]^T$, *define a matrix*

$$\bar{A} = [T^n(n)a \quad T^{n-1}(n)a \quad \cdots \quad T(n)a \quad a]$$

and a function set

$$\mathcal{F}(z) =$$
$$\left\{ f(z) \,\middle|\, f(z) = a_0 + a_1 z + \cdots + a_n z^n + z^{n+1} p(z),\; p(z) = \sum_{i=0}^\infty h_i z^i \right\}.$$

Then the function which belongs to $\mathcal{F}(z)$ *and has minimal* H^∞-*norm is determined by*

$$f(z) = \alpha z^{n+1} \frac{\eta_-(z)}{\xi_+(z)},$$

where

$$\alpha = \bar{\sigma}(\bar{A}), \quad \bar{A}\xi = \bar{\sigma}(\bar{A})\eta, \quad \bar{A}\eta = \bar{\sigma}(\bar{A})\xi .$$

Moreover, $z^{n+1}\eta_-(z)/\xi_+(z)$ is an inner function and $\|f(z)\|_\infty = \alpha$.

Based on Fact 1 and Schur-Takagi-AAK theorem, we obtain the main results of this paper.

Theorem 3. *Assume $u_0 \neq 0$. Then, the unique transfer function which matches the given input-output data (u_0, y_0), (u_1, y_1), $\cdots$, (u_n, y_n) and has the minimal H^∞-norm is given by*

$$f(z) = \alpha \frac{\eta(z)}{\xi(z)},$$

where α is the maximal value which satisfies

$$(\bar{Y}^2 - \lambda^2 \bar{U}^2)\xi' = 0, \qquad \xi' \neq 0,$$

for matrices $\bar{U}$, $\bar{Y}$ defined as

$$\bar{U} = [T^n(n)u \ \ T^{n-1}(n)u \ \ \cdots \ \ T(n)u \ \ u],$$
$$\bar{Y} = [T^n(n)y \ \ T^{n-1}(n)y \ \ \cdots \ \ T(n)y \ \ y],$$

by vectors $u = [u_0 \ u_1 \ \cdots \ u_n]^T$ and $y = [y_0 \ y_1 \ \cdots \ y_n]^T$; and

$$\xi = \bar{U}\xi', \qquad \eta = \frac{1}{\alpha}\bar{Y}\xi' .$$

Moreover, $\eta(z)/\xi(z)$ is inner and $\|f(z)\|_\infty = \alpha$.

Proof. According to Fact 1, a transfer function $G(z) = \sum_0^\infty g_i z^i$ matches the given (u_0, y_0), (u_1, y_1), $\cdots$, (u_n, y_n) if and only if $g = U^{-1}y$, where g, y and U are defined as in Fact 1.

Let matrix

$$\bar{G} = [T^n(n)g \ \ T^{n-1}(n)g \ \ \cdots \ \ T(n)g \ \ g] .$$

Then

$$\begin{aligned}
\bar{G} &= [T^n(n)U^{-1}y \ \ T^{n-1}(n)U^{-1}y \ \ \cdots \ \ T(n)U^{-1}y \ \ U^{-1}y] \\
&= U^{-1}[T^n(n)y \ \ T^{n-1}(n)y \ \ \cdots \ \ T(n)y \ \ y] \\
&= U^{-1}\bar{Y} .
\end{aligned} \tag{7}$$

Assume that α is the maximal singular value of matrix $\bar{G}$, and ξ^0, η^0 is one of the corresponding Schmidt pairs, i.e.,

$$\bar{G}\xi^0 = \alpha\eta^0, \qquad \bar{G}\eta^0 = \alpha\xi^0 . \tag{8}$$

Then, according to Schur-Takagi-AAK Theorem, the unique transfer function which matches (u_0, y_0), (u_1, y_1), $\cdots$, (u_n, y_n) and has the minimal H^∞-norm α is given by

$$f(z) = \alpha z^{n+1} \frac{\eta_-^0(z)}{\xi_+^0(z)}, \qquad (9)$$

and $z^{n+1}\eta_-^0(z)/\xi_+^0(z)$ is inner.

Since $\bar{G}$, $\bar{Y}$ are symmetric, we have

$$\bar{G}^2 = \bar{G}\bar{G}^T = U^{-1}\bar{Y}^2 U^{-T}, \qquad (10)$$

$$U^{-1}\bar{Y} = (U^{-1}\bar{Y})^T = \bar{Y}U^{-T}. \qquad (11)$$

Let

$$\xi' := U^{-T}\xi_0.$$

From (8) and (10), we have

$$\bar{Y}^2\xi' = \alpha^2 UU^T\xi'. \qquad (12)$$

Let

$$T_1 := \begin{bmatrix} 0 & \cdots & 0 & 1 \\ 0 & \cdots & 1 & 0 \\ \vdots & \ddots & \vdots & \vdots \\ 1 & \cdots & 0 & 0 \end{bmatrix}.$$

Then, it is obvious that

$$U^T = T_1\bar{U}, \qquad (13)$$

and

$$UU^T = (T_1\bar{U})^T(T_1\bar{U}) = \bar{U}^2. \qquad (14)$$

Therefore, (12) can be rewritten as

$$(\bar{Y}^2 - \alpha^2\bar{U}^2)\xi' = 0. \qquad (15)$$

Moreover,

$$\begin{aligned}
\xi_+^0(z) &= (U^T\xi')_+(z) = (T_1\bar{U}\xi')_+(z) \\
&= [z^n \; z^{n-1} \; \cdots \; z \; 1]T_1\bar{U}\xi' \\
&= [1 \; z \; \cdots \; z^{n-1} \; z^n]\bar{U}\xi' \\
&= (\bar{U}\xi')(z) = \xi(z).
\end{aligned} \qquad (16)$$

On the other hand, from (8) and (11), we have

$$\eta^0 = \frac{1}{\alpha}\bar{G}\xi^0 = \frac{1}{\alpha}(U^{-1}\bar{Y})(U^T\xi') = \frac{1}{\alpha}\bar{Y}\xi' = \eta, \qquad (17)$$

which implies

$$z^{n+1}\eta_-^0(z) = \eta(z). \qquad (18)$$

According to (9), (16) and (18), we obtain

$$f(z) = \alpha z^{n+1} \frac{\eta_-^0(z)}{\xi_+^0(z)} = \alpha \frac{\eta(z)}{\xi(z)}, \tag{19}$$

and $\eta(z)/\xi(z)$ is inner.

It is obvious that $\|f(z)\|_\infty = \alpha$.   □

We call Theorem 3 the *input-output extrapolation-minimization theorem*. Now, we give an example to illustrate this theorem.

Example 1. Find a transfer function which matches $(1, 0.5)$, $(-1, 1.5)$ and has the minimal H^∞-norm.

According to the definition of $\bar{U}$ and $\bar{Y}$, we have

$$\bar{U} = \begin{bmatrix} 0 & 1 \\ 1 & -1 \end{bmatrix}, \qquad \bar{Y} = \begin{bmatrix} 0 & 0.5 \\ 0.5 & 1.5 \end{bmatrix},$$

and

$$|\bar{Y}^2 - \lambda\bar{U}^2| = \lambda^2 - \frac{9}{2}\lambda + \frac{1}{16}.$$

Therefore, $\alpha = \frac{2+\sqrt{5}}{2}$.

From $(\bar{Y}^2 - \alpha\bar{U}^2)\xi' = 0$, $\xi = \bar{U}\xi'$, $\eta = \frac{1}{\alpha}\bar{Y}\xi'$, we have

$$\xi' = \beta \begin{bmatrix} \sqrt{5} - 1 \\ 1 \end{bmatrix}, \qquad \xi = \beta \begin{bmatrix} 1 \\ \sqrt{5} - 2 \end{bmatrix},$$

$$\eta = \beta \begin{bmatrix} \sqrt{5} - 2 \\ 1 \end{bmatrix}. \qquad (\beta \neq 0)$$

The desirable transfer function is

$$f(z) = \frac{2 + \sqrt{5}}{2} \frac{\beta[\sqrt{5} - 2 + z]}{\beta[1 + (\sqrt{5} - 2)z]} = \frac{(\sqrt{5} + 2)^2}{2} \frac{z + \sqrt{5} - 2}{z + \sqrt{5} + 2}.$$

From Theorem 3, we can immediately derive the following corollary, which is useful in model validation.

Corollary 4. *Assume $u_0 \neq 0$. Then, there exists a transfer function which belongs to $\mathcal{BH}^\infty$ and matches (u_0, y_0), (u_1, y_1), $\cdots$, (u_n, y_n), if and only if $Y^T Y \leq U^T U$, where the matrices U and Y are defined as*

$$U = \sum_0^n u_j T^j(n), \qquad Y = \sum_0^n y_j T^j(n).$$

4 Applications of the Main Results

4.1 Application to Model Validation

The input-output extrapolation-minimization theorem is useful in robust identification and model validation (Poolla et al. [11], Zhou et al. [13, 14, 15]). Here, we discuss its application to model validation.

Assume that the transfer function of a plant is given as

$$G(z) = G_0(z) + \Delta_a(z)W_a(z),$$

where

$$G_0(z) = \frac{a_0 + a_1 z + \cdots + a_p z^p}{1 + b_1 z + \cdots + b_p z^p}, \qquad W_a(z) = \frac{c_0 + c_1 z + \cdots + c_q z^q}{1 + d_1 z + \cdots + d_q z^q},$$

$$\|\Delta_a(z)\|_\infty \leq 1.$$

Moreover, assume that $W_a(z)$ is an unimodular function in $\mathcal{H}^\infty$, the measured input-output data $(u_0, y_0), (u_1, y_1), \cdots, (u_n, y_n)$ are noise free and $u_0 \neq 0$. Then, we have the following theorem.

Theorem 5. *The measured input-output data does not invalidate the given plant model, if and only if*

$$(BY - AU)^T D^T D(BY - AU) \leq U^T B^T C^T CBU,$$

where the matrices A, B, C, D, U, Y are defined as

$$A = \sum_0^n a_j T^j(n), \qquad B = \sum_0^n b_j T^j(n), \qquad C = \sum_0^n c_j T^j(n),$$

$$D = \sum_0^n d_j T^j(n), \qquad U = \sum_0^n u_j T^j(n), \qquad Y = \sum_0^n y_j T^j(n);$$

and $a_j := 0, b_j := 0$, for $j > p$; $c_j := 0, d_j := 0$, for $j > q$; $b_0 := 1, d_0 := 1$.

Proof. It is not difficult to verify that the given plant model is not invalidated by the measured data, if and only if there exists a $\Delta_a(z) \in B\mathcal{H}^\infty$ which matches $(CBu, D(By - Au))$.

According to Corollary 4, this means

$$(BY - AU)^T D^T D(BY - AU) \leq U^T B^T C^T CBU.$$

$\square$

Now, we consider the case where the measured output data are contaminated by additive noise whose magnitude is uniformly bounded by a constant, say, ϵ. That is, $y_i = \tilde{y}_i + v_i$, for all $0 \leq i \leq n$, where $|v_i| \leq \epsilon, 0 \leq i \leq n$; and $\tilde{y}_i, 0 \leq i \leq n$, stands for the noise free response of the plant to the given input $[u_0 \ u_1 \ \cdots \ u_n]$.

Assume that the transfer function of a plant has the same form as that in Theorem 5, through similar arguments, we obtain the following theorem.

Theorem 6. *The measured input-output data does not invalidate the given plant model, if and only if*

$$\min_{|v_i| \leq \epsilon} J_{mv} \leq 1, \quad J_{mv} := \bar{\sigma}[(B(Y-V)-AU)DC^{-1}B^{-1}U^{-1}],$$

where the matrices A, B, C, D, U and Y are the same as those in Theorem 5 and the matrix V is defined as

$$V = \sum_0^n v_j T^j(n) .$$

Since J_{mv} is a convex function of $[v_0 \; v_1 \; \cdots \; v_n]^T$, it is easy to check whether the condition in Theorem 6 is satisfied or not (Dem'Yanov et al. [2]). Similar results can be obtained when the noise is energy bounded.

Using the above approach, analogous results can also be easily derived for multiplicative uncertainty case, coprime factor description uncertainty case, etc.(Poolla et al. [11], Zhou et al. [13]).

4.2 Application to Robust Identification

Now, we consider how to apply the input-output extrapolation-minimization theorem to a robust identification problem.

First, we state the formulation of the identification problem.

Assume: The transfer function of the plant to be identified is represented as

$$G(z) = G_0(z) + \Delta_a(z)W_a(z), \qquad G_0(z) = \frac{a_0 + a_1 z + \cdots + a_p z^p}{1 + b_1 z + \cdots + b_p z^p},$$

where, b_1, b_2, $\cdots$, b_p are prescribed, and $\|\Delta_a(z)\|_\infty \leq 1$, $W_a(z) \in \mathcal{H}^\infty$.

Given:

1. A frequency weight function $W_0(z)$ which is an unimodular function in $\mathcal{H}^\infty$ and represented as

$$W_0(z) = \frac{c_0 + c_1 z + \cdots + c_q z^q}{1 + d_1 z + \cdots + d_q z^q} .$$

2. M sets of noise-free input-output data which correspond to M times identification experiments. Represent them as

$$(u_0^i, y_0^i), (u_1^i, y_1^i), \cdots, (u_{n_i}^i, y_{n_i}^i), \qquad 1 \leq i \leq M,$$

where $u_0^i \neq 0$, $1 \leq i \leq M$.

Find: parameters a_0, a_1, $\cdots$, a_p and a transfer function $W_a(z)$ such that the M sets input-output data are matched by the model set defined by $G_0(z)$ and $W_a(z)$ and the cost function defined as

$$J_{id} = \|W_0(z)W_a(z)\|_\infty$$

is minimized.

For given a_0, a_1, $\cdots$, a_p, we have the following theorem (Zhou et al. [14]).

Theorem 7. *If $W_a(z) = \beta W_0^{-1}(z)$, then there exists $\Delta_i(z) \in \mathcal{BH}^\infty$ such that the transfer function represented as*

$$G_i(z) = G_0(z) + \Delta_i(z)W_a(z)$$

matches the given input-output data (u_0^i, y_0^i), (u_1^i, y_1^i), $\cdots$, $(u_{n_i}^i, y_{n_i}^i)$, $1 \le i \le M$ and the cost function $J_{id} = \|W_0(z)W_a(z)\|_\infty$ is minimized $(J_{id} = \beta)$, where

$$\beta := \max_{1 \le i \le M} \{\bar{\sigma}(U_i^{-1}D_i^{-1}C_i(Y_iB_i - U_iA_i)B_i^{-1})\},$$

and the matrices A_i, B_i, C_i, D_i, U_i and Y_i are defined as

$$A_i = \sum_0^{n_i} a_jT^j(n_i), \qquad B_i = \sum_0^{n_i} b_jT^j(n_i), \qquad C_i = \sum_0^{n_i} c_jT^j(n_i),$$

$$D_i = \sum_0^{n_i} d_jT^j(n_i), \qquad U_i = \sum_0^{n_i} u_j^iT^j(n_i), \qquad Y_i = \sum_0^{n_i} y_j^iT^j(n_i);$$

and $a_j := 0$, $b_j := 0$, for $j > p$; $c_j := 0$, $d_j := 0$, for $j > q$; $b_0 := 1$, $d_0 := 1$.

From convex analysis, the following theorem can be shown.

Theorem 8.

$$\beta(a) = \max_{1 \le i \le M} \{\bar{\sigma}(U_i^{-1}D_i^{-1}C_i(Y_iB_i - U_iA_i)B_i^{-1})\}$$

is a convex function of vector $a := [a_0 \; a_1 \; a_2 \; \cdots \; a_p]^T$.

From the above results, we obtain our identification algorithm.

Step 1. Construct matrices A_i, B_i, C_i, D_i, U_i, Y_i, using the given parameters and input-output data, and leave a_0, a_1, $\cdots$, a_p free;

Step 2. Minimize the function $\beta(a)$ defined in Theorem 8 by means of convex optimization to obtain the nominal model's parameters a_0, a_1, $\cdots$, a_p and the minimal β, say, β^{opt};

Step 3. Let $W_a(z) = \beta^{opt}W_0^{-1}(z)$.

Since convex optimization is performed through linear programming such as cutting plane method, ellipsoid method, etc. and can be solved efficiently up to a thousand variables (Dem'Yanov et al. [2]), the proposed algorithm is linear and computationally tractable.

The above results can be extended to noisy data cases (Zhou et al. [14]).

5 Conclusion

In this paper, we investigated the input-output extrapolation-minimization problem and discussed some of its applications to model validation and robust identification. Although we only discussed the single-input/single-output case here, the results in this paper can be immediately extended to multi-input/multi-output case. Moreover, in addition to unstructured uncertainty, the input-output extrapolation-minimization theorem can also be applied to plants where its transfer function is expressed through nominal model, structured uncertainty and unstructured uncertainty (Zhou et al. [15]).

References

1. V.M.Adamja, D.Z.Arov and M.G.Krein, "Analytic Properties of Schmidt Pairs for a Hankel Operator and the Generalized Schur-Takagi Problem," Math. Sbornik, Vol,15, No.1, (1971), pp.31–73.
2. V.F.Dem'Yanov and L.V.Vasil'ev, "Nondifferentiable Optimization," Optimization Software, Inc. Publications Division, New York, (1985).
3. P.Dorato and R.K.Yedavall, "Recent Advances in Robust Control," IEEE Press, (1990).
4. P.Eykhoff and P.C.Parks, "Special Issue on Identification and System Parameter Estimation," Automatica, Vol.26, No.1, (1990).
5. G.C.Goodwin, B.Ninness and M.E.Saldado, "Quantification of Uncertainty in Estimation," Proc. American Control Conference, San Diego, CA, (1990), pp. 2400–2405.
6. A.J.Helmicki, C.A.Jacobson and C.N.Nett, "Control Oriented System Identification: A Worst-Case/Deterministic Approach in $\mathcal{H}_\infty$," IEEE Trans. AC, Vol.36, No.10, October, (1991), pp.1163–1176.
7. H.Kimura, "On Interpolation-Minimization Problem and System Theory in Hardy Space," Instrument and Control, Vol.24, No.7, (1985), pp.23–32. (In Japanese)
8. H.Kimura, "From $\mathcal{LQG}$ to $\mathcal{H}^\infty$," Instrument and Control, Vol.29, No.2, (1990), pp.111–119. (In Japanese)
9. J.M.Krause and P.P.Khargonekar, "On an Identification Problem Arising in Robust Adaptive Control," Proc. 26th IEEE CDC, Los Angeles, CA, Dec. (1987).
10. L.Ljung, "System Identification: Theory for the User," Englewood Cliffs, NJ, Prentice-Hall, (1987).
11. K.Poolla, P.Khargonekar, A.Tikku, J.Krause and K.Nagpal, "A Time-Domain Approach to Model Validation," Proc. American Control Conference, Chicago, Illinois, (1992), pp. 313–318.
12. R.S.Smith and J.C.Doyle, "Towards a Methodology for Robust Parameter Identification," Proc. American Control Conference, San Diego, CA, (1990), pp. 2395–2399.
13. T.Zhou and H.Kimura, "Minimal H^∞-norm of Transfer Functions Consistent with Prescribed Finite Input-Output Data," SICE'92, Kumamoto, Japan, (1992), pp. 1079–1082.

14. T.Zhou and H.Kimura, "Identification for Robust Control in Time Domain," Submitted for publication.

15. T.Zhou and H.Kimura, "Simultaneous Identification of Nominal Model, Parametric Uncertainty and Unstructured Uncertainty for Robust Control in Time Domain," Submitted for presentation.

Identification of Model Error Bounds in ℓ_1- and $\mathcal{H}_\infty$-Norm

Richard G. Hakvoort and Paul M.J. Van den Hof

Mechanical Engineering Systems and Control Group,
Delft University of Technology, Mekelweg 2, 2628 CD Delft, The Netherlands

Abstract:
A time domain identification procedure is developed which yields upper bounds for the ℓ_1- and $\mathcal{H}_\infty$-norm of the model error for a given nominal model. In the procedure use is made of measurement data and a priori information, consisting of a time domain bound on the noise and information about the decay-rate of the pulse response of the model error. The upper bounds are determined by parametrizing the model error and calculating the worst-case model error by solving a set of linear programming problems.

1 Introduction

Robust control theory is able to cope with a system representation consisting of a nominal model and a bound on the model error. In $\mathcal{H}_\infty$-optimal feedback design a bound on the $\mathcal{H}_\infty$-norm of the model error is required (Maciejowski [11]) and in ℓ_1-optimal feedback design a bound on the ℓ_1-norm of the model error is required (Khammash and Pearson [8]). Consequently in literature attention has been paid to the problem of how to derive such upper bounds from measurement data. The problem of estimating $\mathcal{H}_\infty$-bounds on the model error is for example considered in Wahlberg and Ljung [16] where the least-squares algorithm is used to derive frequency domain error bounds. Lau et al. [9] and Younce and Rohrs [17] consider a combined parametric-nonparametric uncertainty approach. Goodwin et al. [3] and Ninness and Goodwin [14] adopt a stochastic approach to the $\mathcal{H}_\infty$-identification problem. Finally in Van den Boom et al. [15] and De Vries and Van den Hof [2] frequency domain error bounds are derived assuming the noise is bounded in the frequency domain. The problem of identification of model error bounds in ℓ_1-norm is considered by Jacobson and Nett [7], Mäkilä [12] and Chen et al. [1].

In this paper the problems are considered of identifying an upper bound for the ℓ_1-norm and an upper bound for the $\mathcal{H}_\infty$-norm of the model error for a given nominal model. For both problems the same setting is adopted. A general uncertainty description is considered, weighted additive uncertainty, which includes the multiplicative uncertainty description. The information assumed available is one or more sequences of time domain measurement data and prior information consisting of a bound on the pulse response of the model error and a time domain bound on the noise. There are no restrictions on the experimental conditions, i.e. the shape of the input signal. No model order assumption on the true system is made, nor any statistical properties of the noise are assumed. In order to calculate the upper bounds the model error is parametrized after which the identification problems under consideration are posed as constrained optimization problems.

For the problem of worst-case identification in ℓ_1-norm bounds on the pulse response of the model error are calculated using linear programming techniques from which an upper bound on the ℓ_1-norm of the model error can readily be derived. For the problem of worst-case identification in $\mathcal{H}_\infty$-norm a frequency dependent bound on the model error is calculated by again solving a set of linear programming problems from which an upper bound on the $\mathcal{H}_\infty$-norm of the model error can easily be derived. This upper bound is shown to be non-conservative under certain conditions, which means that no smaller bound for the $\mathcal{H}_\infty$-norm of the model error can be derived from the information available. All calculated error bounds are hard bounds, which means that the true model error is guaranteed to be smaller than the calculated upper bound, provided the prior information that is used is correct.

The outline of the paper is as follows. In the next section the a priori information assumed available is summarized. In Sect. 3 it is described how this information, including the measurement data, is processed, such that it can be used in the worst-case analysis. Then in Sect. 4 a solution is given for the problem of identifying an upper bound for the ℓ_1-norm of the model error. Next in Sect. 5 a solution is given for the problem of identifying an upper bound for the $\mathcal{H}_\infty$-norm of the model error. Then in Sect. 6 an example is given of the calculation of these upper bounds. The paper ends with conclusions.

2 A Priori Knowledge

We consider a discrete time, asymptotically stable, linear, time-invariant, causal SISO system $G_0(q) = \sum_{k=0}^{\infty} g_0(k)q^{-k}$ (where q is the forward shift operator) with additive bounded output noise. The input-output behaviour of the plant is assumed to be given by the equation

$$y(t) = G_0(q)u(t) + H(q)e(t), \ e(t) \in [e_l(t), e_u(t)], \tag{1}$$

where $u(t)$ is the measured input signal, $y(t)$ is the measured output signal, $e(t)$ is the noise, only known to be bounded by $e_l(t)$, $e_u(t)$. $H(q)$ is some (a priori given) noise model, that can be used to bring in a priori knowledge about the frequency distribution of the noise.

For identification purposes we need measurements of the input signal $u(t)$ and the output signal $y(t)$ acting on the system, $t = 1, 2, \ldots, N$. There are no prior restrictions whatsoever on $u(t)$, it may for example be generated in closed loop. Here the situation is considered that one data sequence is available. It is however straightforward to extend to the case that more measurement sequences are available.

We consider the uncertainty configuration

$$G_0(q) = \hat{G}(q) + \Delta_{\hat{G}}(q)W(q), \tag{2}$$

where $\hat{G}(q)$ is some a priori given nominal model, constructed by any identification or modelling procedure and $W(q)$ is an a priori specified fixed weighting function. The system $G_0(q)$ and the model error $\Delta_{\hat{G}}(q)$ are unknown. The aim is to find an upper bound for the model error in ℓ_1- and $\mathcal{H}_\infty$-norm. All transfer functions in (2) are assumed to be stable. We require the weight $W(q)$ and the noise model $H(q)$ to be minimum-phase. Without loss of generality we further require $W(q)$ to be biproper. Any transfer function can be represented by its (possibly infinite) pulse response sequence, which will be denoted by the corresponding lower-case character. Hence (2) defines the pulse responses g_0, $\hat{g}$, $\delta_{\hat{G}}$ and w.

Next we assume to know an $M > 0$ and $\rho > 1$ such that

$$|\delta_{\hat{G}}(k)| \leq M\rho^{-k}, \, \forall k \geq 0 . \tag{3}$$

If more information is available about the pulse response of the model error this may be included as well. It is for example possible to consider an interval bound on $\delta_{\hat{G}}(k)$ independent of the bound for other values of k.

Finally we assume to know upper bounds on past (unmeasured) data $u(t)$ and $y(t)$,

$$|u(t)| \leq \bar{u}, \, |y(t)| \leq \bar{y}, \, \forall t \leq 0 . \tag{4}$$

If the system is at rest at $t = 0$, $\bar{u}$ can be chosen to be equal to 0.

Notice that no model order assumption has been imposed on the system $G_0(q)$, neither any statistical properties have been assumed for the noise $e(t)$.

The information obtained so far does not uniquely determine the system $G_0(q)$ and the model error $\Delta_{\hat{G}}(q)$. There is a set of systems $G(q)$ and corresponding model errors $\Delta(q)$ consistent with the data and the prior information. We accordingly define

$$\Delta_{\hat{G},C} = \{\Delta(q) \mid (1), t = 1, \ldots, N, (2), (3) \text{ and } (4) \text{ are satisfied}\},$$

with the property $\Delta_{\hat{G}}(q) \in \Delta_{\hat{G},C}$.

3 Processing the Information

The idea now is to formulate the identification problems under consideration as constrained optimization problems. However the set $\Delta_{\hat{G},C}$ has a complicated (implicit) structure and is not well suited for numerical optimization techniques. In this section it is described how a set $\Delta_{\hat{G},L}$ consisting of a number of linear constraints, can be obtained, that is a close outer approximation of the set $\Delta_{\hat{G},C}$. This set of linear constraints will then be used in numerical optimization techniques, or more specifically linear programming, in order to calculate upper bounds on the ℓ_1- and $\mathcal{H}_\infty$-norm of the model error.

In order to obtain a finite dimensional optimization problem the number of unknowns in $\Delta_{\hat{G}}(q)$ has to be reduced to a finite number $n+1$. For that reason the model error $\Delta_{\hat{G}}(q)$ is split into two parts:

$$\Delta_{\hat{G}}(q) = \tilde{\Delta}_{\hat{G}}(q) + \bar{\Delta}_{\hat{G}}(q),$$

$$\tilde{\Delta}_{\hat{G}}(q) = \sum_{k=0}^{n} \delta_{\hat{G}}(k)q^{-k}, \quad \bar{\Delta}_{\hat{G}}(q) = \sum_{k=n+1}^{\infty} \delta_{\hat{G}}(k)q^{-k}, \tag{5}$$

where n is a design variable which influence will be discussed later on.

We substitute (2) into (1), divide by $H(q)$ and introduce $\tilde{H}(q) = H^{-1}(q)$, $\tilde{G}(q) = H^{-1}(q)\hat{G}(q)$ and $\tilde{W}(q) = H^{-1}(q)W(q)$, yielding

$$\tilde{H}(q)y(t) = \tilde{G}(q)u(t) + \Delta_{\hat{G}}(q)\tilde{W}(q)u(t) + e(t), \quad e(t) \in [e_l(t), e_u(t)] \ . \tag{6}$$

When calculating various signals a distinction will be made between the known part of a signal and the unknown part. Bounds will be calculated for these unknown parts and their influence will be captured in the bounded output noise. Using (4) in order to calculate the worst-case influence of the initial conditions, we write for the terms appearing in (6),

$$\tilde{H}(q)y(t) = x(t) + a(t), \quad x(t) = \sum_{k=0}^{t-1} \tilde{h}(k)y(t-k),$$

$$|a(t)| \le \bar{a}(t) = \sum_{k=t}^{\infty} |\tilde{h}(k)|\bar{y}, \quad t = 1, \ldots, N, \tag{7}$$

$$\tilde{G}(q)u(t) = v(t) + b(t), \quad v(t) = \sum_{k=0}^{t-1} \tilde{g}(k)u(t-k),$$

$$|b(t)| \le \bar{b}(t) = \sum_{k=t}^{\infty} |\tilde{g}(k)|\bar{u}, \quad t = 1, \ldots, N,$$

$$\tilde{W}(q)u(t) = w(t) + c(t), \quad w(t) = \sum_{k=0}^{t-1} \tilde{w}(k)u(t-k),$$

$$|c(t)| \le \bar{c}(t) = \sum_{k=t}^{\infty} |\tilde{w}(k)|\bar{u}, \ t = -n+1, \ldots, N,$$

with the additional property that

$$\bar{c}(-t) = \ldots = \bar{c}(0) \ge \bar{c}(1) \ge \ldots \ge \bar{c}(t), \ w(-t) = 0, \ \forall \, t \ge 0 \ .$$

In fact $\bar{a}(t)$, $\bar{b}(t)$ and $\bar{c}(t)$ are decreasing functions of t.

We now obtain

$$\Delta_{\hat{G}}(q)\tilde{W}(q)u(t) = (\tilde{\Delta}_{\hat{G}}(q) + \bar{\Delta}_{\hat{G}}(q))(w(t) + c(t)) = \tilde{\Delta}_{\hat{G}}(q)w(t) + d(t) + f(t),$$

$$d(t) = \bar{\Delta}_{\hat{G}}(q)c(t), \ f(t) = \bar{\Delta}_{\hat{G}}(q)(w(t) + c(t)),$$

where, using (3), $d(t)$ can be bounded by

$$|d(t)| \le \sum_{k=0}^{n} |\delta_{\hat{G}}(k)||c(t-k)| \le \sum_{k=0}^{n} M\rho^{-k}\bar{c}(t-k) = \bar{d}(t), \ t = 1, \ldots, N,$$

which is also a decreasing function of t, and $f(t)$ can be bounded by

$$|f(t)| \le \sum_{k=n+1}^{\infty} |\delta_{\hat{G}}(k)|(|w(t-k)| + |c(t-k)|)$$

$$\le \sum_{k=n+1}^{\infty} M\rho^{-k}(|w(t-k)| + \bar{c}(t-k))$$

$$= \sum_{k=n+1}^{t-1} M\rho^{-k}(|w(t-k)| + \bar{c}(t-k)) + \sum_{k=t}^{\infty} M\rho^{-k}\bar{c}(0)$$

$$= \sum_{k=n+1}^{t-1} M\rho^{-k}(|w(t-k)| + \bar{c}(t-k)) + M\rho^{-t+1}(\rho-1)^{-1}\bar{c}(0)$$

$$= \bar{f}(t), t = 1, \ldots, N,$$

that will generally not vanish for increasing t, especially due to the contribution of $|w(t-k)|$.

With these results equation (6) can be written as

$$x(t) + a(t) = v(t) + b(t) + \tilde{\Delta}_{\hat{G}}(q)w(t) + d(t) + f(t) + e(t), \ t = 1, \ldots, N,$$

$$e(t) \in [e_l(t), e_u(t)], \ |a(t)| \le \bar{a}(t), \ |b(t)| \le \bar{b}(t), \ |d(t)| \le \bar{d}(t), \ |f(t)| \le \bar{f}(t) \ . \tag{8}$$

If we now introduce extended noise bounds

$$n_l(t) = x(t) - v(t) - e_u(t) - \bar{a}(t) - \bar{b}(t) - \bar{d}(t) - \bar{f}(t),$$
$$n_u(t) = x(t) - v(t) - e_l(t) + \bar{a}(t) + \bar{b}(t) + \bar{d}(t) + \bar{f}(t),$$

then (3) and (8) yield a set of (linear inequality) constraints for the unknown pulse response parameters $\delta_{\hat{G}}(k)$, $\Delta_{\hat{G}}(q) \in \Delta_{\hat{G},L}$, where

$$\Delta_{\hat{G},L} = \left\{ \Delta(q) \left| \begin{array}{l} n_l(t) \leq \sum_{k=0}^{n} \delta(k)w(t-k) \leq n_u(t), \ t=1,\ldots,N \\ -M\rho^{-k} \leq \delta(k) \leq M\rho^{-k}, \ k=0,\ldots,\infty \end{array} \right. \right\} .$$

The set $\Delta_{\hat{G},L}$ has been constructed in such a way that $\Delta_{\hat{G},C} \subseteq \Delta_{\hat{G},L}$ as each element satisfying the constraints of the first set also satisfies the constraints of the latter set. In general the set $\Delta_{\hat{G},L}$ will be a fairly tight approximation of the set $\Delta_{\hat{G},C}$ provided n has been chosen large enough, as in that case the signals $\bar{a}(t)$, etc. remain small (compared to the noise level $e_l(t)$, $e_u(t)$). The set $\Delta_{\hat{G},L}$ is even identical to $\Delta_{\hat{G},C}$ if $\bar{u} = 0$, $H(q) = 1$ and $n \geq N - 1$ as in that case the signals $\bar{a}(t)$, $\bar{b}(t)$, $\bar{c}(t)$, $\bar{d}(t)$ and $\bar{f}(t)$ are zero $\forall t$. At the price of some conservatism we thus have obtained a set of linear constraints, which contains information from the data and the prior information, that is well suited for usage in numerical optimization techniques.

4 Identification of an Upper Bound for the ℓ_1-norm of the Model Error

In this section we focus on the problem of estimating an upper bound for the ℓ_1-norm of the model error, $\|\Delta_{\hat{G}}\|1$.

Applying the orthotopic outer bounding procedure of Milanese and Belforte [13] we calculate

$$\delta_l(k) = \min_{\Delta \in \Delta_{\hat{G},L}} \delta(k), \quad \delta_u(k) = \max_{\Delta \in \Delta_{\hat{G},L}} \delta(k), \quad k = 0,\ldots,n,$$

which requires solving $2(n+1)$ linear programming problems for $n+1$ unknowns subject to $2(N+n+1)$ linear inequality constraints. This can be done using standard linear programming software available. See Luenberger [10] for an extensive treatment of the linear programming problem and numerical algorithms to solve it. Here we notice the price for choosing a large value of n in (5) as in that case more and larger linear programming problems have to be solved.

We thus obtain the outer bounding box description

$$\Delta_{\hat{G},B} = \left\{ \Delta(q) \left| \begin{array}{l} \delta_l(k) \leq \delta(k) \leq \delta_u(k), \ k=0,\ldots,n, \\ -M\rho^{-k} \leq \delta(k) \leq M\rho^{-k}, \ k=n+1,\ldots,\infty \end{array} \right. \right\},$$

which has the property $\Delta_{\hat{G},L} \subseteq \Delta_{\hat{G},B}$. The orthotope $\Delta_{\hat{G},B}$ is tight in the sense that there does not exist an orthotope (with the same orientation) of smaller size which contains the set $\Delta_{\hat{G},L}$. The set $\Delta_{\hat{G},B}$ has been calculated on the basis of the set $\Delta_{\hat{G},L}$ which has in turn been constructed from the data and the prior information. Hence the quality of the data and prior information directly influences the size of $\Delta_{\hat{G},B}$. If e.g. many measurements are available with a low noise level, the set $\Delta_{\hat{G},B}$ will be relatively small.

The simple box structure of $\Delta_{\hat{G},B}$ will be utilized now to solve the ℓ_1-identification problem posed. The following theorem gives an upper bound for the ℓ_1-norm of the model error.

Theorem 1.

$$\|\Delta_{\hat{G}}\|1 = \sum_{k=0}^{\infty} |\delta_{\hat{G}}(k)| \leq \max_{\Delta \in \Delta_{\hat{G},C}} \|\Delta\|1$$

$$\leq \max_{\Delta \in \Delta_{\hat{G},L}} \|\Delta\|1 \leq \max_{\Delta \in \Delta_{\hat{G},B}} \|\Delta\|1$$

$$= \sum_{k=0}^{n} \max\{-\delta_l(k), \delta_u(k)\} + \frac{M\rho^{-n}}{\rho - 1} \ .$$

Proof. The first equality is the definition of the ℓ_1-norm. The inequalities are direct implications of the construction of the sets involved, where $\Delta_{\hat{G}}(q) \in \Delta_{\hat{G},C} \subseteq \Delta_{\hat{G},L} \subseteq \Delta_{\hat{G},B}$. The last equality finally follows from the fact that for any $\Delta(q) \in \Delta_{\hat{G},B}$,

$$\|\Delta\|1 = \sum_{k=0}^{\infty} |\delta(k)| = \sum_{k=0}^{n} |\delta(k)| + \sum_{k=n+1}^{\infty} |\delta(k)|$$

$$\leq \sum_{k=0}^{n} \max\{-\delta_l(k), \delta_u(k)\} + \sum_{k=n+1}^{\infty} M\rho^{-k},$$

which yields the desired result by noting that there is a worst-case model error in the set $\Delta_{\hat{G},B}$ for which the ℓ_1-norm equals this upper bound. □

The upper bound obtained in this way has been calculated using information from the data and the prior information. In general however it is not the minimal upper bound that can (theoretically) be derived from this information. By definition this smallest upper bound is given by $\max_{\Delta \in \Delta_{\hat{G},C}} \|\Delta\|1$, which generally can not be calculated. In Hakvoort [4] an approach has been proposed to calculate $\max_{\Delta \in \Delta_{\hat{G},L}} \|\Delta\|1$ although in that paper this set has been defined slightly different. If that procedure is carried out in general a tighter bound for $\|\Delta_{\hat{G}}\|1$ will be obtained than given by Theorem 1, however the procedure is computationally very involved. In Sect. 6 an example is given of the calculation of an upper bound according to Theorem 1 which shows that this can be carried out with some computational effort.

5 Identification of an Upper Bound for the $\mathcal{H}_\infty$-norm of the Model Error

In this section we consider the problem of estimating an upper bound for the $\mathcal{H}_\infty$-norm of the model error, $\|\Delta_{\hat{G}}\|\infty$. To derive an upper bound we will again use the fact that $\Delta_{\hat{G}} \in \Delta_{\hat{G},C} \subseteq \Delta_{\hat{G},L}$ and hence

$$\|\Delta_{\hat{G}}\|\infty = \sup_{\omega \in [0,\pi]} |\Delta_{\hat{G}}(e^{i\omega})| \leq \max_{\Delta \in \Delta_{\hat{G},C}} \|\Delta\|\infty \leq \max_{\Delta \in \Delta_{\hat{G},L}} \|\Delta\|\infty . \quad (9)$$

In this section we will focus on deriving a bound for the last term appearing in this formula.

In the sequel we will need the following lemma.

Lemma 2. *Consider the function* $f_m(x) : \mathbb{C} \to \mathbb{R}$, *defined by*

$$f_m(x) = \max_{k=1,\dots,m} Re(c_k x),$$

with $c_k = e^{2\pi \frac{k}{m} i}$, $k = 1, 2, \dots, m \geq 3$, *where* $Re(\cdot)$ *denotes the real part of* $\cdot$, *then*

(i) $f_m(x) \leq |x| \leq \dfrac{f_m(x)}{\cos\left(\frac{\pi}{m}\right)}$,

(ii) $\lim_{m \to \infty} f_m(x) = |x|$.

Proof. For any x and any c_k with $|c_k| = 1$, $Re(c_k x) \leq |c_k x| \leq |c_k||x| = |x|$, which proves the left-hand inequality of (i). Further for any x there exist an integer l and $\delta \in [-\frac{\pi}{m}, \frac{\pi}{m})$ such that $x = |x|e^{(2\pi \frac{l}{m} + \delta)i}$, yielding

$$Re(c_k x) = Re\left(e^{2\pi \frac{k}{m} i}|x|e^{(2\pi \frac{l}{m} + \delta)i}\right)$$

$$= |x|Re\left(e^{(2\pi \frac{k+l}{m} + \delta)i}\right) = |x| \cos\left(2\pi \frac{k+l}{m} + \delta\right) .$$

If we now choose $k = k^*$ such that $k^* + l = nm$ for some integer n, we obtain

$$Re(c_{k^*} x) = |x| \cos(2\pi n + \delta) = |x| \cos(\delta) \geq |x| \cos\left(\frac{\pi}{m}\right) \Leftrightarrow |x| \leq \dfrac{Re(c_{k^*} x)}{\cos\left(\frac{\pi}{m}\right)},$$

which proves the right-hand inequality of part (i). Finally part (ii) immediately follows from part (i) for $m \to \infty$.   □

Moreover it is easy to show that the bounds in (i) are tight in the sense that there exists an x such that the underbound becomes equality, and there is an x such that the upper bound becomes equality. The lemma in fact says that the amplitude of a complex number can be calculated approximately by checking a number of different directions in the complex plane.

We will derive bounds for the model error by evaluating $|\Delta(e^{i\omega})|$ for a finite number of frequencies. The behaviour between the frequencies will later on be estimated by means of a worst-case interpolation argument. The set of frequencies is given by

$$\Omega = \{\omega_1, \dots, \omega_l\}, \ 0 \leq \omega_1 < \cdots < \omega_l \leq \pi .$$

By definition for each frequency ω_j the model error is bounded by

$$\Delta_{\hat{G}}(e^{i\omega_j}) \in \left\{ \Delta(e^{i\omega_j}) \,\middle|\, \Delta(q) \in \Delta_{\hat{G},L} \right\} .$$

Now for each frequency ω_j, $j = 1, \ldots, l$ this frequency domain uncertainty set will be evaluated using the tool provided in Lemma 2. With the c_k as defined in Lemma 2 we calculate

$$\mu_k(\omega_j) = \max_{\Delta \in \Delta_{\hat{G},L}} \operatorname{Re}\left(c_k \tilde{\Delta}(e^{i\omega_j}) \right) = \max_{\Delta \in \Delta_{\hat{G},L}} \operatorname{Re}\left(c_k \sum_{k'=0}^n \delta(k') e^{-ik'\omega_j} \right)$$

$$= \max_{\Delta \in \Delta_{\hat{G},L}} \sum_{k'=0}^n \delta(k') \operatorname{Re}\left(c_k e^{-ik'\omega_j} \right), \quad k = 1, \ldots, m, \; j = 1, \ldots, l,$$

(10)

with $\mu_{k+m} = \mu_k$, which requires solving ml linear programming problems for $n + 1$ unknowns subject to $2(N + n + 1)$ linear inequality constraints (see Luenberger [10] for details about linear programming). In this way for each frequency ω_j a convex polytope $\mathbf{P}_{m,j}$ in the complex plane is determined,

$$\mathbf{P}_{m,j} = \left\{ \tilde{\Delta}(\omega_j) \,\middle|\, \operatorname{Re}\left(c_k \tilde{\Delta}(\omega_j) \right) \leq \mu_k(\omega_j), \; k = 1, \ldots, m \right\} .$$

The convex polytope has vertices $v_k(\omega_j)$, $k = 1, \ldots, m$ which satisfy

$$\operatorname{Re}\left(c_k v_k(\omega_j) \right) = \mu_k(\omega_j), \quad \operatorname{Re}\left(c_{k+1} v_k(\omega_j) \right) = \mu_{k+1}(\omega_j), \quad k = 1, \ldots, m . \quad (11)$$

This set has the property that it contains the exact uncertainty set for the truncated model error,

$$\left\{ \tilde{\Delta}(e^{i\omega_j}) \,\middle|\, \Delta(q) \in \Delta_{\hat{G},L} \right\} \subseteq \mathbf{P}_{m,j},$$

and hence $\tilde{\Delta}_{\hat{G}}(e^{i\omega_j}) \in \mathbf{P}_{m,j}$. For the case $m = 4$ it simply means that for each frequency ω_j the minimum and maximum of the real and imaginary part of the model error $\tilde{\Delta}(\omega_j)$ are calculated, which yields a box in the complex plane. Notice that the model error is evaluated in the frequency domain without transforming the measurement data to the frequency domain.

The following lemma establishes the bound obtainable in this way.

Lemma 3. *Let $\mu_k(\omega_j)$ be defined by (10) and $v_k(\omega_j)$ by (11), then*

(i) $\displaystyle \max_k \mu_k(\omega_j) \leq \max_{\Delta \in \Delta_{\hat{G},L}} |\tilde{\Delta}(e^{i\omega_j})| \leq \max_k |v_k(\omega_j)| \leq \max_k \frac{\mu_k(\omega_j)}{\cos(\frac{\pi}{m})}$,

(ii) $\displaystyle \lim_{m \to \infty} \max_k \mu_k(\omega_j) = \max_{\Delta \in \Delta_{\hat{G},L}} |\tilde{\Delta}(e^{i\omega_j})| = \lim_{m \to \infty} \max_k |v_k(\omega_j)|$.

Proof. The first inequality in part (i) directly follows from the definition of $\mu_k(\omega_j)$ and Lemma 2. The second inequality follows from the definition of $v_k(\omega_j)$. The bound defined by the second inequality is the tightest bound that can be derived from the set $\mathbf{P}_{m,j}$. Noting that the last expression appearing in the first part above is also an upper bound according to Lemma 2, we conclude that the third inequality must hold, which proves part (i). Next the statement in part (ii) is proven by noting that the left-hand expression in (i) converges to the right-hand expression for $m \to \infty$. $\square$

In this way we have established a frequency dependent bound on the model error $\tilde{\Delta}_{\hat{G}}(q)$ for a finite number of frequencies. Taking into account the worst-case influence of the tail $\bar{\Delta}_{\hat{G}}(q)$ and of the intersample behaviour of the model error, an upper bound for $\|\Delta_{\hat{G}}\|\infty$ can readily be derived. Introduce the parameter λ_j for the intersample frequency distance

$$\lambda_j = \max\{\omega_j - \omega_{j-1}, \omega_{j+1} - \omega_j\}, \quad j = 1, \ldots, l \ . \tag{12}$$

Now a bound on the $\mathcal{H}_\infty$-norm of the model error is established in the following theorem.

Theorem 4. *Let $\beta(n, \Omega, m)$ be defined by*

$$\beta(n, \Omega, m) = \max_j \left\{ \max_k |v_k(\omega_j)| + \frac{1}{2}\lambda_j \frac{M\rho}{(\rho-1)^2} \right\} + \frac{M\rho^{-n}}{\rho-1},$$

with $v_k(\omega_j)$ defined by (11) and λ_j by (12) then

(i) $\|\Delta_{\hat{G}}\|\infty \leq \max\limits_{\Delta \in \Delta_{\hat{G},L}} \|\Delta\|\infty \leq \beta(n, \Omega, m)$,

(ii) $\lim\limits_{\Omega \to [0,\pi]} \lim\limits_{n,m \to \infty} \beta(n, \Omega, m) = \max\limits_{\Delta \in \Delta_{\hat{G},L}} \|\Delta\|\infty$.

Proof. The first inequality in part (i) has been established in (9). Next (5) yields

$$\|\Delta_{\hat{G}}\|\infty \leq \|\tilde{\Delta}_{\hat{G}}\|\infty + \|\bar{\Delta}_{\hat{G}}\|\infty,$$

and using the fact that the ℓ_1-norm upper bounds the $\mathcal{H}_\infty$-norm,

$$\max_{\Delta \in \Delta_{\hat{G},L}} \|\bar{\Delta}\|\infty \leq \max_{\Delta \in \Delta_{\hat{G},L}} \|\bar{\Delta}\|_1 = \sum_{k=n+1}^{\infty} M\rho^{-k} = \frac{M\rho^{-n}}{\rho-1} \ .$$

In De Vries and Van den Hof [2] it has been shown that assumption (3) implies that

$$\left| \frac{d\,|\Delta_{\hat{G}}(e^{i\omega})|}{d\omega} \right| \leq \left| \frac{d\,\Delta_{\hat{G}}(e^{i\omega})}{d\omega} \right| \leq \frac{M\rho}{(\rho-1)^2}, \ \forall \omega \ .$$

Worst-case interpolation considerations now give

$$|\tilde{\Delta}_{\hat{G}}(e^{i\omega})| \leq |\tilde{\Delta}_{\hat{G}}(e^{i\omega_j})| + \frac{M\rho}{(\rho-1)^2}|\omega_j - \omega|, \ \forall \omega, \omega_j,$$

which in combination with Lemma 3 yields the desired result (i). Next, using Lemma 3, we obtain

$$\lim_{\Omega \to [0,\pi]} \lim_{n \to \infty} \max_{j,k} \mu_k(\omega_j) \leq \max_{\Delta \in \Delta_{\partial,L}} \|\Delta\|_\infty \leq \lim_{\Omega \to [0,\pi]} \lim_{n \to \infty} \beta(n, \Omega, m)$$

$$= \lim_{\Omega \to [0,\pi]} \lim_{n \to \infty} \max_{j,k} |v_k(\omega_j)|$$

$$\leq \lim_{\Omega \to [0,\pi]} \lim_{n \to \infty} \max_{j,k} \frac{\mu_k(\omega_j)}{\cos(\frac{\pi}{m})},$$

which proves part (ii) by noting that the right-hand side expression converges to the left-hand side expression for $m \to \infty$.   □

We see that the total error in (i) consists of three parts. The first contribution stems from the fundamental uncertainty in the data. This can basically only be reduced by using more informative data or a priori information (besides a small reduction obtainable by choosing a large value for m). The second contribution to the error in (i) comes from the fact that only a finite number of frequencies is evaluated and the intersample behaviour is analyzed by some worst-case interpolation argument. This error can be reduced by evaluating more frequencies (which can be done for the same data set) or doing a more careful interpolation analysis, see De Vries and Van den Hof [2]. However it should be emphasized that often this interpolation contribution is much too conservative and that generally a more realistic indication of the model error is obtained by simply linearly interpolating the computed model error between two subsequent frequencies. The third contribution follows from the truncation in the parametrization of the model error and reduction of this error requires an increase in the computational effort as the size of the linear programming problems increases. However this can also be performed for the same data set.

6 Example

In this section an example is presented which shows the applicability of the theory developed. For a given nominal model we evaluate the model error both in ℓ_1-norm and in $\mathcal{H}_\infty$-norm.

In Fig. 1 a Bode diagram is given of the 5th order system $G_0(q)$, the 3rd order nominal model $\hat{G}(q)$ and the 7th order weighting function $W(q)$ that have been chosen (quite randomly). Their transfer functions are given by

$$G_0(q) = \frac{0.703\,q^5 - 0.893\,q^4 + 0.240\,q^3 + 0.524\,q^2 - 0.902\,q + 0.401}{q^5 - 2.474\,q^4 + 2.891\,q^3 - 1.981\,q^2 + 0.834\,q - 0.181},$$

$$\hat{G}(q) = \frac{0.806\,q^3 - 1.106\,q^2 - 0.0230\,q + 0.322}{q^3 - 2.280\,q^2 + 1.945\,q - 0.622},$$

$$W(q) = \frac{3.15\,q^7 - 2.23\,q^6 + 1.62\,q^5 - 2.01\,q^4 + 1.64\,q^3 - 0.202\,q^2 + 0.104\,q - 0.258}{q^7 - 1.16\,q^6 + 0.567\,q^5 - 0.336\,q^4 + 0.216\,q^3 - 0.194\,q^2 + 0.151\,q - 0.0521}.$$

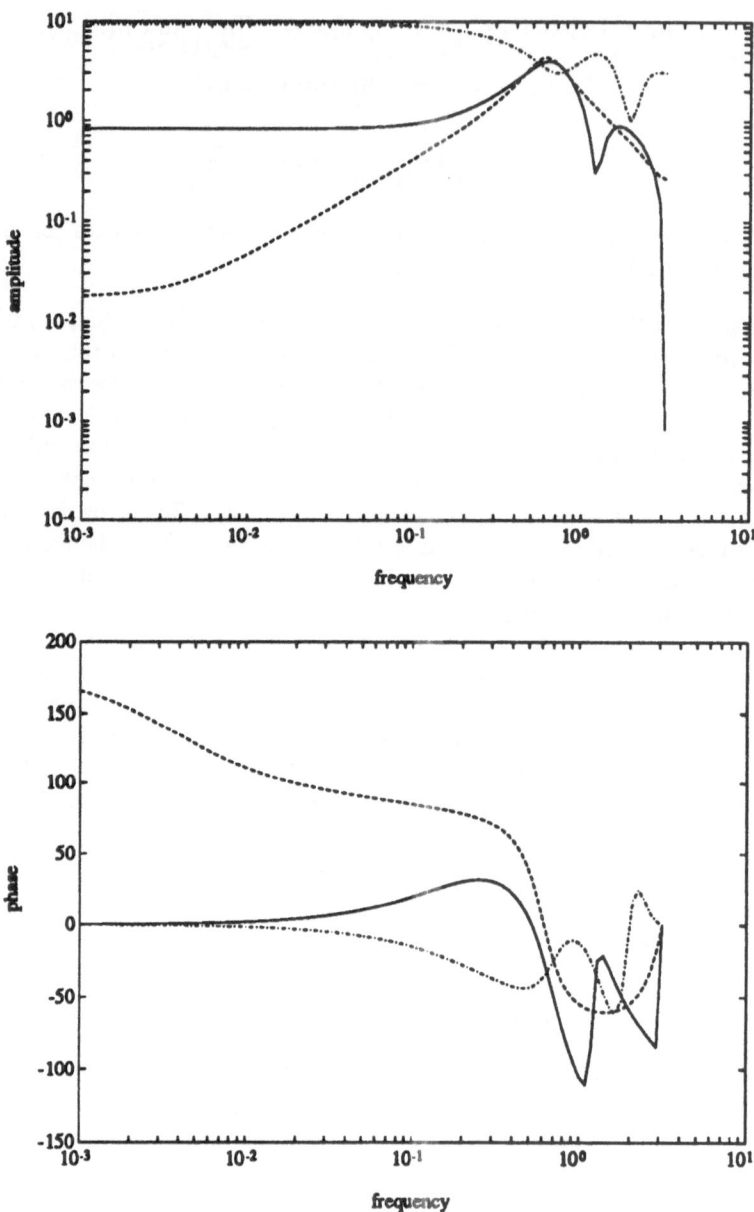

Fig. 1. Bode diagram system $G_0(q)$ (solid), nominal model $\hat{G}(q)$ (dashed) and weight $W(q)$ (dash-dotted)

Starting from zero initial conditions ($\bar{u} = 0$), a simulation experiment has been performed with a Gaussian white noise input signal (variance 1) and a uniformly distributed additive output noise ($e(t) \in [-0.3, 0.3]$, $H(q) = 1$, hence the choice of $\bar{y}$ is irrelevant as $\bar{a}(t) = 0$, $\forall t$ in (7)). We chose $M = 2$, $\rho = 1.1$, which is a very conservative choice. We used 1000 samples for identification purposes.

First we calculated an upper bound for the ℓ_1-norm of the model error applying the procedure of Sect. 4. We chose $n = 80$, which means that to obtain the set $\Delta_{\hat{G},B}$ 162 linear programming problems had to be solved for 81 unknowns subject to 2162 linear inequality constraints. This has been done on a VAX workstation 3100 using the linear programming software in the Fortran NAG library. Solving one such linear programming problem takes about 4 minutes CPU time. The result is shown in Fig. 2, where the calculated upper and lower bound of the pulse response sequence of the model error are plotted together with the pulse response of the true model error. The worst-case ℓ_1-norm found in this way is 1.88, a factor 2 larger than the ℓ_1-norm of the true model error $\Delta_{\hat{G}}$ defined by (2), which is equal to 0.90.

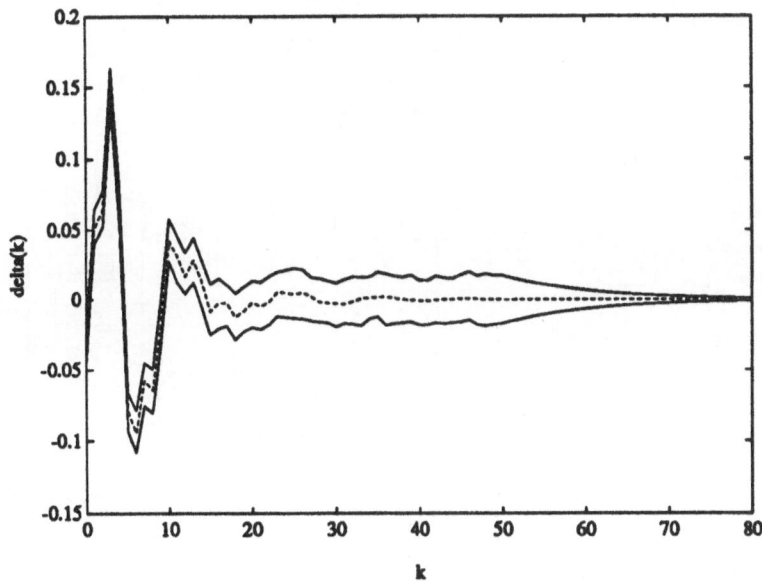

Fig. 2. Pulse response true model error $\delta_{\hat{G}}(k)$ (dashed) and identified lower bound $\delta_l(k)$ and upper bound $\delta_u(k)$ (solid)

Next we calculated an upper bound for the $\mathcal{H}_\infty$-norm of the model error applying the procedure of Sect. 5. Now we chose $n = 130$ (which implies that the contribution of $\bar{\Delta}(q)$ to the model error is almost zero) and $m = 4$. We calculated the worst-case model error for 100 frequencies logarithmically distributed between 0.01 and π. This means that 400 linear programming problems had to be solved for 131 unknowns subject to 2262 linear inequality constraints. The outcome of a linear programming problem for a certain frequency can be used as initial estimate when solving a linear programming problem for a next frequency. If done so solving one such linear programming problem takes about 4 minutes CPU time. The result is shown in the Nyquist plot of Fig. 3, where the calculated uncertainty regions are shown, together with the true model error $\Delta_{\hat{G}}(q)$, which is of course inside the uncertainty regions. In Fig. 4 the upper bound on the amplitude of the model error is plotted in a Bode diagram together with the amplitude of the true model error. The influence of the interpolation contribution to the worst-case model error in Theorem 4 has not been taken into account here. The $\mathcal{H}_\infty$-norm of the worst-case model error is 0.65 whereas the $\mathcal{H}_\infty$-norm of the true model error is 0.57. We conclude that a tight error bound has been obtained.

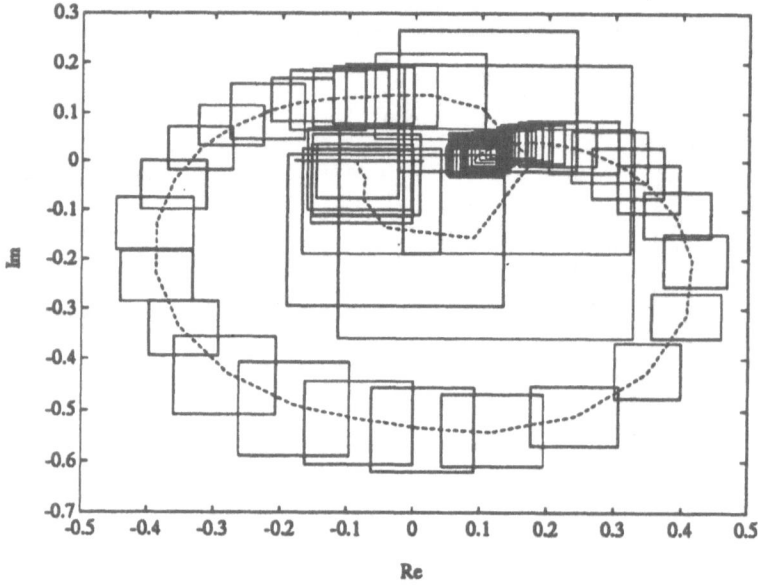

Fig. 3. Nyquist diagram frequency dependent uncertainty regions $\mathbf{P}_{m,j}$ (boxes) and true model error $\Delta_{\hat{G}}(q)$ (dashed)

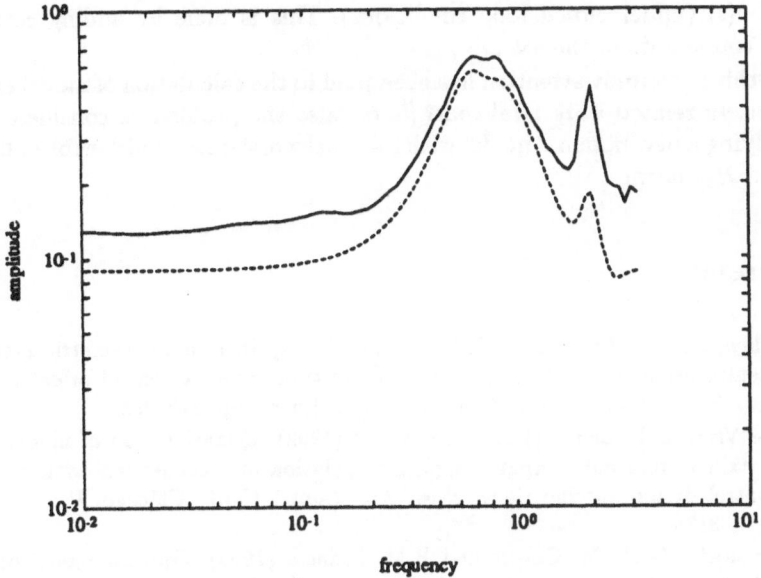

Fig. 4. Upper bound on the amplitude of the model error $\beta(n, \Omega, m)$ (solid) and amplitude of the true model error $\Delta_{\hat{G}}(q)$ (dashed)

7 Conclusions

An identification procedure has been developed which yields for a given nominal model $\hat{G}(q)$ and weight $W(q)$ upper bounds for the ℓ_1-norm and the $\mathcal{H}_\infty$-norm of the model error, starting from measurement data and some a priori information. This prior information consists of a time domain bound on the noise and information about the decay rate of the pulse response of the model error. There are no restrictions on the shape of the input signal. Numerically it requires solving a set of linear programming problems. The bounds are guaranteed upper bounds on the model error. Moreover under certain conditions the $\mathcal{H}_\infty$-bound is the smallest upper bound on the $\mathcal{H}_\infty$-norm of the model error that can be computed using the information available in the data and the priors. However the ℓ_1-bound is conservative in general.

In an example the procedure has been illustrated. It appeared that tight error bounds were obtainable with some computational effort. However other simulation studies show that the calculated upper bounds may be much larger than the true model error. This is basically due to the worst-case character of the analysis, the noise is always assumed to take the worst possible value. Smaller upper bounds on the model error may be obtained by using more data and more (and more accurate) prior information. Current investigations show that it is possible to utilize the fact that the noise $e(t)$ is uncorrelated to the input signal $u(t)$ (under open loop conditions) or some external reference signal $r(t)$ (under closed loop conditions). This is done by adding certain linear constraints to the set $\Delta_{\hat{G},L}$.

In this paper only attention has been paid to the calculation of model error bounds. In related work (Hakvoort [5, 6]) also the problem is considered of identifying a new nominal model with minimal worst-case model error in both ℓ_1- and $\mathcal{H}_\infty$-norm.

References

1. Chen, J., C.N. Nett and M.K.H. Fan (1992). Optimal non-parametric system identification from arbitrary corrupt finite time-series: a control oriented approach. *Proc. Am. Contr. Conf.*, Chicago, U.S.A., pp. 279–285.

2. De Vries, D.K. and P.M.J. Van den Hof (1992). Quantification of model uncertainty from data: input design, interpolation and connection with robust control design specifications. *Proc. Am. Contr. Conf.*, Chicago, U.S.A., pp. 3170–3175.

3. Goodwin, G.C., M. Gevers and B.M. Ninness (1990). Optimal model order selection and estimation of model uncertainty for identification with finite data. *Proc. Conf. Dec. Contr.*, Honolulu, U.S.A., pp. 285–290.

4. Hakvoort, R.G. (1991). Identification of an upper bound for the ℓ_1-norm of the model uncertainty. *Sel. Topics in Id., Mod. and Contr.*, Vol. 3, eds. O.H. Bosgra and P.M.J. Van den Hof, Delft Univ. Press, pp. 51–58.

5. Hakvoort, R.G. (1992a). Worst-case system identification in ℓ_1: error bounds, optimal models and model reduction. *Proc. Conf. Dec. Contr.*, Tucson, U.S.A.

6. Hakvoort, R.G. (1992b). Worst-case system identification in $\mathcal{H}_\infty$: error bounds and optimal models. *Sel. Topics in Id., Mod. and Contr.*, Vol. 5, eds. O.H. Bosgra and P.M.J. Van den Hof, Delft Univ. Press.

7. Jacobson, C.A. and C.N. Nett (1991). Worst case system identification in ℓ_1: optimal algorithms and error bounds. *Proc. Am. Contr. Conf.*, Boston, U.S.A., pp. 3152–3157.

8. Khammash, M. and J.B. Pearson (1991). Performance robustness of discrete-time systems with structured uncertainty. *IEEE Trans. Autom. Contr.*, Vol. AC-36, pp. 398–412.

9. Lau, M., R. Kosut and S. Boyd (1990). Parameter set estimation of systems with uncertain nonparametric dynamics and disturbances. *Proc. Conf. Dec. Contr.*, Honolulu, U.S.A., pp. 3162–3167.

10. Luenberger, D.G. (1984). *Linear and Nonlinear Programming*, Addison Wesley, U.K., 2nd edition.

11. Maciejowski, J.M. (1989). *Multivariable Feedback Design*, Addison Wesley, U.K.

12. Mäkilä, P.M. (1991). Robust identification and Galois-sequences. Techn. rep., Proc. Contr. Lab., Dept. Chem. Eng. University of Åbo, Finland.

13. Milanese, M. and G. Belforte (1982). Estimation theory and uncertainty intervals evaluation in presence of unknown but bounded errors: linear families of models and estimators. *IEEE Trans. Autom. Contr.*, Vol. AC-27, pp. 408–414.

14. Ninness, B.M. and G.C. Goodwin (1992). Robust frequency response estimation accounting for noise and undermodelling. *Proc. Am. Contr. Conf.*, Chicago, U.S.A., pp. 2847–2851.

15. Van den Boom, T., M. Klompstra and A. Daamen (1991). System identification for $\mathcal{H}_\infty$-robust control design. *Proc. 9th IFAC Symp. Ident. and Syst. Par. Est.*, Budapest, Hungary, pp. 1431–1436.

16. Wahlberg, B. and L. Ljung (1991). On estimation of transfer function error bounds. *Proc. European Contr. Conf.*, Grenoble, France, pp. 1378–1383.

17. Younce, R.C. and C.E. Rohrs (1992). Identification with nonparametric uncertainty. *IEEE Trans. Autom. Contr.*, Vol. AC-37, pp. 715–728.

Asymptotic Worst-Case Identification with Bounded Noise

Munther A. Dahleh*

Lab. for Information & Decision Systems, M.I.T., Cambridge MA 02139

Abstract:
This paper presents an overview of the problem of asymptotic worst-case identification in the presence of bounded noise.

1 Introduction

Recently, there has been an increasing interest among the control community in the problem of identifying plants for control purposes. This generally means that the identified model should approximate the plant as it operates on a rich class of signals, namely signals with bounded norm, since this allows for the immediate use of robust control tools for designing controllers [6, 15]. This problem is of special importance when the data are corrupted with bounded noise. The case where the objective is to optimize prediction for a fixed input was analyzed by many researchers in [18, 15, 17, 18, 19, 22]. The problem is more interesting when the objective is to approximate the original system as an operator, a problem extensively discussed in [31]. For linear time invariant plants, such approximation can be achieved by uniformly approximating the frequency response (in the $\mathcal{H}_\infty$-norm) or the impulse response (in the ℓ_1 norm). In $\mathcal{H}_\infty$ identification, it was shown that robustly convergent algorithms can be furnished, when the available data is in the form of a corrupted frequency response, at a set of points dense on the unit circle [8, 9, 10, 11, 12]. When the topology is induced by the ℓ_1 norm, a complete study of asymptotic identification was given in [27, 28, 30] for arbitrary inputs, and the question of optimal input design was addressed. Related work on this problem was also reported in [7, 13, 14, 16, 20, 21].

The work of Tse et al. [27, 28, 30] and [3] allows for the analysis of large classes of systems including nonlinear fading memory systems. The study is done in two steps. The first step is concerned with obtaining tight upper and lower bounds on the optimal achievable error, for a given fixed experiment.

---

* Research Supported by AFOSR under grant AFOSR-91-0368 and by NSF under grant 9157306-ECS.

The second step is then to study these bounds and characterize the inputs that will minimize them. The upper and lower bounds are characterized under some mild topological assumptions on the model set through the diameter of the worst-case uncertainty set, a concept borrowed from Information Based Complexity [25, 26]. This characterization is valid for any input, and allows for closed loop identification. The papers include a detailed analysis of this diameter for different classes of systems, including LTI stable systems, unstable systems, and nonlinear time-invariant fading memory systems. Conditions on the input set are derived to guarantee finite (optimal) worst-case errors.

Another issue of importance in the context of worst-case identification is *Complexity*. It turns out that it is generally much harder to devise experiments that can guarantee small worst-case errors in the presence of bounded noise. This problem has been extensively analyzed in [4, 23].

It is important to caution at this point that the terminology "worst-case" does not mean that one can furnish guarantees on the worst-case error. Clearly, any result we obtain is a function of prior assumptions (which are not verifiable in general), and thus the results hold only when these assumptions are valid. This is no different from the traditional stochastic approach for system identification.

In this paper, we will give an overview of the results we have developed in the context of this problem. Although we will not present proofs, the material is presented in a unified fashion. The objective is to study:

1. The fundamental limitations of worst-case identification.
2. The impact this has on input design.
3. The impact this has on Algorithms.

Details of these results can be found in [3, 4, 27].

2 Problem set-up

There are three basic ingredients for any identification problem. These are

1. *Prior knowledge and assumptions.* This includes assumptions/knowledge of the model set, disturbances, noise, etc.
2. *Data.* This includes specifications on input design, knowledge of open loop or closed loop data, and so on.
3. *Error criterion.* This dictates the sense in which the identified model approximates the actual process.

Worst case identification in the presence of bounded noise refers to a problem with specific ingredients. These are discussed below.

2.1 Prior Knowledge

Let $\mathcal{X}$ be the class of all causal, single-input single-output, time-invariant, discrete-time systems. Let $\mathcal{M} \subset \mathcal{X}$ be the *model set* which is assumed to contain the unknown plant h to be identified. The set $\mathcal{M}$ captures the experimenter's a priori knowledge about h. Some examples of $\mathcal{M}$ are the set of all stable linear time-invariant systems, the set of stable systems with a bound on the decay rate of the impulse response, the set of all finite-dimensional systems with a bound on the order, space of stable nonlinear systems with fading memory, etc.

The process generating the outputs is assumed to have the form:

$$y = h(u) + d$$

where u, y are the inputs and the outputs of the process respectively, and d is the disturbance. The disturbance is assumed to be unknown but bounded:

$$\|d\|_{\infty} = \sup_{k} |d(k)| \leq \delta \ .$$

In this setting, there are no statistical assumptions made on the noise. Such assumptions have dominated the traditional approaches in system identification. Deterministic assumptions are not new in the estimation/identification literature, for instance see [24].

2.2 Data

For any experiment, only a finite set of data is available:

$$[(y_k, u_k) \quad | \quad k = 0, 1, \ldots, n]$$

where $u \in \mathcal{U}$: the set of all input sequences that can be used in the identification experiments. Typically, $\mathcal{U}$ is a norm-bounded set, to reflect physical limitations, power restrictions, safety, or to maintain the validity of the linear model of the plant. An *experiment* is conducted by either choosing or measuring an input sequence $u \in \mathcal{U}$ and measuring the output sequence y, related to u by

$$y = h(u) + d \tag{1}$$

for a length of time n. If a number of experiments are performed, we use a vector of inputs $\mathbf{u}$ and outputs $\mathbf{y}$.

2.3 Error Criterion

An *identification algorithm* is a mapping ϕ which generates, at each time instant n, an estimate

$$\hat{h}^{(n)} \equiv \phi(P_n \mathbf{u}, P_n \mathbf{y}) \in \mathcal{X}$$

given the finite sequence of inputs and outputs. Here, P_n is the truncation operator, defined by $P_n x = (x_0, x_1, \ldots, x_n)$ for each infinite sequence x. Its use signifies that the algorithm ϕ generates at each time instant an estimate based only on the input-output data it has seen so far. Generally, we will assume that the algorithm has access to what the model set $\mathcal{M}$ is and also the value of δ, the bound on the disturbance. In the terminology of Helmicki et. al. [12], the algorithm is *tuned*. However, in some cases, we will be able to give stronger results using algorithms which are untuned to the value of δ.

The error criterion is defined through a metric $\rho_{\mathcal{X}}$. This metric can be chosen to be the induced operator norm over the spaces ℓ_∞, ℓ_2 and so on. We would like to evaluate the error $\rho_{\mathcal{X}}(h, \hat{h}_n)$. Of course this is difficult to evaluate since h is not known. The terminology worst-case identification is in fact motivated from looking at the maximum value this error criterion takes over all plants in the model set.

To explain the error criterion, consider the set of behavior $\mathcal{L}$:

$$\mathcal{L} = \{y = h(u) + d \mid d \in \|d\|_\infty \leq 1\} \ .$$

So, for any plant $h \in \mathcal{M}$, $u \in \mathcal{U}$, this set consists of all possible outputs that can be generated by disturbances with $\|d\|_\infty \leq \delta$. The estimates based on these outputs can be different. The worst-case error criterion simply says that the distance between the estimate taken after a long experiment and the plant h, for any possible output in the behavior set, and for any $h \in \mathcal{M}$ should be minimized. The error is defined as:

$$e_\infty(\phi, \mathcal{M}, \mathbf{u}, \delta) \equiv \sup_{h \in \mathcal{M}} \ \sup_{\|\mathbf{d}\|_\infty \leq \delta} \ \limsup_{n \to \infty} \rho_{\mathcal{X}}(\phi(P_n\mathbf{u}, P_n(\mathbf{u} * h + d)), h) \ .$$

In the above definition of the worst-case asymptotic error, although convergence of the estimates to within $e_\infty(\phi, \mathcal{M}, \mathbf{u}, \delta)$ is guaranteed for all admissible plants and disturbance sequences, the *rate* of convergence may be arbitrarily slow for some plants and some disturbances. The worst-case asymptotic error is said to be *uniform* if the rate of convergence is uniform over all admissible plants and disturbance sequences. If the convergence is uniform, the worst-case asymptotic error defined above is the same as the limit of the worst-case error taken at each finite time n, i.e.

$$e_\infty(\phi, \mathcal{M}, \mathbf{u}, \delta) = \limsup_{n \to \infty} \sup_{h \in \mathcal{M}} \ \sup_{\|\mathbf{d}\|_\infty \leq \delta} \ \rho_{\mathcal{X}}(\phi(P_n\mathbf{u}, P_n(\mathbf{u} * h + d)), h) \ .$$

This allows one to a priori determine the experiment length required to guarantee that *any* plant in the model set can be identified to a prescribed accuracy. It is the notion of convergence considered by Helmicki et al. in their framework [12].

Demanding uniform convergence is too restrictive a formulation for a general theory of fundamental limitations of worst-case identification. Although such uniform convergence is certainly desirable, it is impossible to achieve for many interesting model sets. In fact, for many inherently infinite-dimensional model sets, the worst-case error at each finite time is always infinite, while the worst-case asymptotic error can be made small using an appropriate identification algorithm and inputs. Our formulation thus allows us to discuss optimal worst-case identification and optimal inputs for a much broader class of model sets. Besides, in some applications of identification, such as adaptive control, uniform convergence of estimates is not necessary to fulfill the desired objectives. However, because of the special importance of uniform convergence, we will give additional conditions on the model set for this to take place. It will be seen that these conditions are quite strong and essentially require the model set to be finite-dimensional. It is worthwhile to note that the model set considered in [9, 10] satisfies these conditions.

The optimal worst-case asymptotic error $E_\infty(\mathbf{u}, \mathcal{M}, \delta)$ is defined as the smallest error achievable by any algorithm:

$$E_\infty(\mathbf{u}, \mathcal{M}, \delta) \equiv \inf_\phi e_\infty(\phi, \mathcal{M}, \mathbf{u}, \delta) .$$

3 General Results

Given the above set-up of the worst-case identification problem, there are general results that can be derived to characterize the optimal error $E_\infty(\mathbf{u}, \mathcal{M}, \delta)$. The following presentation is a summary of results reported in [3, 4, 27].

An important concept in such a characterization is the concept of an *Uncertainty Set*. Simply, the uncertainty set at time n is the set of all plants consistent with the data, i.e.,

$$S_n(\mathcal{M}, \mathbf{u}, \mathbf{y}, \delta) = \{g \in \mathcal{M} : \|P_n(g(\mathbf{u}) - \mathbf{y})\|_\infty \leq \delta\}$$

and the infinite-horizon uncertainty set is the set of all plants consistent with the infinite-horizon input and output, i.e.,

$$S_\infty(\mathcal{M}, \mathbf{u}, \mathbf{y}, \delta) = \{g \in \mathcal{M} : \|g(\mathbf{u}) - \mathbf{y}\|_\infty \leq \delta\} .$$

3.1 Lower Bounds on $E_\infty(\mathbf{u}, \mathcal{M}, \delta)$

For a given input, and a plant in the set $\mathcal{M}$, any output in the set of behaviors $\mathcal{L}$ is a possible output, which in turn results in a specific sequence of uncertainty sets. At time n, all the plants in S_n are indistinguishable, and any estimator will result in an error which is closely related to the diameter of this set. Finally, since the actual process is unknown, we need to look at the maximum diameter generated for different choices of $h \in \mathcal{M}$. This motivates the next definitions.

For any set $A \subset \mathcal{X}$, define the diameter of the set A as

$$\mathrm{diam}(A) = \sup_{g,h \in A} \rho_\mathcal{X}(g, h) .$$

Next, we define a quantity known as the *Diameter of Information*.

Definition 3.1 *Given a choice of the inputs* u, *define the infinite-horizon diameter of information* $D(u, \mathcal{M}, \delta)$ *the diameter of the largest possible uncertainty set :*

$$D(u, \mathcal{M}, \delta) \equiv \sup_{h \in \mathcal{M}} \sup_{\|d\|_\infty \leq \delta} \text{diam}(S_\infty(\mathcal{M}, u, u * h + d, \delta)) .$$

In information-based complexity terminology , these quantities correspond to the *diameter of information* for the infinite horizon problem where the information available is the entire infinite output sequence. The quantity $D(u, \mathcal{M}, \delta)$ is the largest distance between two plants for which there are admissible disturbances such that the plants give exactly the same outputs. It turns out that it is precisely this quantity that characterizes the optimal worst-case asymptotic errors. First we show that half the infinite-horizon diameter of information is a lower bound to the optimal asymptotic error.

Proposition 1 *Let* $\mathcal{M}$ *be any model set,* u *be any vector of inputs and* $\delta \geq 0$. *Then*

$$e_\infty(\phi, \mathcal{M}, u, \delta) \geq D(u, \mathcal{M}, \delta)/2$$

for any algorithm ϕ.

3.2 Upper Bound

Under some mild assumptions on the model set, the *Diameter of Information* is an upper bound on the best achievable error. Such assumptions are related to σ-compactness of the model set. A model set $\mathcal{M}$ is σ-compact if it is a countable union of compact, nested subsets, i.e., $\mathcal{M} = \cup_i \mathcal{M}_i$. A model set is separable if it is the closure of a σ-compact set. The following result gives the upper bound. In here ρ_χ indicates some induced norm.

Theorem 2. *Suppose that the model set* $\mathcal{M}$ *is* σ-compact, *or separable in the* ρ_χ *topology, there is an identification algorithm* ϕ^* *such that*

$$e_\infty(\phi^*, \mathcal{M}, u, \delta) \leq D(u, \mathcal{M}, \delta)$$

for all u *and* $\delta \geq 0$.

This theorem can be interpreted as follows: For model sets with such topological structure, and for any input, there exists an algorithm such that the asymptotic worst-case error is bounded by the diameter of information. Such an algorithm is based on the Occam's Razor principle: Pick the estimate in the smallest set consistent with the data. This algorithm requires the knowledge of δ, and thus it is a *tuned* algorithm.

It should be noted that by an elementary result in information-based complexity theory, the optimal worst-case error achievable when the algorithm has *full* access to the entire *infinite* input-output sequences is also bounded between the infinite-horizon diameter of information and half the diameter of information. Our two results (Proposition 1 and Theorem 2) are of an entirely different nature: they assert that the optimal worst-case asymptotic error achievable when the algorithm has access to *finite* but arbitrarily long data records also satisfies the same bounds. The assumed topological conditions are crucial for the validity of Theorem 2.

3.3 Input Design

From the above discussion, the inputs should be designed to minimize the diameter of information, i.e.,

$$\inf_{u \in \mathcal{U}} D(u, \mathcal{M}, \delta) \ .$$

If this quantity is infinite, then accurate worst-case identification is not possible. For certain model sets, such a minimization can be performed and optimal inputs can be furnished. In the sequel, we will present a few such examples.

4 Application to Specific model Sets

The above general results, can be applied to specific model sets, to derive optimal inputs that guarantee accurate identification.

4.1 Stable LTI Systems

Here $\mathcal{X}$ is the space ℓ_1. This is the space of BIBO linear time-invariant, causal operators on ℓ_∞. The metric $\rho_\mathcal{X}$ can be either the ℓ_1 norm, or the $\mathcal{H}_\infty$. The set $\mathcal{M}$ will be any balanced (i.e. if $h \in \mathcal{M}$ then $-h \in \mathcal{M}$) and convex closed subset of $\mathcal{X}$ (with diameter larger than 2δ).

Since the space ℓ_1 is separable (with respect to both the ℓ_1 topology and $\mathcal{H}_\infty$ topology), then there is an identification algorithm ϕ^* such that

$$\frac{D(u, \mathcal{M}, \delta)}{2} \leq e_\infty(\phi^*, \mathcal{M}, u, \delta) \leq D(u, \mathcal{M}, \delta)$$

for all u and $\delta \geq 0$.

To estimate the diameter of information, we first notice that

$$D(u, \mathcal{M}, \delta) \geq 2\delta \ .$$

In fact, one can show that there exists an input u^*, such that the diameter of information is equal to 2δ.

Theorem 3. *Assume $\mathcal{M}$ is balanced and convex and contains only stable plants. If u^* contains all finite sequences of 1's and −1's, then*

$$D(u^*, \mathcal{M}, \delta) \leq 2\delta \ .$$

In the above theorem, the diameter is computed with respect to either the ℓ_1 norm or $\mathcal{H}_\infty$ norm. The results are the same. This, combined with the earlier result, shows that there exists a single input u^*, such that

$$E_\infty(u^*, \mathcal{M}, \delta) \leq 2\delta .$$

Hence, to identify a plant accurately in the limit, it is enough to know a priori that it is stable; no additional information, such as bounds on decay rate and gain, is necessary. The achievable accuracy varies continuously with the noise bound δ for small δ; thus, identification can be performed robust to measurement noise.

Uniform Convergence. Next, we look at the issue of uniform convergence. For the model set ℓ_1, it can at once be seen that although convergence to a small asymptotic error is possible, such convergence cannot be uniform. To guarantee uniform convergence, we need to look at compact model sets.

Proposition 4 *Let $\mathcal{M} \subset \ell_1$ be a compact set (in the ℓ_1-topology) or a subset of a compact set in ℓ_1. For the single input u^* which contains all finite sequences of 1's and -1's, there is an algorithm the estimates of which converge, uniformly for all $h \in \mathcal{M}$ and all $\|d\|_\infty \leq \delta$, to an ℓ_1 ball of radius 2δ around the true plant. Moreover the algorithm does not require the knowledge of the value of δ to compute its estimates.*

Common examples of such compact model sets are the uniformly stable ones, of the form $M_s(g) \equiv \{h : |h_i| \leq |g_i| \text{ for all } i\}$ where g is any stable plant. The specific model sets considered in [9] and [10] belong to this class. In the particular case when g_i is taken to be 0 for all i larger than some given M, we get the model set of finite-impulse- response of length M. For this model set the near-optimal algorithm ϕ^* is given by

$$\phi^*(P_n u, P_n y) = \arg \min_{|h_i| \leq |g_i|, i=0,1,\dots M} \|P_n(y - u * h)\|_\infty$$

which is computable by linear programming.

Sample Complexity. The input designed to achieve a radius of information equal to 2δ is quite a rich input. This suggests that achieving accurate identification when only stability is assumed can be quite difficult in practice. To study this problem in a more precise fashion, let $\mathcal{M}_N$ denote the subset of ℓ_1 that contains systems with finite impulse response of length N. Let $\mathcal{U}_n$ be the set of all infinite real sequences $\{u_i\}_{i=1}^\infty$ such that $|u_i| \leq 1$ for all i, and $u_i = 0$ for $i > n$. Any element of $\mathcal{U}_n$ will be called an *input of length* n. Let $D_{N,n}^*$ be the smallest diameter of information over the class of inputs in $\mathcal{U}_n$, i.e.,

$$D_{N,n}^* = \inf_{u \in \mathcal{U}_n} D(\mathcal{M}_N, u_n, \delta) .$$

We have shown earlier that

$$\lim_{n \to \infty} D_{N,n}^* = 2\delta \ .$$

The sample (time) complexity of such problems is captured in the length of n necessary to get within a factor of 2δ. The next result shows that this length is exponential in N, and thus can be quite unrealistic for large N.

Theorem 5. *Fix some $K > 1$ and let*

$$n^*(N, K) = \min\{n \mid D_{N,n}^* \le 2K\delta\}$$

Then

1. $n^*(N, K) \ge 2^{Nf(1/K)-1} - N$
2. $\lim_{N \to \infty} \frac{1}{N} \log n^*(N, K) = f(1/K)$.

Here, $f : (0, 1) \mapsto \Re$ is the function defined by

$$f(\alpha) = 1 + \left(\frac{1-\alpha}{2}\right) \log \left(\frac{1-\alpha}{2}\right) + \left(\frac{1+\alpha}{2}\right) \log \left(\frac{1+\alpha}{2}\right) \ . \qquad (2.11)$$

Notice that the function f defined by (2.11) satisfies $f(\alpha) = 1 - H((1 - \alpha)/2)$, where H is the binary entropy function. In particular, f is positive and continuous for $\alpha \in (0, 1)$.

This theorem furnishes a lower bound on the minimum time required to get within a factor of the optimum solution, which is exponential in N. This lower bound is asymptotically tight.

4.2 Nonlinear Systems with Fading Memory

The set $\mathcal{X}$ contains causal functions from $\mathcal{U}$ to $\Re^\infty$; these are discrete-time systems, possibly non-linear, which take as input a sequence in $\mathcal{U}$ to give an output sequence in $\Re^\infty$. The input and the output at time n will be denoted by u_n and $h_n(u)$. Assume that they further satisfy the following properties

1. $h_n(u)$ depends continuously on $u_0, \ldots, u_{n-1}$.
2. h has equilibrium-initial behavior:

$$h_{n+1}(0u) = h_n(u) \text{ for all } n$$

where $0u$ is the input $0, u_0, u_1, \ldots$.

The model sets $\mathcal{M}$ we shall consider will be *balanced* and *convex* subsets of this class of functions $\mathcal{X}$. In general, we will use the notation vw for concatenation, i.e. first apply the finite sequence v, then w. Since we are dealing with causal systems, we shall slightly abuse the notation and write $h_n(w)$ to mean $h_n(u)$, where u is any infinite sequence the first n elements of which is the finite sequence w.

The metric $\rho_{\mathcal{X}}$ will be taken to be the operator-induced norm:

$$\|h\| = \sup_{u \in \mathcal{U}} \|h(u)\|_\infty \ .$$

This is the natural norm to consider for robust control applications.

Definition of Fading Memory Systems.

Definition 4.1 *An operator h has fading memory (FM) if for each $\varepsilon > 0$ there is some $T = T(\varepsilon)$ such that: for every k, every $t \geq T$ and every finite sequences $v \in [-1, 1]^k$, $w \in [-1, 1]^t$,*

$$\|h_{t+k}(vw) - h_t(w)\| < \varepsilon .$$

It can easily be seen that fading memory systems satisfying properties (1) and (2) have bounded operator-induced norms.

Examples of Fading Memory Systems.

Example 2. (Stable LTI systems) For each $h \in \ell_1$ define the input/output map $u \mapsto u * h$ by convolution. It is clear that these systems satisfy the above conditions. The operator-induced norm in this case is just the ℓ_1 norm.

Example 3. (Hammerstein Systems) These are systems which are formed by composition of a LTI system followed by a memoryless nonlinear element:

$$y_n = g((u * h)_n)$$

for some $h \in \ell_1$ and some continuous function $g : \Re \rightarrow \Re$. It is easy to verify that these systems satisfy the first two conditions above. Since $|(h * u)_n| \leq \|h\|_1$ then g is uniformly continuous on $[-\|h\|_1, \|h\|_1]$. Hence h has fading memory. Any stable system with fading memory can be approximated arbitrarily closely by a Hammerstein system.

Proposition 6 *The class of all fading memory systems is separable.*

A σ-compact set is constructed as follows: for every $n \geq 0$, consider all Hammerstein systems, in which h is FIR of length n. The set of all stable systems with fading memory is the closure of the countable union of such systems.

This means that when we consider fading memory system, we can apply the general results, and reduce the analysis of asymptotic optimal error to the analysis of infinite-horizon diameter, i.e.,

$$\frac{D(\mathbf{u}, \mathcal{M}, \delta)}{2} \leq e_\infty(\phi^*, \mathcal{M}, \mathbf{u}, \delta) \leq D(\mathbf{u}, \mathcal{M}, \delta)$$

for all u and $\delta \geq 0$.

It is evident that for any $\mathcal{M}$ in $\mathcal{X}$, balanced and convex (with diameter larger than 2δ), $\mathcal{M}$ satisfies:

$$D(u, \mathcal{M}, \delta) \geq 2\delta .$$

The existence of an experiment that results in equality is established below.

Theorem 7. *Let the model set $\mathcal{M}$ be some subset of the set of fading memory systems. Let W be any countable dense subset of $[-1, 1]$ and consider any input $u^* \in [-1, 1]^\infty$ which contains all possible finite sequences of elements of W. Then*

$$D(u^*, \mathcal{M}, \delta) \leq 2\delta$$

Complexity. Consider the class of p memory systems. These are systems that operate on the last p component of the input, and are given by continuous functions on $[-1, 1]^p$. Let g be such a function. It can be easily seen that in general, the time needed to identify a system to a prescribed accuracy grows exponentially as the order of the system, even when there is no noise. For example, if we assume a certain Lipschitz condition on the order p memory function g, such as $|g(x) - g(y)| < M\|x - y\|$, then to identify the function up to accuracy ϵ (in the $\|\cdot\|_\infty$ norm), the number of data points needed is at least the minimum number of ϵ-balls to cover $[-1, 1]^p$. Since the volume of an ϵ-ball is $O(\epsilon^p)$, it is clear that this minimum number is $\Omega((\frac{1}{\epsilon})^p)$, and hence so is the experiment length. This means that if p is large, the experiment length will be very long if we make no further assumption on the unknown plant.

It is interesting to compare this situation with the problem of identifying *linear* finite impulse response systems. For nonlinear systems the time complexity is exponential of the order, whether or not there is noise. For the linear case, while it takes only linear time to identify a FIR system *exactly* when there is no noise, we have shown that the time complexity immediately becomes exponential once we introduce any unknown but bounded noise. Moreover, it has been demonstrated that if we are willing to put a probability distribution on the noise, polynomial time complexity can often be obtained [29]. These facts show that while in the nonlinear case, the plant uncertainty determines the time complexity of the identification, in the linear case, the complexity is sensitive to how the noise is modeled.

4.3 Untuned algorithms

So far, all the algorithms devised to deliver accurate identification are *tuned*. As we discussed, such algorithms are based on Occam's Razor in which the simplest model that explains the data is picked. That is, if $\mathcal{M} = \cup_i \mathcal{M}_i$, where the $\mathcal{M}_i$'s are nested compact sets, and S_n is the uncertainty set at time n, then the estimate $\hat{h}_n$ can be any element in the intersection of S_n and $\mathcal{M}_j$, where j is the smallest integer such that this intersection is not empty. To implement such an algorithm, the sets S_n need to be computed. This requires the knowledge of the bound on the disturbance, δ.

Notice that these results are derived for any choice of inputs in $\mathcal{U}$. It is possible, however, to derive *untuned* algorithms for specific experiments. In particular, this is possible when the input used has the property that it minimizes the diameter of information. Such results can be found in [3, 20] and will not be discussed here.

5 Summary

Finally, I would like to discuss some of the issues that this line of research has highlighted.

1. Choose the simplest estimate to explain the data. This is a consequence of the Occam's Razor principle [1]. This implies that the algorithm should have enough information to make such a decision, which typically translates into the knowledge of an upper bound on the magnitude of the noise. Estimates that fit the data very well generally fit the noise as well, and result in non-convergent estimates.

2. For nonlinear fading memory systems, we need local interpolants to fit the nonlinearities. This can be accomplished by basis functions such as "Linear Splines", or what is known as "Gaussian radial" functions, or any other local interpolant. Polynomial interpolants generally are not appropriate. This is due to the fact that if the interpolant in some neighborhood depends on far away points, noise can be amplified and the algorithm may not be convergent.

3. Inputs have to be quite rich. Of course this depends on the amount of prior information. This can have major impact on closed loop identification, where the inputs are not arbitrary. On the other hand, for fairly understood models (small model sets), worst-case identification can be quite easy, see for example [13].

4. It may be too "hard" in practice to give guarantees under such noise assumptions. This follows from the sample complexity results discussed earlier. These results capture the fundamental limitations of worst-case identification in the presence of bounded noise. Also, this suggests alternate formulations in which the error criterion is a worst-case criterion, however, noise is assumed to be stochastic. It is hinted in [23, 29] that experiments can be much shorter in such formulations.

5. More work is needed in the area of algorithms and particularly recursive algorithms.

References

1. A. Blumer, A. Ehrenfeucht. D. Haussler, M. Warmuth, "Occam's Razor", *Information Processing Letters 24*, pp.377-380, 1987.

2. M.A. Dahleh and M.H. Khammash, "Controller Design in the Presence of Structured Uncertainty," to appear in *Automatica* special issue on robust control.

3. M.A. Dahleh, E.D. Sontag, D.N. Tse and J.N. Tsitsiklis, "Worst-Case identification of nonlinear fading memory systems," to appear in *ACC Proc.*, 1992.

4. M.A. Dahleh, T. Theodosopoulos, and J.N. Tistsiklis, "The sample complexity of worst-case identification of F.I.R. Linear systems," To appear in Systems and Control Letters.

5. J. C. Doyle, "Analysis of feedback systems with structured uncertainty," *IEEE Proceedings* 129, 242-250,1982.

6. E. Fogel and Y. F. Huang, " On the value of information in system identification–bounded noise case," *Automatica*, vol.18, no.2, pp.229-238, 1982.

7. G.C. Goodwin, M. Gevers and B. Ninness, "Quantifying the error in estimated transfer functions with application to model order selection," *IEEE Trans. A-C*, Vol 37, No. 7, July 1992.

8. G. Gu, P.P. Khargonekar and Y. Li, "Robust convergence of two stage nonlinear algorithms for identification in $\mathcal{H}_\infty$," *Systems and Control Letters*, Vol 18, No. 4, April 1992.

9. G. Gu and P.P. Khargonekar, "Linear and nonlinear algorithms for identification in $\mathcal{H}_\infty$ with error bounds," *IEEE Trans. A-C*, Vol 37, No. 7, July 1992.

10. A.J. Helmicki, C.A. Jacobson and C.N. Nett, "Identification in H_∞: A robust convergent nonlinear algorithm," Proceedings of the 1989 International Symposium on the Mathematical Theory of Networks and System, 1989.

11. A.J. Helmicki, C.A. Jacobson and C.N. Nett, "Identification in H_∞: Linear Algorithms," Proceedings of the 1990 American Control Conference, pp 2418-2423.

12. A.J. Helmicki, C.A. Jacobson and C.N. Nett, "Control-oriented System Identification: A Worst-case/deterministic Approach in H_∞," *IEEE Trans. A-C*, Vol 36, No. 10, October 1991.

13. C.A. Jacobson and C.N. Nett, "Worst-case system identification in ℓ_1: Optimal algorithms and error bounds," in *Proc. of the 1991 American Control Conference*, June 1991.

14. J.M. Krause, G. Stein, P.P. Khargonekar, "Robust Performance of Adaptive Controllers with General Uncertainty Structure," Proceedings of the 29th Conference on Decision and Control, pp. 3168-3175, 1990.

15. R. Lozano-Leal and R. Ortega, "Reformulation of the parameter identification problem for systems with bounded disturbances," *Automatica*, vol.23, no.2, pp.247-251, 1987.

16. M.K. Lau, R.L. Kosut, S. Boyd, "Parameter Set Estimation of Systems with Uncertain Nonparametric Dynamics and Disturbances," Proceedings of the 29th Conference on Decision and Control, pp. 3162-3167, 1990.

17. M. Milanese and G. Belforte, "Estimation theory and uncertainty intervals evaluation in the presence of unknown but bounded errors: Linear families of models and estimators," *IEEE Trans. Automatic Control*, AC-27, pp.408-414, 1982.

18. M. Milanese and R. Tempo, "Optimal algorithm theory for robust estimation and prediction," *IEEE Trans. Automatic Control*, AC-30, pp. 730-738, 1985.

19. M. Milanese, "Estimation theory and prediction in the presence of unknown and bounded uncertainty: a survey," in *Robustness in Identification and Control*, M. Milanese, R. Tempo, A. Vicino Eds, Plenum Press, 1989.

20. P.M. Makila, "Robust Identification and Galois Sequences," Technical Report 91-1, Process Control Laboratory, Swedish University of Abo, January, 1991.

21. P.M. Makila and J.R. Partington, "Robust Approximation and Identification in H_∞," Proc. 1991 American Control Conference, June, 1991.

22. J.P. Norton, "Identification and application of bounded-parameter models," *Automatica*, vol.23, no.4, pp.497-507, 1987.

23. K. Poolla and A. Tikku, "On the time complexity of worst-case system identification," preprint, 1992.

24. F.C. Schweppe. *Uncertain dynamic systems*, Prentice-Hall, Englewood cliffs, N.J., 1973.

25. J.F. Traub and H. Wozniakowski, *A General Theory of Optimal Algorithms*, Academic Press, New York, 1980.

26. J.F. Traub, G. Wasilkowski and H. Wazniakowski, *Information-Based Complexity*, Academic Press, 1988.

27. D. Tse, M.A. Dahleh and J.N. Tsitsiklis. "Optimal Asymptotic Identification under bounded disturbances," To appear in *IEEE Trans. Automat. Contr.*.

28. D.N.C. Tse, M.A. Dahleh, J.N. Tsitsiklis, "Optimal and Robust Identification in the ℓ_1 norm," in *Proc. of the 1991 American Control Conference*, June 1991.

29. D.N.C. Tse and J.N. Tsitsiklis, "Sample complexity of system identification under a high probability formulation," in preparation.

30. D.N.C. Tse, "Optimal and robust identification under bounded disturbances," Master's thesis, Dept. of Elec. Eng. and Comp. Sci., M.I.T., February, 1991.

31. G. Zames, "On the metric complexity of casual linear systems: ϵ-entropy and ϵ-dimension for continuous-time," *IEEE Trans. on Automatic Control*, Vol. 24, April 1979.

Sequential Approximation of Uncertainty Sets via Parallelotopes*

Antonio Vicino[1] *and Giovanni Zappa*[2]

[1] Università di L'Aquila, Dipartimento di Ingegneria Elettrica
   67040 Poggio di Roio, L'Aquila, Italy

[2] Università di Firenze, Dipartimento di Sistemi e Informatica
   Via di Santa Marta, 3 - 50139 Firenze, Italy

1 Introduction

Recent years have witnessed a growing renewed interest in system identification ([1, 2]). Research activity has been mainly stimulated by the growing need for techniques providing the basic information required by advanced robust and adaptive control schemes developed in the past decade [3]. Both soft (stochastic) and hard (deterministic) bound settings have been widely investigated (see e.g., [3, 4] for soft bounds mixed parametric/nonparametric approaches, [5, 6] for H_∞ or l_1 nonparametric techniques, [7, 8, 9] for hard bound purely parametric approaches and [10, 11, 16] for hard bound mixed parametric/nonparametric approaches). Mixed parametric/nonparametric approaches appear very promising for providing the necessary information for applicability of the techniques recently devised in the robust control field for structured and unstructured uncertainties [13, 14, 15, 16].

The present paper is embedded in a hard bound setting, where knowledge about disturbances and a priori information is given in terms of deterministic bounds. A fixed order model and a possible block accounting for unmodeled dynamics are allowed. The contribution of this paper is in the spirit of [10, 16]. The main distinguishing feature is that instead of constructing adaptive ellipsoidic approximations for the parameter uncertainty set, i.e. the set of parameters compatible with the disturbance bounds and the a priori knowledge on the unmodeled dynamics, it proposes recursive approximations of orthotopic or parallelotopic shape. Beyond the intrinsic interest from a theoretical standpoint, the main practical motivation for this different characterization of the parameter uncertainty set estimates may be found in the recent results on robust control of plants subject to parametric or mixed parametric/nonparametric perturbations. Most of these contributions refer to uncertainty regions in plant parameter space of hyperrectangular or polytopic

* This work was partially supported by funds of the Ministero della Università e della Ricerca Scientifica e Tecnologica and CNR.

shape [13, 17, 14]. The main purpose of these references is to characterize extremal subsets of the uncertainty region providing worst case properties of the uncertain system from the stability or performance viewpoint. The interesting feature of polytopic regions is that it is possible to find very 'small' subsets (made of vertices or edges) providing the 'worst case' information contained in the whole uncertainty set.

In this paper, an adaptive algorithm is provided for constructing recursively an outer bounding parallelotopic approximation of the parameter uncertainty set. The procedure represents a counterpart to the algorithm originally proposed in [18] and improved in [19] as suggested in [18]. It can be employed both in a purely parametric or in a mixed parametric/nonparametric setting of the identification problem. Though the computational burden of the algorithm is comparable to that in [18, 19], it is a good candidate to provide better approximations of the parameter uncertainty set, on the ground that the family of approximating parallelotopes is parameterized according to a larger number of degrees of freedom than ellipsoids.

The paper is structured as follows. Section 2 introduces notation and problem formulation. Section 3 presents basic results for optimal approximation of the uncertainty set at a fixed step of the adaptive procedure, while the adaptive algorithm is discussed in Sect. 4. Concluding remarks are reported in Sect. 5.

2 Notation and Problem Formulation

Consider the linear regression equation

$$y(k) = \phi'(k)\theta + e(k), \quad k = 1, 2, \ldots \tag{1}$$

where $y(k)$ is the k-th scalar measurement on the system under investigation, $\theta = [\theta_1, \ldots, \theta_n]'$ is the model parameter vector, $\phi(k) = [\phi_1(k), \ldots, \phi_n(k)]'$ is the regressor and $e(k)$ represents an error term such that

$$|e(k)| \leq r(k), \quad k = 1, 2, \ldots \tag{2}$$

where $r(k) > 0$ is a known sequence of error bounds. Notice that, as will be better specified at the end of this section, θ may include both parameters of a fixed order nominal model and parameters describing the unmodelled dynamics possibly associated with the nominal model (see e.g. [16]). Denote by $\Theta(k)$ the *uncertainty parameter set* at time k, i.e. the set of θ consistent with the model equation (1) and the error bound (2) up to the k-th measurement, i.e.

$$\Theta(k) = \{\bigcap_{i=1}^{k} \Sigma_i\}, \tag{3}$$

where Σ_i is the set of parameters consistent with the i-th measurement

$$\Sigma_i = \{\theta \in R^n : |y(i) - \phi'(i)\theta| \leq r(i)\} .$$

A set in R^n defined as Σ_i will be called a 'strip'. It is easy to check that $\Theta(k)$ is a convex polytope. We assume that $\Theta(k)$ is nonempty for any k.

Since the main objective of next sections is to approximate $\Theta(k)$ through simple shaped regions like parallelotopes, we introduce a description of such regions. Denote by $\mathcal{R}(\theta^c)$ the unit ball in the l_∞ norm centered at θ_c

$$\mathcal{R}(\theta^c) = \{\theta : \max_{i=1,\dots,n} |\theta_i - \theta_i^c| \leq 1\} .$$

A parallelotope can be defined through $\mathcal{R}(\tilde{\theta}^c)$ and a nonsingular transformation $T \in R^{n,n}$

$$\mathcal{P}(T, \theta^c) = \{\theta : \theta = T\tilde{\theta}, \ \tilde{\theta} \in \mathcal{R}(\tilde{\theta}^c)\} = \{\theta : \|P(\theta - \theta^c)\|_\infty \leq 1\}$$

where $P = T^{-1}$. We denote by $t_j, j = 1,\dots,n$ and $p_i', i = 1,\dots,n$ the columns and rows of matrices T and P, respectively. Of course, the parallelotope $\mathcal{P}(T, \theta^c)$ can be expressed as the intersection of n normalized strips in parameter space, each centered around θ^c

$$\mathcal{P}(T, \theta^c) = \{\bigcap_{i=1}^n S_i\} ,$$

where

$$S_i = \{\theta : |p_i'\theta - c_i| \leq 1\} . \tag{4}$$

with $c_i = p_i'\theta^c$ (observe that $\tilde{\theta}^c = [c_1, \dots, c_n]'$). Since we look for 'optimal', in the sense of minimal volume, outer approximations of $\Theta(k)$, we choose as 'measure' μ of a parallelotope in R^n its volume

$$\mu(\mathcal{P}(T, \theta^c)) = \text{vol}\,(\mathcal{P}(T, \theta^c)) .$$

We recall the relationship between the volumes of a unit ball $\mathcal{R}(\theta^c)$ and $\mathcal{P}(T, \theta^c)$

$$\begin{aligned} \mu(\mathcal{R}(\theta^c)) &= 2^n \\ \mu(\mathcal{P}(T, \theta^c)) &= \mu(\mathcal{R}(\theta^c))|\det\,(T)| \end{aligned} \tag{5}$$

Hence, the requirement of minimal volume for a parallelotopic domain is equivalent to one of minimum determinant magnitude for the matrix T defining the parallelotope.

Now, we can formulate the problem solved in the forthcoming section.

Consider the linear regression model (1) with error bounds given by (2). Let an outer estimate of $\Theta(k)$ be given at time k in the form of a parallelotope $\mathcal{P}(T, \theta^c)$. Suppose that an additional measurement at time $k + 1$ becomes available. The problem is to use the new information to update in an optimal way the parallelotopic estimate. More precisely, denoting by $\mathcal{P}_k$ the parallelotopic estimate of $\Theta(k)$ at time k, i.e.

$$\mathcal{P}_k = \mathcal{P}(T(k), \theta^c(k)) ,$$

we want to find the minimal volume parallelotope $\mathcal{P}_{k+1}$ consistent with the previous estimate $\mathcal{P}_k$, the new measurement $y(k+1)$ and the corresponding error bound $r(k+1)$. Of course, a priori information on the system is assumed available in order to determine a suitable initial estimate $\mathcal{P}_0$, while a priori information on the data is necessary for evaluating the error bounds $r(k)$.

Before ending this section, we remark that the above problem formulation includes both purely parametric model estimation, classical in the set membership uncertainty community (see e.g. [7, 8, 9]), and mixed parametric and nonparametric identification in a hard bound context (see e.g. [11, 16, 10]). The major requirement is a linear parameterization of the model. Hence, ARMA models can be dealt with in an equation error approach. Output error models where the parametric part is a FIR model or a linear combination of orthogonal filters (like Laguerre or Kautz filters) ([16, 20, 21]) can be tackled equally well. When dealing with purely parametric models, the a priori information generally consists in an initial uncertainty parallelotope for the parameters and a measurement error bound. When mixed parametric/nonparametric models are of concern, a priori information on the non-parametric part of the model becomes of crucial importance and it requires suitable techniques to translate it into an initial parallelotopic estimate $\mathcal{P}_0$. A good example of this can be found in [16], where the nonparametric part is modelled via a FIR model cascaded to a suitable shaping filter and the corresponding a priori information is mapped into an ellipsoid in the FIR parameter space. The a priori information which can be assumed in the context of parallelotopic approximations may be given like in [16] in terms of

- a hard bound on the tail contribution of the nonparametric part of the model (this is equivalent to assuming a certain rate of decay of the impulse response of the nonparametric part)..
- hard bounds on the errors between the first n samples of the 'true' impulse response samples and the FIR model parameters.
- hard bounds on discrepancies between the frequency response magnitude of the nonparametric part and the truncated approximation.

3 Set Estimation via Minimal Volume Parallelotopes

In this section, we provide a solution to the following problem:

Given the parallelotope $\mathcal{P}_k$ and the strip Σ_{k+1} provided by the $(k+1)$-th measurement, find the minimal volume parallelotope $\mathcal{P}_{k+1}$ containing the polytope $\mathcal{V}_k \doteq \mathcal{P}_k \cap \Sigma_{k+1}$.

Notice that $\mathcal{V}_k$ is the intersection of $n+1$ strips in the parameter space, each bounded by a pair of parallel hyperplanes. Clearly, some of these hyperplanes may not be tangent to $\mathcal{V}_k$; moreover, the intersection of a tangent hyperplane with an n-th dimensional polytope is an l-dimensional face, with $0 \leq l \leq n-1$.

Elementary geometrical considerations show that $\mathcal{V}_k$ has m faces of dimension $n-1$, $n+1 \le m \le 2(n+1)$, belonging to at least n different strips.

In order to compute the optimal outbounding paralleloptope, we first proceed to detecting the bounding hyperplanes of the various strips which are non tangent to $\mathcal{V}_k$. For a given strip, if both of the bounding hyperplanes are not tangent to $\mathcal{V}_k$, then the strip does not provide any new information and therefore can be discarded. In this case, the problem is trivially solved by taking $\mathcal{P}_{k+1}$ as the intersection of the remaining n strips. Conversely, if for a given strip, only one hyperplane is tangent to $\mathcal{V}_k$, then the strip need be 'tightened' by replacing the non tangent hyperplane by a new parallel tangent hyperplane. Iterating this tightening procedure for all the $n+1$ strips defining $\mathcal{V}_k$ leads to the following expression for $\mathcal{V}_k$

$$\mathcal{V}_k = \left\{ \bigcap_{i=1}^{n+1} S_i \right\} \tag{6}$$

where S_i are defined as in (4) and all the strips are tightened, i.e. all the hyperplanes

$$c_i - p_i'\theta = \pm 1, \quad i = 1, \ldots, n+1$$

are tangent to $\mathcal{V}_k$.

Notice that the result of the tightening process is independent of the order according to which the strips are tightened. A technique for solving this problem will be presented in the next section.

The next lemma which provides a parameterization of a generic strip containing $\mathcal{V}_k$, follows easily from the tightened representation of $\mathcal{V}_k$ given above.

Lemma 1. *Any strip $\bar{S}_i$ outbounding $\mathcal{V}_k$ can be expressed as*

$$\bar{S}_i \doteq \{\theta : |\bar{c}_i - \bar{p}_i\theta| \le 1\} \tag{7}$$

where $\bar{p}_i$ and $\bar{c}_i$ are given by

$$\bar{p}_i = \sum_{j=1}^{n+1} a_{ij}p_j, \qquad \bar{c}_i = \sum_{j=1}^{n+1} a_{ij}c_j,$$

with

$$a_{ij} \ge 0, \qquad \sum_{j=1}^{n+1} a_{ij} \le 1 . \tag{8}$$

In order to find the minimal volume parallelotope, define the quantities

$$P_j \doteq \begin{bmatrix} p_1' \\ \vdots \\ p_{j-1}' \\ p_{j+1}' \\ \vdots \\ p_{n+1}' \end{bmatrix} \in R^{n,n}, \qquad C_j \doteq \begin{bmatrix} c_1 \\ \vdots \\ c_{j-1} \\ c_{j+1} \\ \vdots \\ c_{n+1} \end{bmatrix} \in R^n, \qquad j = 1, \ldots, n+1 . \tag{9}$$

We can now state the following theorem which is the main result of the paper.

Theorem 2. *The minimal volume parallelotope $\mathcal{P}_{k+1}$ outbounding $\mathcal{P}_k \cap \mathcal{S}_{k+1}$ is given by*

$$T(k+1) = (P_{j\bullet})^{-1}, \qquad \theta^c(k+1) = (P_{j\bullet})^{-1} C_{j\bullet}.$$

where

$$j^* = \arg \max_{j=1,\ldots,n+1} \{|det\, P_j|\} .$$

Proof. (see [22]).

Remark. The previous theorem implies that the hyperplanes defining the minimal volume parallelotope are parallel to n out of the $n+1$ hyperplanes defining $\mathcal{V}_k$.

4 Sequential Parameter Uncertainty Set Estimation

We begin this section by giving a result which allows one to tighten a given strip Σ with respect to a parallelotope $\mathcal{P}(T, \theta^c)$ (explicit dependence of ϕ, y, r, $\mathcal{P}$, etc. on k is dropped here for notational convenience). A strip Σ

$$\Sigma = \{\theta \in R^n : |y - \phi'\theta| \le r\} .$$

is tight with respect to $\mathcal{P}(T, \theta^c)$ if

$$\{\theta \in \mathcal{P}(T, \theta^c) : y - \phi'\theta = r\} \neq \emptyset \text{ and } \{\theta \in \mathcal{P}(T, \theta^c) : y - \phi'\theta = -r\} \neq \emptyset .$$

The above definition means that Σ is tight with respect to $\mathcal{P}(T, \theta^c)$ if both of its bounding hyperplanes are tangent to the intersection set $\Sigma \cap \mathcal{P}(T, \theta^c)$.

Let us associate the following quantities to Σ and $\mathcal{P}(T, \theta^c)$

$$\bar{r} \doteq \phi'\theta^c + \sum_{i=1}^{n} |\phi't_i|$$

$$\underline{r} \doteq \phi'\theta^c - \sum_{i=1}^{n} |\phi't_i|$$

$$r^+ \doteq \min(y + r, \bar{r}) \tag{10}$$

$$r^- \doteq \max(\underline{r}, y - r) .$$

We can now state the following result

Theorem 3. *The strip*

$$\mathcal{S} = \{\theta : |p'\theta - c| \le 1\}$$

with

$$p = 2\phi/(r^+ - r^-), \quad c = (r^+ + r^-)/(r^+ - r^-) .$$

is tight with respect to the parallelotope $\mathcal{P}(T, \theta^c)$.

Proof. (see [22]).

Remark. Observe that if $\bar{r} \geq y + r$ and $\underline{r} \leq y - r$, then the strip Σ itself is tight with respect to $\mathcal{P}(T, \theta^c)$. In this case, the strip S provided by Theorem 3 represents the normalized expression for Σ needed in order to apply Theorem 2.

Based on the above result and Theorem 2 of the preceding section, we can build a recursive algorithm for outbounding the parameter uncertainty set via parallelotopes. The input of the algorithm at time $k + 1$ is the estimate $\mathcal{P}_k$ at time k and the strip Σ_{k+1} representing the $k + 1$-th measurement.

Step 1
Compute a description of $\mathcal{V}_k \doteq \mathcal{P}_k \cap \Sigma_{k+1}$ in terms of $n + 1$ tightened strips S_i like in (6).

Step 2
Form the $n + 1$ matrices P_j and vectors C_j defined in (9)

Step 3
Solve

$$j^* = \arg \max_{j=1,\ldots,n+1} \{|\det P_j|\}$$

and compute the corresponding n optimal strips S_i^* defined by the rows of P_j^* and C_j^*.

Step 4
Construct $\mathcal{P}_{k+1}$ as

$$\mathcal{P}_{k+1} = \bigcap_{i=1}^{n} S_i^* .$$

As a comment on the algorithm, we notice that in order to perform Step 1, Theorem 3 must be applied to each of the $n + 1$ strips determining $\mathcal{V}_k$, where the reference parallelotope is defined by the remaining n strips.

Observation A simplified version of the recursive algorithm can be employed for deriving orthotopic approximations of the parameter uncertainty set. For this problem, only Step 1 of the algorithm delineated above needs to be performed. In fact, orientations of the hyperplanes bounding the approximating parallelotope are not free in this case, being fixed by the orthotopic shape assumption.

5 Numerical Example

The algorithm presented in the previous section was exploited for the set membership identification of the parameters of the ARX model

$$y(k) = -1.3y(k-1) - 0.4y(k-2) + u(k-1) + 0.8u(k-2) + w(k) \quad (11)$$

where the disturbance $w(k)$ is independently uniformly distributed (i.u.d.) in $[-1, 1]$, while the input $u(k)$ is i.u.d. in $[0, 1]$. Notice that the same example was considered in [18] and [19]. Results of a typical simulation experiment are illustrated in Figs. 1 and 2. Figure 1 shows the behavior of the components of the center $\theta^c(k)$ of the outbounding parallelotope $\mathcal{P}_k$. Figure 2 shows the estimated parallelotope volume $2^n|\det T(k)|$ (in logarithmic scale) as a function of the number of samples processed.

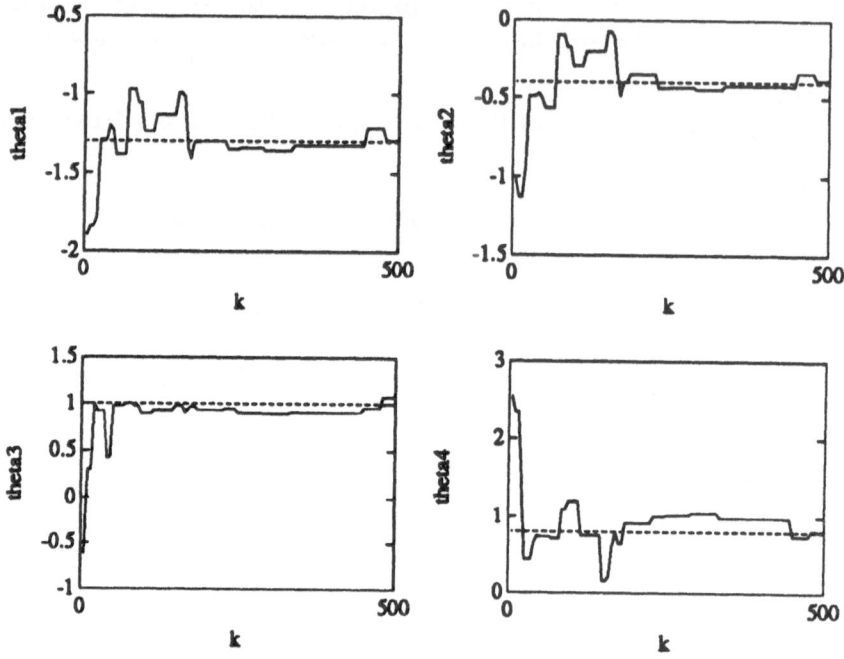

Fig. 1. Behavior of central parameter estimates $\theta_i^c, i = 1, \ldots, 4$ (dashed line - true value; solid line - estimated value)

A preliminary analysis of these results shows that the proposed algorithm behaves much better (by several orders of magnitude) than the Fogel-Huang ellipsoidic algorithm [18] and compares favourably with respect to the modified ellipsoidic algorithm used in [19].

6 Concluding Remarks

In this paper an algorithm has been proposed for sequential estimation of the parameter uncertainty set in a linear regression model. A hard bound setting

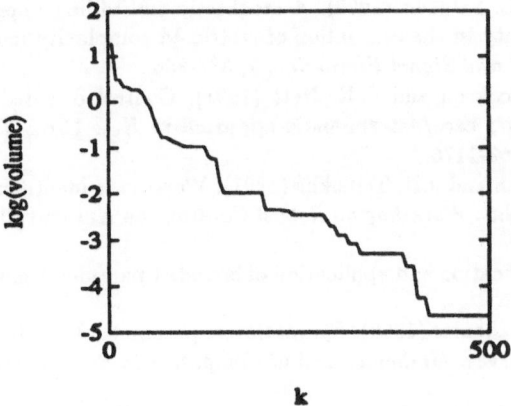

Fig. 2. Log-volume of the outbounding parallelotope

of the underlying identification problem has been considered. The procedure provides an outer approximation of the uncertainty set alternative to commonly used ellipsoidic bounds. Several ramifications connected with the proposed algorithm are under investigation. Finding conditions under which a new measurement at time $k + 1$ allows for a reduction of the parallelotope volume at time k is an important issue from a convergence standpoint. Devising suitable techniques for mapping different kinds of assumed prior knowledge on the unmodelled dynamics and disturbances into prior uncertainty estimates is another significant topic. Moreover, numerical efficiency and robustness of different techniques for implementing the algorithm, which involves matrix determinant computations, need to be investigated and compared with the ellipsoidic case. Finally, an analysis of the achievable estimate accuracy as related to the number of measurements processed at the same time at each step of the algorithm, would be useful for establishing a tradeoff between allowable computational burden and desired accuracy of the uncertainty approximation in practical applications.

References

1. Ljung, L. (1987). *System Identification: Theory for the User*, Prentice-Hall, Englewood Cliffs, N.J. .
2. Ljung L., B. Wahlberg and H. Hjalmarsson (1991). Model quality: the roles of prior knowledge and data information. *Proc. 30-th CDC*, Brighton (UK), 273-278.
3. Gevers, M. (1991). Connecting identification and robust control: a new challenge. *Proc. 9-th IFAC Symp. on Identification and Sys. Parameter Estimation*, Budapest (H), 1-10.

4. Goodwin, G.C. and M.E. Salgado (1990). A stochastic embedding approach for quantifying uncertainty in the estimation of restricted complexity models. *Int. J. Adaptive Control and Signal Processing*, **3**, 333-356.

5. Helmicki, A.J., C.A. Jacobson and C.N. Nett (1991). Control oriented system identification: a worst case/deterministic approach in H_∞. *IEEE Trans. Automat. Contr.*, **36**, 1163-1176.

6. Tse, D.N.C., M.A. Dahleh and J.N. Tsitsiklis (1991). Worst case identification for robust control. *Proc. Int. Workshop on Robust Control*, San Antonio (USA), CRC Press, 311-328.

7. Norton, J. (1987). Identification and application of bounded-parameter models. *Automatica*, **23**, 497-507.

8. Walter, E. and H. Piet-Lahanier (1990). Estimation of parameter bounds from bounded-error data: a survey. *Mathematics and Computers in Simulation*, **32**, 449-468.

9. Milanese M. and A. Vicino (1991). Optimal estimation theory for dynamic systems with set membership uncertainty: an overview. *Automatica*, **27**, 997-1009.

10. Kosut R.L., M.K. Lau and S.P. Boyd (1991). System identification for robust control design. *Proc. European Control Conference*, Grenoble (F), 1384-1388.

11. Younce, R.C. and C.E. Rohrs (1990). Identification with non-parametric uncertainty. *Proc. 29th CDC*, Honolulu (USA), 3154-3161.

12. Wahlberg, B. and L. Ljung (1991). On estimation of transfer function error bounds. *Proc. European Control Conference*, Grenoble (F), 1378-1383.

13. Chapellat, H., M. Dahleh, and S. P. Bhattacharyya (1990). Robust stability under structured and unstructured perturbations. *IEEE Trans. on Automat. Contr.*, **35**, 1100-1108.

14. Dahleh, M., A. Tesi and A. Vicino (1992). An overview on extremal properties for robust control of interval plants. *Automatica*, to appear.

15. Doyle, J. C. (1982). Analysis of feedback systems with structured uncertainties. *IEE Proc., Part D*, **129**, 242-250.

16. Doyle, J. C., J. E. Wall and G. Stein (1982). Performance and robustness analysis for structured uncertainty. *Proc. 21st IEEE CDC*, Orlando (USA), 629-636.

17. Bartlett, A. C., C. V. Hollot and L. Huang (1988). Root locations of an entire polytope of polynomials: it suffices to check the edges. *Mathematics of Contr., Sign. and Syst.*, **1**, 61-71.

18. Fogel, E. and F. Huang (1982). On the value of information in system identification-Bounded noise case. *Automatica*, **18**, 229-238.

19. Belforte, G., B. Bona and V. Cerone (1990). Parameter estimation algorithms for set membership description of uncertainty. *Automatica*, **26**, 887-898.

20. Wahlberg, B. (1986). System identification using Laguerre models. *IEEE Trans. Automat. Contr.*, **36**, 551-562.

21. Goodwin, G., M. Gevers and B. Ninness (1991). Optimal model order selection and estimation of model uncertainty for identification with finite data. *Proc. 30-th CDC*, Brighton (UK), 285-290.

22. A. Vicino and G. Zappa (1992). "Adaptive approximation of uncertainty sets for linear regression models", *Int. Res. Rep., Dipartimento di Ingeneria Elettrica, Università di L'Aquila*.

A Robust Ellipsoidal-Bound Approach to Direct Adaptive Control

Tung-Ching Tsao and Michael G. Safonov

Department of Electrical Engineering — Systems,
University of Southern California, Los Angeles, CA 90089-2563, U.S.A.

1 Introduction

There are two main approaches to adaptive control. The most widely known is the indirect approach based on a separation of estimation and control. In this approach a plant model identification scheme is combined with a controller design scheme (see, for example, [1] and the references therein). In the direct approach there is no separation. Instead, the controller parameters which achieve the desired control performance objectives are directly identified from real-time input-output data (e.g., [2]). An adaptive control design method is said to be robust if it can tolerate the inevitable residual plant modeling error. Early approaches to robust adaptive control focused on the direct approach [3]. A characteristic of the early robust adaptive control methods was the assumption of a priori bounds on residual plant uncertainty; i.e., the measurement data is not used to develop refined estimates of effects of modeling uncertainty on control performance.

Motivated in part by desire to produce robust indirect adaptive control designs in which information about model accuracy contained in real time data is used to improve the control design, recent research efforts have focused on a form of control-oriented system identification in which plant input-output data is used to refine ellipsoidal bounds on plant parameters. Used as part of an indirect adaptive control scheme, this approach to identification offers the distinct advantage of providing plant uncertainty bounds suitable for use with robust control synthesis methods such as H^∞ control. However, there is no separation theorem to ensure the optimality of this, or any other, indirect approach to adaptive control. For example, since the ellipsoidal bounds provided by control-oriented identification methods are not tight and not unique and since achievable control performance depends on these bounds, there is an inevitable coupling between control and identification within this framework. Attempts to artificially enforce the separation between control and identification by choosing a particular form or shape for the ellipsoidal bound in the identification phase can be expected to lead to unduly conservative designs. Nevertheless, the ellipsoidal bound approach has been until

now the only adaptive control approach to date that takes explicit account
of the input-output data in assessing the effects of plant uncertainty.

In this paper we show how to overcome these deficiencies. We show
how the mathematical methods used in ellipsoidal-bounding approaches to
control-oriented identification (e.g., [4, 5]) can be adapted to enable one to
do direct robust adaptive control by directly identifying controller parameters which achieve specified closed-loop performance objectives. Our main
result is a formulation of the direct robust adaptive control problem within
the ellipsoidal bound framework. A key idea is the introduction of a fictitious
reference signal, which facilitates direct identification of the controller (DIC)
to meet specifications of performance using measurements of plant input and
output. This enables us to bypass identification of the plant and its uncertainties as would be required in an indirect adaptive control approach based
on the separation principle.

The applicability of the proposed method of direct identification of controller parameters indicates that the use of the separation approach, in which
controllers are designed based on the identified nominal models of plant and
uncertainty information, is not a necessity in the design of robust adaptive
controllers.

Throughout the paper the generic model structure for closed-loop systems
depicted in Fig. 1 will be used.

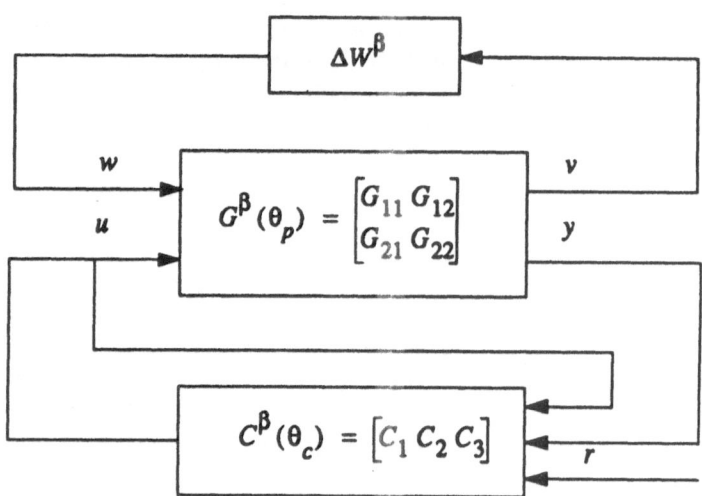

Fig. 1. Generic model structure for closed-loop systems.

In the figure, Δ is an operator with $\|\Delta\|_{2i} < 1$, where $\|.\|_{2i}$ is L_2-induced
operator norm, W^β is an appropriate weighting function, θ_p is plant parameter, θ_c is controller parameter, $G^\beta(\theta_p)$ is nominal plant model, $C^\beta(\theta_c)$ is

controller, β is an index used to indicate that the triple $(W^\beta, G^\beta, C^\beta)$ forms an element of some set, v, w are internal signals of the plant, u and y are plant input and output, r is reference input signal.

We will show in later sections that both plant and controller parameter identification problems can be treated as constrained optimization problems of the form:

> minimize β
>
> subject to

$$1.\ (W^\beta, g^\beta) \in \mathcal{H} \tag{1}$$
$$2.\ \exists\ \text{operator } \Delta \text{ with } \|\Delta\|_{2i} < 1 \text{ and } e = \Delta W^\beta h$$
$$3.\ g^\beta(\theta, v, w, u, y, r) = 0$$

where β is a cost index for elements in $\mathcal{H}$, with the smallest β indicating the highest preference, θ is the parameter to be identified, W^β is a weighting function, and equation $g^\beta = 0$ defines a possible interconnection of the closed-loop system. In the case of plant parameter identification, e and h are just internal signals w and v; in the DIC case they are functions of the measurable signals u, y and r.

Given a nominal model G of a plant and a weighting function W, we are not guaranteed the existence of an operator Δ of unit norm that describes the plant input-output behavior observed. Most of the time we need to either change the complexity of the nominal model by increasing its order, say, or iterate on the weighting functions. This is why we use β and the set $\mathcal{H}$ in (1). This β-iteration of identification is similar in spirit to the γ-iteration of robust control design of [6]. The same arguments hold in the DIC case, which will be described in more detail in Sect. 3.

The uncertainty operator Δ of (1) can be either a linear time invariant or an arbitrary operator. When it is confined to be linear time invariant, several frequency domain estimators have been developed for the identification of plant parameters with such uncertainty assumption, e.g. [5, 7, 8], or one can adopt time domain approach using an algorithm that combines features of balanced stochastic truncation (BST) and Hankel SVD [9], or using a model validation method of [10]. However, for robust controller design purpose and also for our proposed DIC approach Δ can be regarded as an arbitrary operator on L_2 space of unit operator norm even for linear time invariant systems. In this case, condition 2 of (1) is equivalent to

$$\|e\|_{t2} < \|W^\beta h\|_{t2}, \quad \forall t > 0 \tag{2}$$

where $\|.\|_{t2}$ is the truncated L_2 norm, $\|x\|_{t2}^2 = \int_0^t x^T(\tau)x(\tau)\,d\tau$. When e and h in (1) are linear in the parameters to be identified, (2) defines a quadrics, i.e. an ellipsoid or hyperboloid etc., at each time t. This time domain approach leads to a simple quadratic formulation – the unknown-but-bounded formulation, or in some places called set-membership or ellipsoidal algorithms.

Unknown-but-bounded method is a deterministic approach for identification [11], which deals with the calculation of the intersection of ellipsoids. Because the exact intersection of ellipsoids is difficult to describe using simple expressions, many successful results have been established in determining the outer approximate of the intersection, such as the recursive schemes of [12], [13] and [14], and the minimax approach of [15]. Although improvements are required in finding tighter estimate of the intersection, the simplicity of the quadratic inequality (2) makes time-domain set-membership identification a very attractive approach.

The organization of the paper is as follows. In Sect. 2, we discuss the concept of consistent estimates. Section 3 contains the main contribution of this paper, in which we describe the approach, introduce our direct identification control (DIC) approach to robust direct adaptive control and prove its achievement of performance specifications. Section 4 concludes the paper.

2 Identifying the Consistent Sets

The design of robust controllers requires information about a nominal model and its associated model accuracy. Traditional system identification techniques, which only produce a nominal model without providing information about accuracy, are therefore inadequate for the purpose of robust controller design. Motivated by the demands from robust control design, many control-oriented identification methodologies have been developed. One of the important features of the new methods is the concept of consistent or unfalsified model [16] which results in a set estimate instead of the point estimate of the traditional identification. In the following we will review the definition of consistent sets [17], which is required in Sectoin 3.

Definition 1 Consistent set at time t Θ_t^*. The parameter estimate that satisfies the constraints in (1) up to time t is called a consistent estimate at time t. The set formed by all consistent estimates at time t is called a consistent set at time t, denoted Θ_t^*. When the preference level β needs to be emphasized, we use $\Theta_t^{\beta*}$. Also current consistent set $\tilde{\Theta}_t^*$ means the intersection of the consistent set at current time and the starting set Θ_0, that is $\tilde{\Theta}_t^* = \Theta_t^* \cap \Theta_0$.

Definition 2 Consistent set Θ^*. The consistent set Θ^* is the following infinite intersection

$$\Theta^* = \bigcap_{t>0} \Theta_t^* .$$

The consistency requirement for parametric identification of plant parameter defined by (1) is usually, see Fig. 1,

$$1. \ (\beta W, G) \in \mathcal{H} \tag{3}$$

2. $w = \beta \Delta W\, v$ for some Δ with $\|\Delta\|_{2i} < 1$

3. $\begin{bmatrix} -I & G_{11}(\theta_p) & G_{12}(\theta_p) & 0 \\ 0 & G_{21}(\theta_p) & G_{22}(\theta_p) & -I \end{bmatrix} \begin{bmatrix} v \\ w \\ u \\ y \end{bmatrix} = 0$

that is $W^\beta = \beta W$ for some fixed W and $G^\beta = G$, and r and C^β do not take part in the identification of plant parameter. The identification of plant parameter becomes the process of minimization of β under the consistency constraint (3). The following examples demonstrate this idea.

Example 4. Assume that the plant is described by

$y = Gu$ with $G = \dfrac{(\theta_1 + \beta \Delta_2)s + \theta_2}{s^2 + \theta_3 s + \theta_4}(1 + \beta \Delta_1)$ and $\|\Delta_1\|_{2i} < 1, |\Delta_2| < 1$.

Let $\theta_p^T = [\theta_1, \theta_2, \theta_3, \theta_4]$ be the plant parameter vector to be identified.

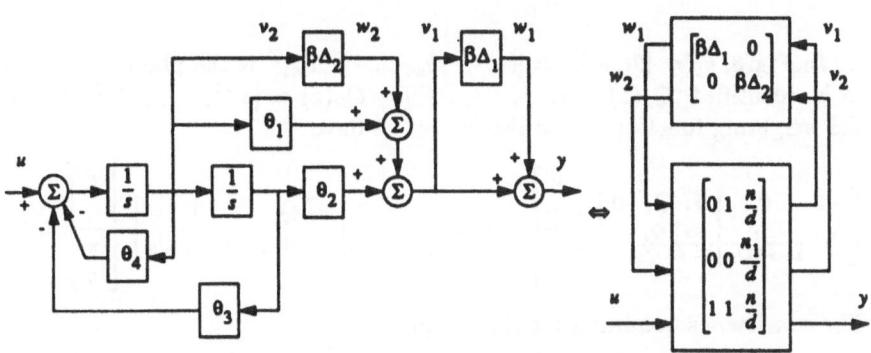

Fig. 2. Example 1

In Fig. 2, $n(\theta_p) = \theta_1 s + \theta_2$, $d(\theta_p) = s^2 + \theta_3 s + \theta_4$ and $n_1(\theta_p) = \theta_1 s$. From the figure condition 3 of (3) becomes

$$\begin{bmatrix} -1 & 0 & 0 & 1 & \frac{n}{d} & 0 \\ 0 & -1 & 0 & 0 & 0 & \frac{n_1}{d} \\ 0 & 0 & 1 & 1 & \frac{n}{d} & -1 \end{bmatrix} \begin{bmatrix} v_1 \\ v_2 \\ w_1 \\ w_2 \\ u \\ y \end{bmatrix} = 0 \Rightarrow \begin{bmatrix} -1 & 0 & 0 & 0 & \frac{n}{d} & 0 \\ 0 & -1 & 0 & 0 & 0 & \frac{n_1}{d} \\ 0 & 0 & -1 & 0 & -\frac{n}{d} & 1 \end{bmatrix} \begin{bmatrix} v_1 - w_2 \\ v_2 \\ w_1 + w_2 \\ w_2 \\ u \\ y \end{bmatrix} = 0 \,.$$

From which the consistency condition for identifying θ_p becomes:

$\exists w_2 \in L_{2e}(0, \infty)$ and operators Δ_i with $\|\Delta_i\|_{2i} < 1, i = 1, 2$

and $\begin{cases} -\frac{n}{d}u + y - w_2 = \beta \Delta_1 \left(\frac{n}{d}u + w_2 \right) \\ w_2 = \beta \Delta_2 \left(\frac{n_1}{d}u \right) \end{cases}$

Example 5. [4]. In this example, the space under discussion is l_2. Assume that the plant is described by

$$y = Gu \text{ with } G = \frac{b_1 z^{-1} + \cdots + b_m z^{-m}}{1 + a_1 z^{-1} + \cdots + a_n z^{-n}}(1 + \Delta W_G) \text{ and } \|\Delta\|_{2i} < 1.$$

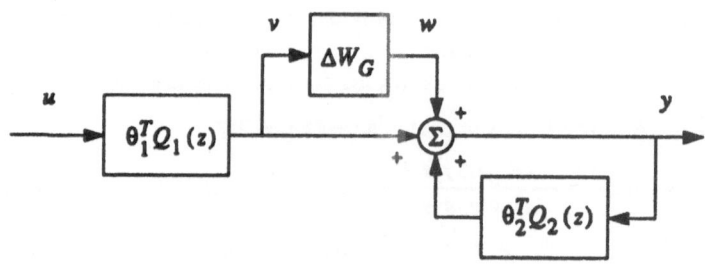

Fig. 3. Example 2

In Fig. 3, $\theta_p = [\theta_1, \theta_2]^T = [b_0, \cdots, b_n, a_0, \cdots, a_n]^T$ is the plant parameter to be identified, $Q_1(z) = [z^{-1}, \cdots, z^{-m}]^T$, $Q_2(z) = [z^{-1}, \cdots, z^{-n}]^T$, W_G is the weighting function. From the figure, we have

$$\begin{bmatrix} -1 & 0 & \theta_1^T Q_1 & 0 \\ 0 & \frac{-1}{1-\theta_2^T Q_2} & \frac{-\theta_1^T Q_1}{1-\theta_2^T Q_2} & 1 \end{bmatrix} \begin{bmatrix} v \\ w \\ u \\ y \end{bmatrix} = 0 \Rightarrow \begin{bmatrix} -1 & 0 & \theta_1^T Q_1 & 0 \\ 0 & -1 & -\theta_1^T Q_1 & 1 - \theta_2^T Q_2 \end{bmatrix} \begin{bmatrix} v \\ w \\ u \\ y \end{bmatrix} = 0 .$$

The consistency condition for this example is:

$$\exists \text{ an operator } \Delta \text{ with } \|\Delta\|_{2i} < 1 \text{ and}$$
$$y - \theta_1^T Q_1 u - \theta_2^T Q_2 y = \beta \Delta W_G(\theta_1^T Q_1 u) .$$

Example 6. [18]. Assume that the plant input u and output y are described by

$$y = Gu \text{ with } G = \frac{b_1 s^{n-1} + b_2 s^{n-2} + \cdots + b_n + \beta \Delta_1}{s^n + a_1 s^{n-1} + \cdots + a_n + \beta \Delta_2} \text{ and } \|\Delta_1\|_{2i}, \|\Delta_2\|_{2i} < 1 .$$

θ_p is the same as in Example 2 and $Q(s) = [\frac{1}{s+p_1}, \cdots, \frac{1}{s+p_n}]^T$ for some positive p_i's. From Fig. 4, we can write

$$\begin{bmatrix} -1 & 0 & 0 & 0 & 1 & 0 \\ 0 & -1 & \frac{1}{1-\theta_2^T Q} & \frac{1}{1-\theta_2^T Q} & \frac{\theta^T Q}{1-\theta_2^T Q} & 0 \\ 0 & 0 & \frac{1}{1-\theta_2^T Q} & \frac{1}{1-\theta_2^T Q} & \frac{\theta^T Q}{1-\theta_2^T Q} & -1 \end{bmatrix} \begin{bmatrix} v_1 \\ v_2 \\ w_1 \\ w_2 \\ u \\ y \end{bmatrix} = 0$$

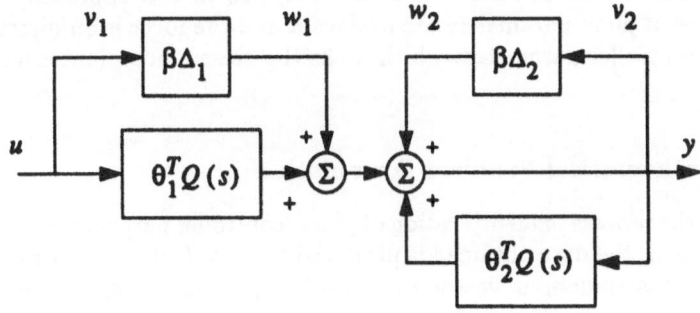

Fig. 4. Example 3

$$\Rightarrow \begin{bmatrix} -1 & 0 & 0 & 0 & 1 & 0 \\ 0 & -1 & 0 & 0 & \frac{\theta_1^T Q}{1-\theta_2^T Q} & 0 \\ 0 & 0 & -1 & 0 & -\theta_1^T Q & 1-\theta_2^T Q \end{bmatrix} \begin{bmatrix} v_1 \\ v_2 - \frac{1}{1-\theta_2^T Q}(w_1 + w_2) \\ \hline w_1 + w_2 \\ \hline w_2 \\ \hline u \\ y \end{bmatrix} = 0 .$$

The consistency condition is

$$\exists w_2 \in L_{2e}(0, \infty) \text{ and operators } \Delta_i \text{ with } \|\Delta_i\|_{2i} < 1, i = 1, 2$$

$$\text{and } \begin{cases} -\theta_1^T Q u + (1 - \theta_2^T Q) y - w_2 = \beta \Delta_1 u \\ w_2 = \beta \Delta_2 (y) \end{cases} .$$

Or one can add the above two equations to form:

$$-\theta_1^T Q u + (1 - \theta_2^T Q) y = \beta [\Delta_1, \Delta_2] \begin{bmatrix} u \\ y \end{bmatrix} . \tag{4}$$

3 Direct Identification of Controller Parameters

Controller parameters should be determined by the performance requirements. There are basically two approaches to determine controller parameters: the direct and indirect methods. Indirect method is an application of
the widely used separation principle, in which plant parameters are identified
first either off-line or on-line and then controllers are designed based on the
identified parameters to fulfill the performance requirements. But no one has
yet shown the use of separation principle to be optimal. On the contrary,
set estimation and robust controller design might benefit from being coupled
together as pointed out in [4]. In the following we combine the set estimation
and robust controller design in such a way that the entire process of plant
parameter identification is no longer needed. Instead we directly identify the
controller parameters that are able to produce the specified performance from

the measured data of plant inputs and outputs. In this approach, errors in estimates of plant parameters are irrelevant and the focus is on directly computing controller parameters which make the closed-loop plant error signals small.

3.1 Mathematical Problem Formulation

During the process of identification of plant/controller parameters in a closed-loop system, the data obtained is produced by a controller with time-varying and perhaps undesired vector of controller parameters θ_c. Since, at first glance, what happened has happened and it does not seem to be possible to formulate an expression for the controller parameter vector that will yield desired performance in the future without first obtaining the plant parameters. But if one regards the obtained data as being produced by some fixed desired θ_c and an appropriate fictitious signal $\tilde{r}$ acting as reference input signal and sets up an equation based on the performance specification with actual reference signal r replaced by $\tilde{r}$, an expression for the controller parameter is obtained. The question remains whether the θ_c satisfying the expression will produce the desired performance. The answer is positive for a large class of reference signals as will be shown in Sect. 3.2.

We now give the explicit formulation. From Fig. 1, we have

$$u = C_1(\theta_c(t))u + C_2(\theta_c(t))y + C_3(\theta_c(t))r \ . \tag{5}$$

Assume that u in (5) is produced by some fixed θ_c, the actual plant output y and some fictitious reference signal $\tilde{r}$, i.e. $u = C_1 u + C_2 y + C_3 \tilde{r}$, so that we have

$$\left[C_1(\theta_c) - I \ C_2(\theta_c) \ C_3(\theta_c) \right] \begin{bmatrix} u \\ y \\ \tilde{r} \end{bmatrix} = 0 \ . \tag{6}$$

When determining the performance specifications for a control design, we usually need to trade off between performance and cost. The performance can be expressed in terms of the allowed error level at different frequencies, which can be represented by different weighting functions. The cost can be represented by the complexity of the controller structure. So we can set up a family of pairs of weighting functions and controller structures and order the pairs according to our preferences on the combination of the two. In other words, we can define the family to be $\mathcal{H} \triangleq \{(W_s^\beta, C^\beta(\theta_c))\}$ where W_s^β is weighting function, C^β represents controller structure and β indicates the preference level. The design of controller becomes a process of choosing a best possible element from $\mathcal{H}$, or equivalently it is the solution of the following optimization problem:

minimize $\quad \beta$

subject to

1. $(W_s^\beta, C^\beta) \in \mathcal{H}$ $\hfill (7)$

2. $\exists$ operator Δ with $\|\Delta\|_{2i} < 1$ and $\tilde{r} - y = \Delta(W_s^\beta(\tilde{r}))$

3. $\left[C_1^\beta(\theta_c) - I \; C_2^\beta(\theta_c) \; C_3^\beta(\theta_c) \right] \begin{bmatrix} u \\ y \\ \tilde{r} \end{bmatrix} = 0$

This is of the same form as (1). When solving the problem using a time domain approach, condition 2 of (7) can be replaced by the L_2-norm inequality:

$$\|\tilde{r} - y\|_{t2} < \|W_s^\beta(\tilde{r})\|_{t2}, \quad \forall t > 0 \ . \hfill (8)$$

In (8), performance specification is directly placed on the plant input and output, not on the model-based sensitivity function. This is desirable since the sensitivity function of a linear time invariant system with model uncertainty is close to the actual sensitivity function only at frequencies where the complementary sensitivity is close to one [19].

In the following we give an example to show that for a particular structure of the controller, the relation (8) results in a quadratic inequality of θ_c which is of the same form as that for the plant parameter case in [4].

Example 7. Assume u, y and r are of the same dimension, and that the controller are of the structure shown in Fig. 5.

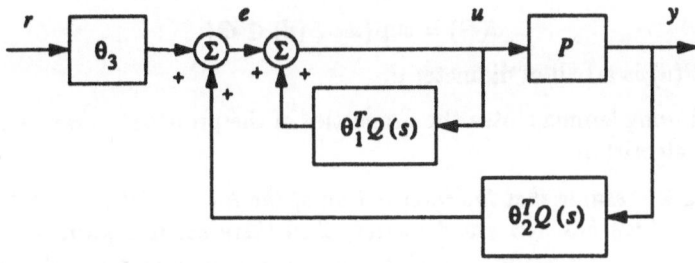

Fig. 5. Direct identification of controller parameters

The constraints in (7) for time domain approach now become

$$\|\tilde{r} - y\|_{t2} < \|W_s^\beta(\tilde{r})\|_{t2}, \quad \forall t > 0 \hfill (9)$$

$$\text{and } \left[\theta_1^T Q - I \; \theta_2^T Q \; \theta_3 \right] \begin{bmatrix} u \\ y \\ \tilde{r} \end{bmatrix} = 0 \ .$$

$Q = [\frac{1}{s+p_1}, \cdots, \frac{1}{s+p_m}]^T$ for some positive p_i's and some $m > 0$. After simple algebraic manipulations, inequality (9) can be put into the form

$$\|g - \theta^T \phi\|_{t2} < \|W_s(g - \theta^T \psi)\|_{t2}, \quad \forall t > 0 \qquad (10)$$

where

$$g = u, \quad \theta = \begin{bmatrix} \theta_1 \\ \theta_2 \\ \theta_3 \end{bmatrix}, \quad \phi = \begin{bmatrix} Qu \\ Qy \\ y \end{bmatrix}, \quad \psi = \begin{bmatrix} Qu \\ Qy \\ 0 \end{bmatrix}.$$

Equation (10) is of the same form as that in [4]. It can be shown that when $\|W_s\|_{2i} \leq 1$ the solution for θ_1 and θ_2 defined by (10) at each fixed value of θ_3 is a set formed by intersection of ellipsoids. Note that one can also specify performance for each channel independently, under this condition the corresponding Δ in (7) will be of diagonal structure, and (10) becomes a system of L_2-norm inequalities.

3.2 Performance

We first introduce an identification algorithm that will produce consistent parameter estimates within finite amount of time. We shall refer to an approach to robust direct adaptive control in which controller parameters are selected to satisfy the inequality (8) as *direct identification control (DIC)*.

We will show that the DIC method yields the desired performance after finite amount of time.

The notion of the diameter of a set is needed in the statement of our results.

Definition 3 Diameter of a set $d(\Theta)$. The diameter of a set Θ is defined to be

$$d(\Theta) \triangleq \sup\{d : B(d) \subset \Theta\}$$

where $B(d)$ is a ball of diameter d.

The following lemma states the properties of the proposed parameter identification algorithm.

Lemma 1 *Assume that the intersection of the the consistent set Θ^* and the starting set Θ_0 has nonzero diameter, then there exists a finite $T > 0$ such that the parameter estimate $\hat{\theta}(t)$ becomes a constant vector and lies within $\Theta^* \cap \Theta_0$ after $t > T$ if the following parameter update law is used:*

$$\hat{\theta}(t) = \begin{cases} unchanged, & when \ \hat{\theta}(t) \in \tilde{\Theta}_t^* \\ \theta_t \in \tilde{\Theta}_t^* \ s.t. \ dist(\theta_t, \partial\tilde{\Theta}_t^*) > \alpha d(\tilde{\Theta}_t^*), & when \ \hat{\theta}(t) \notin \tilde{\Theta}_t^* \end{cases} \qquad (11)$$

where α is a fixed constant, $0 < \alpha < 1$, $\tilde{\Theta}_t^* = \Theta_t^* \cap \Theta_0$, $\partial\Theta$ denotes the boundary set of Θ and

$$dist(\theta_t, \partial\Theta) = \inf\{\|\theta_t - \theta\| : \theta \in \partial\Theta\} .$$

$$(12)$$

Proof of lemma. Let $S = \{t_i\}$, $i = 1, 2, \cdots$, denote the sequence of instances when the switching occurs. We will prove by contradiction. Suppose S is infinite. Let $T_1 = t_1$, from the definition of the parameter update law and the fact

$$\Theta_{t_1}^* \supseteq \Theta_{t_2}^*, \quad \forall t_1 \geq t_2$$

there exists $t_n \in S$ for some n such that

$$d(\tilde{\Theta}_{t_n}^*) \leq (1 - \alpha) d(\tilde{\Theta}_{T_1}^*) .$$

Define $T_2 = t_n$. Similarly, $\exists t_m \in S$ for some m s.t.

$$d(\tilde{\Theta}_{t_m}^*) \leq (1 - \alpha) d(\tilde{\Theta}_{T_2}^*) .$$

and so on. So, we have for all $i \geq 1$ that

$$d(\tilde{\Theta}_{T_i}^*) \leq (1 - \alpha)^{i-1} d(\tilde{\Theta}_{T_1}^*) .$$

The above inequality implies that

$$d(\Theta^* \cap \Theta_0) = d((\cap_{i,i=1}^\infty \Theta_{t_i}^*) \cap \Theta_0) = \lim_{i \to \infty} d(\tilde{\Theta}_{t_i}^*) = \lim_{i \to \infty} d(\tilde{\Theta}_{T_i}^*) = 0 .$$

since $0 < \alpha < 1$, which contradicts with the assumption of the diameter of $\Theta^* \cap \Theta_0$. $\qquad\Box$

Theorem 1 *Assume $\mathcal{H}$ contains only one element, Θ_i^* is defined by (7) and that the intersection of the consistent set and the starting set $\Theta^* \cap \Theta_0$ has nonzero diameter. Assume also that $\hat{\theta}_c(t)$ is updated by using (11) and $\exists$ an arbitrary $\delta > 0$ s.t. given any t_1 there exists $t_3 > t_2 > t_1$ such that $\int_{t_2}^{t_3} \|W_s r(\tau)\|^2 \, d\tau > \delta$. Then $\exists T > 0$ s.t.*

$$\|r - y\|_{t2} \leq \|W_s(r)\|_{t2}, \quad \forall t > T .$$

Remark. When $\mathcal{H}$ contains more than one element, we need to do β-iteration. In this case, we require that there exists at least one β such that $\Theta^{\beta*} \cap \Theta_0$ has nonzero diameter and that $d(\Theta^{\beta*} \cap \Theta_0) > 0$ when it is not an empty set. With these modifications the result of Theorem 1 still holds.

Remark. For the regulation problems, i.e. the situation that $r(t) = 0$ after $t > T$ for some $T > 0$ or $r(t)$ goes to zero asymptotically , it can be shown that, with same assumptions on $\Theta^* \cap \Theta_0$ and parameter update law, output y goes to zero asymptotically.

Remark. Parameter update law (11) can also be used in controller designs based on separation approach. One advantage of using (11) is that each time the controller is redesigned one regards the nominal model as determined by a single estimate instead of considering the whole current consistent set as nominal model set as used in [20]. Although the design is easier for single estimate, the transient behavior is not clear.

Proof of theorem. By Lemma 1, $\exists$ finite T_1 s.t. $\hat{\theta}_c(t) \in \Theta^* \cap \Theta_0$, and becomes a constant vector $\forall t > T_1$. After $t > T_1$, the equations governing u, y, $\tilde{r}$ in (7) become the same as those for u, y, r, and both can be regarded as driven by the same u. Hence, $\tilde{r}(t) = r(t)$ for $t > T_1$. Or we can write

$$\tilde{r}(t) = r(t) + r_1(t), \quad \forall t > 0$$

for some $r_1(t)$ of finite duration with $r_1(t) = 0$ when $t > T_1$.

Since $\|\tilde{r} - y\|_{t2} < \|W_s(\tilde{r})\|_{t2}$, $\forall t > 0$, there exists Δ with $\|\Delta\|_{2i} < 1$ s.t. $\tilde{r} - y = \Delta(W_s(\tilde{r}))$. Let $\epsilon > 0$ be such that $\|\Delta\|_{2i} = 1 - \epsilon$, then we have

$$\|\tilde{r} - y\|_{t2} \leq (1 - \epsilon)\|W_s(\tilde{r})\|_{t2}, \forall t > 0$$

$$\Rightarrow \|r - y\|_{t2} - \|r_1\|_{t2} \leq \|r + r_1 - y\|_{t2} \leq (1 - \epsilon)(\|W_s(r)\|_{t2} + \|W_s(r_1)\|_{t2}),$$
$$\forall t > 0$$

$$\Rightarrow \|r - y\|_{t2} \leq (1 - \epsilon)\|W_s(r)\|_{t2} + (1 - \epsilon)\|W_s(r_1)\|_{t2} + \|r_1\|_{t2}, \forall t > 0 .$$

Because the last two terms are bounded, from the assumption of $r(t)$, there exists $T > 0$ such that their sum becomes less than $\epsilon\|W_s(r)\|_{t2}$ when $t > T$, so we obtain following inequality:

$$\|r - y\|_{t2} \leq (1 - \epsilon)\|W_s(r)\|_{t2} + \epsilon\|W_s(r)\|_{t2} = \|W_s(r)\|_{t2}, \forall t > T .$$

$\square$

Remark. The time T specified in the theorem depends on ϵ and the magnitude of the reference signal $r(t)$. If the performance requirement is not too demanding for the controller structure and the reference signal, ϵ will not be small, so T will not be too large when $r(t)$ is not small. This fact is shown in the simulation results given in Sect. 3.3.

Effects of Initial Conditions and Disturbances. One of the key issues of Theorem 1 is the assumption of nonzero diameter of $\Theta^* \cap \Theta_0$. Because the operator Δ is assumed to have zero initial conditions so that (7) and (8) are equivalent, when initial conditions are not zero, the consistent set defined by

$$\|y_m - r\|_{t2} < \|W_s(r)\|_{t2}, \forall t > 0$$

may be an empty set, where $y_m = y + y_i$ is the measured output, y is the output without the effect of initial condition, and y_i is produced by initial conditions. Hence, the assumption of Theorem 1 is violated. To deal with this situation, we can assume that there exists stabilizing θ_c with some stability margin so that the effects of nonzero initial conditions decay exponentially when such θ_c is employed. With this assumption, $\|y_i\|_{t_2}$ in the following inequality is uniformly bounded,

$$\|y_m - r\|_{t2} \leq \|y - r\|_{t2} + \|y_i\|_{t2} \leq (1 - \epsilon)\|W_s(r)\|_{t2} + \|y_i\|_{t2}$$

where ϵ is as defined in the proof of Theorem 1. So, from the assumption of $r(t)$, there exists $T_0 > 0$ such that the consistent set defined by

$$\|y_m - r\|_{t2} < \|W_s(r)\|_{t2}, \ \forall t > T_0$$

has nonzero diameter so that the result of Theorem 1 still holds if we start the identification process after T_0. If one does not have an estimate of T_0, one way to solve the problem is to include an iteration with respect to the starting time of identification precess in the β-iteration.

Situations when nonzero deterministic disturbances exist can also be dealt with in a similar fashion.

3.3 Simulation Results

The plant and controller structure are the same as the one given in Example 7 of Sect. 3.1 with P and Q in Fig. 5 defined as:

$$P(s) = \frac{1}{s+1}\left(1 + \frac{(0.8)(100)}{s^2 + 10s + 100}\right), \quad Q(s) = \frac{1}{s+1} .$$

$\theta_1, \theta_2, \theta_3$ are scalars here. For simplicity, we assume $\mathcal{H}$ contains only one element with $W_s(s) = \frac{s+3.5}{s+35}$ and the controller as defined in Fig. 5. The initial conditions are assumed to be zero and there is no disturbance. Initial parameters are set to be $\hat{\theta}(0) = (20, 0, 15)$, which results in an unstable closed-loop system with poles at $19, -5 \pm 8.66j$ and -1. Parameter update law (11) and the time domain approach (8) are used to adjust the parameter; the update law starts at $t = 0.1$ sec. and is executed every 0.01 sec. When switching is required, the estimate is switched to a point around the geometry center of the current consistency set. The reference signal is $r(t) = \sin(t) + \sin(3t) + 0.5\sin(5t)$.

The a priori information on the parameter set is that $d(\Theta^* \cap \Theta_0) \neq 0$. The starting set Θ_0 is the union of five discs in θ_1-θ_2 plane of radius 80 centered at $(20, 0, 15)$, $(-300, -100, 200)$, $(-500, -100, 300)$, $(-700, -100, 400)$ and $(-900, -100, 500)$, respectively.

From the response plots, we see that the undesirable transients appearing in most conventional adaptive control schemes do not exist. The magnitude of resulting transfer function from r to $y - r$ denoted $\|T_{y-r,r}\|$ is also plotted; although it deviates from the expected weighting function the performance specification is still achieved after about $t = 7$ sec. The explanation for this is that for the particular reference signal, the desired performance specified by L_2-norm inequality can be achieved without requiring closed-loop system to acquire the specified sensitivity.

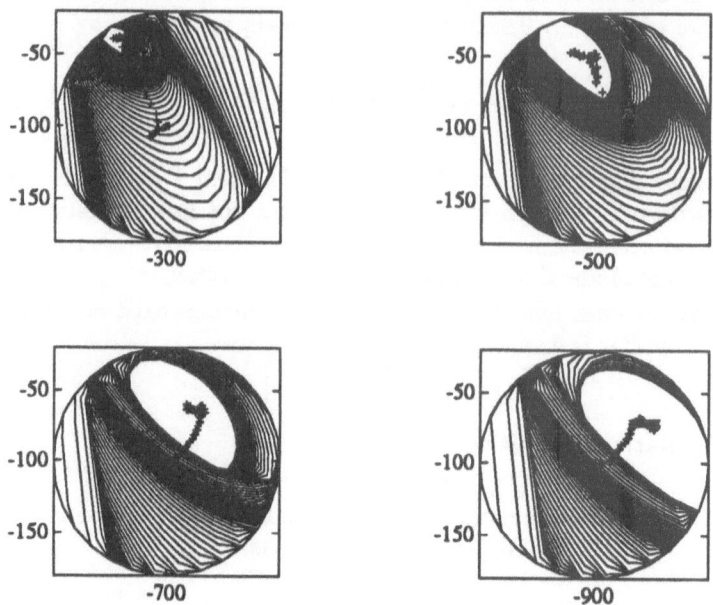

Fig. 6. Simulation results: evolution of consistent sets and time and frequency domain responses: + indicates trajectory of the centers

4 Conclusion

Our direct identification control (DIC) approach is essentially a *set theoretic* approach to robust adaptive control. The DIC approach of direct controller parameter identification can be employed as an alternative to the separation principle in the synthesis of robust adaptive control laws. The proposed method, DIC method, is developed with the mathematical framework of ellipsoidally-bounded parameter estimation used in the control-oriented identification literature [4, 5, 17, 18, 20]. The difference here is that we have reformulated the problem so that the parameters to be identified are controller gains rather than plant parameters and the identification error criterion is based on worst-case closed-loop control system performance rather than parameter estimation error. Our DIC controller synthesis approach is shown to be able to achieve performance specifications after finite amount of time, and is able to handle nonzero initial conditions and deterministic disturbances.

Simulation results indicate good performance of the proposed DIC approach. However, in performing the simulation, difficulties are encountered in obtaining the a priori information, i.e. the starting set Θ_0, and the current consistent set $\tilde{\Theta}_i^*$. An ad hoc approach has been devised to record the

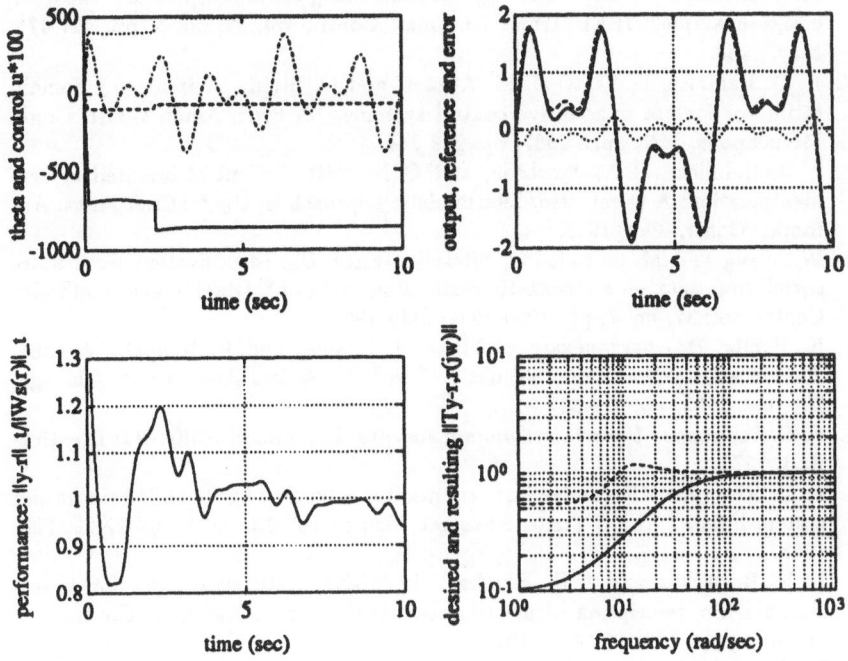

Fig. 7. Simulation results: evolution of consistent sets and time and frequency domain responses: time domain responses and $\|T_{y-r,r}\|$

boundary of $\tilde{\Theta}_t^*$. Since the DIC approach requires good estimate of the current consistent set, one of the important issues of future research is the finding of a tighter approximate of $\tilde{\Theta}_t^*$.

References

1. G. C. Goodwin and K. S. Sin, *Adaptive filtering prediction and control.* Englewood Cliffs, NJ: Prentice-Hall, 1984.
2. K. S. Narendra, Y. H. Lin, and L. S. Valavani, "Stable Adaptive controller design, Part II: Proof of stability," IEEE Trans. Automat. Contr., vol. AC-25, no. 3, pp. 440-448, June 1980.
3. P. A. Ioannou and P. V. Kokotović, "Robust redesign of adaptive control," IEEE Trans. Automat. Contr., vol. AC-29, no. 3, pp. 202-211, Mar. 1984.
4. R. L. Kosut, M. K. Lau, and S. P. Boyd, "Set-membership identification of systems with parametric and nonparametric uncertainty," IEEE Trans. Automat. Contr., vol. 37, no. 7, pp. 929-941, July 1992.
5. R. C. Younce and C. E. Rohrs, "Identification with nonparametric uncertainty," in Proc. IEEE Conf. Decision Contr., Honolulu, HI, Dec. 1990, pp. 3154-3161.

6. M. G. Safonov and R. Y. Chiang, "CACSD using the state-space L^∞ theory–a design example," IEEE Trans. Automat. Contr., vol. 33, no. 5, pp. 477-479, May. 1988.

7. R. O. LaMaire, L. Valavani, M. Athans, and G. Stein, "A frequency-domain estimator for use in adaptive control systems," in Proc. Amer. Contr. Conf., Minneapolis, MN, June 1987, pp. 238-244.

8. A. J. Helmicki, C. A. Jacobson, and C. N. Nett, " Control oriented system identification: A worst-case/deterministic approach in H_∞," IEEE Trans. Automat. Contr., Oct. 1991.

9. W. Wang and M. G. Safonov, "Relative–error H_∞ identification from auto-correlation data — a stochastic realization method," IEEE Trans. Automat. Contr., vol. 37, no. 7, pp. 1000-1004, July 1992.

10. K. Poolla, P. Khargonekar, A. Tikku, J. Krause, and K. Nagpal, "A time-domain approach to model validation," in Proc. Amer. Contr. Conf., Chicago, IL, June 1992, pp. 313-317.

11. F. C. Schweppe, *Uncertain dynamic systems.* Englewood Cliffs, NJ: Prentice-Hall, 1973.

12. F. C. Schweppe, "Recursive state estimation: unknown but bounded errors and system inputs," IEEE Trans. Automat. Contr., vol. 13, no. 1, pp. 22-28, Feb. 1968.

13. D. P. Bertsekas and I. B. Rhodes, "Recursive state estimation for a set-membership description of uncertainty," IEEE Trans. Automat. Contr., vol. 16, no. 2, pp. 117-128, Apr. 1971.

14. E. Fogel and Y. F. Huang, "On the value of information in system identification–bounded noise case," Automatica, vol. 18, no. 2, pp. 229-238, 1982.

15. M. Milanese and R. Tempo, "Optimal algorithms theory for robust estimation and prediction, " IEEE Trans. Automat. Contr., vol. 30, no. 8, pp. 730-738, Aug. 1985.

16. B. Wahlberg and L. Ljung, "Hard frequency-domain model error bounds from least-squares like identification techniques," IEEE Trans. Automat. Contr., vol. 37, no. 7, pp. 730-738, July 1992.

17. J. J. Krause, G. Stein, and P. P. Khargonekar, "Robust Performance of adaptive controllers with general uncertainty structure," in Proc. IEEE Conf. Decision Contr., Honolulu, HI, Dec. 1990, pp. 3168-3175.

18. J. J. Krause and P. P. Khargonekar, "Parameter identification in the presence of nonparametric dynamic uncertainty," Automatica, vol. 26, no. 1, pp. 113-123, 1990.

19. M. G. Safonov, A. J. Laub, and G. L. Hartmann, "Feedback properties of multivariable system; The role and use of the return difference matrix," IEEE Trans. Automat. Contr., vol. 26, no. 1, pp. 47-62, Feb. 1981.

20. M. K. Lau, S. Boyd, R. L. Kosut, and G. F. Franklin, "Robust control design for ellipsoidal plant set," in Proc. IEEE Conf. Decision Contr., Brighton, UK, Dec. 1991, pp. 291-296.

On Line Model Uncertainty Quantification: Hard Upper Bounds and Convergence*

Er-Wei Bai and Sundar Raman

Dept. of Electrical and Computer Engineering, University of Iowa,
Iowa City, IA 52242

Abstract:
   This paper considers the problem of on line uncertainty bound quantification in identification of restricted complexity models. Algorithms are presented, which provide hard and tight upper bound on the unknown model uncertainty in H_2, H_∞ and pointwise sense respectively. The algorithms proposed are very simple, on line and recursive. This allows robust control and adaptive identification to be combined.

1 Introduction and Problem Statement

The work of this paper is motivated by the recent research on robust control and adaptive control. Robust control and adaptive control are two of the most promising and most active research areas in systems and control theory. Robust control methods allow for the inclusion of model uncertainty into design. The key here is the availability of a hard and tight upper bound on the unknown uncertainty. While a loose bound would result in an overly conservative design, a soft bound would make the design dangerously optimistic. However, the problem of obtaining this bound is not solved by robust control designs. Adaptive identification, on the other hand, emphasizes the use of on line knowledge which would be useful to extract information on the bound of uncertainty.

   Use of identification techniques to determine a nominal model (or estimate) as well as a bound on the model uncertainty for robust control design is an emerging research area and is referred to as system identification for control design. Since its appearance in the late 1980's and the early 1990's, research in this area has led to a burst of activity and frequent breakthroughs due to many contributors. The role of system identification for control design is to reduce the plant uncertainty in such a way so as to enable the design of robust control schemes which can achieve the desirable performance specifications for the control system in the presence of model uncertainty. This requires that the system identification methods yield an explicit upper bound

---

* This work was supported in part by NSF ECS-9011359

on the uncertainty, corresponding to the identified model (estimate) or the nominal model. Moreover, this bound must be in a form compatible with the robust control methods. Only when all of these requirements are satisfied, can the desired controller be systematically designed through the combined use of system identification and robust control design. Note that most robust control designs require an estimate of the true, but unknown plant plus an explicit worst case deterministic upper bound on the uncertainty. This is an apparent shortcoming of the traditional identifiers in that the traditional identifiers only produce an estimate without providing enough information on the bound of the uncertainty so that the robust control design can be carried out.

The issue of system identification for control design addressed today is more realistic than it was a couple of years ago. The progress, largely due to the need from robust control design, has resulted in a deeper understanding of adaptive systems and a reexamination of existing identification methods. Further evidence can be found in numerous nicely written papers, e.g. by Glover, Lam and Partington [3], Goodwin et al [4], Gu and Khargonekar [6], Helmicki, Jacobson and Nett [7], Kosut et al [9], Makila [11,12], Parker and Bitmead [13], Partington [14], Smith and Doyle [15], and Wahlberg [16].

In this paper, we follow along the line of Goodwin et al [4] and our earlier works [1,2], focusing on on-line identification with restricted complexity models. On line identification has certainly several advantages over off line identification. Perhaps, the primary reason is that on line information can be used to successively generate more appropriate models and uncertainty bounds, thus refining the controller. In practice, for various reasons, it is often preferable to restrict the complexity of the estimate (or nominal model) i.e. to assume the estimate to be a low order transfer function. It is then inevitable that this simplified mathematical representation be characterized by modeling errors or model uncertainty and it is well known that the effect of the model uncertainty has a major impact on the design of robust controllers. In order to apply robust control design for the inclusion of model uncertainty, a necessary ingredient is the availability of a hard and tight upper bound on the uncertainty. Apparently, the traditional identifiers can not fulfill this task. We propose in this paper a new identifier which consists of two parts: plant identifier and uncertainty bound identifier. While the plant identifier provides an optimal estimate in some sense, the uncertainty bound which can be in H_2, H_∞ or pointwise sense, is characterized by the uncertainty identifier. The purpose of the uncertainty identifier is to extract uncertainty bound information and not to model the undermodeling. We shall use adaptive rational approximation method as in our earlier work, but with a completely different philosophy. Each member of the sequence is a hard upper bound of the uncertainty and moreover this sequence converges to the actual bound of the uncertainty. In other words, the uncertainty bound is approximated on line by a sequence of approximants which converge to actual uncertainty bound monotonically from above. The proposed on line algorithms are very simple and recursive, and basically of least squares type, even though the quantification of uncertainty is done in H_2, H_∞ or pointwise sense.

In theory, the topology that is used to quantify the uncertainty bound could be any one depending on the designs. In reality, the choice of topology should be in such a way that it leads to a practically solvable algorithm and is compatible with the robust control designs. In this paper, we consider uncertainty bound quantification in H_2, H_∞ and pointwise sense. If the intended use is LQR, quantification in H_2 sense may be useful. If the control design is in the framework of H_∞, quantification of uncertainty bound in H_∞ or pointwise sense may be preferable.

A closely related idea that can possibly be used for quantification of uncertainty is model reduction. First apply identification techniques in traditional sense and the order of the estimate is set to be high enough so we can assume that the estimate is close to the true system. Then by model reduction, a low order approximation of the estimate is obtained. The difference between the high order model and its low order approximation can be considered as an estimate of the uncertainty. Compared to this approach, we believe that the methods proposed in this paper are superior in several aspects. The first reason is that the methods presented in this paper quantify the actual difference between the estimate and the true system, while the model reduction approach gives a difference between the high order estimate and its low order approximation. How close between the high order estimate and the true system can not be answered by the traditional identification techniques. Secondly, high-order system identification is usually not a easy task. It has a slow convergence rate. In contrast, the methods proposed in this paper separate plant identification and uncertainty bound quantification and therefore each of them is of relatively low order.

To facilitate the description of the class of systems, we introduce the following notation.

$g(s)$ the unknown true system
$\hat{g}(s)$ the estimate
$\Delta(s) = g(s) - \hat{g}(s)$ the model uncertainty

It is assumed throughout the paper that

(A1) The unknown plant is a SISO LTI system with an unknown stable and strictly proper transfer function of possibly infinite dimension.

(A2) The estimate is a finite order, stable and strictly proper transfer function. It can be known a priori or obtained by an on line identification method. The order of the estimate is fixed.

By using this class of description, the unknown plant can be represented by

$$g(s) = \hat{g}(s) + \Delta(s)$$

which is consistent with the results of robust control designs. Our goal in this paper is to quantify the upper bound on $|\Delta(j\omega)|$ in H_2, H_∞ and pointwise sense, although $\Delta(j\omega)$ is unknown. Only single input single output systems are discussed in the paper, however extensions of the results to multivariable systems are straightforward but tedious.

2 Quantification of Uncertainty Bounds in H_2 Sense

2.1 Theory for H_2 Quantification

In this section, we consider the quantification of uncertainty bound in 2 norm in Hardy space H_2. The Hardy space H_2 consists of all analytic functions in the open right half plane which are square integrable on the $j\omega$ axis. The H_2 norm is defined as

$$\|f(j\omega)\|_2 = \left\{ \frac{1}{2\pi} \int_{-\infty}^{\infty} |f(j\omega)|^2 d\omega \right\}^{1/2}$$

for every $f \in H_2$.

Let $\Delta(s) = g(s) - \hat{g}(s)$ be the uncertainty. It is assumed that $\Delta(s)$ is a rational function which is stable, strictly proper and possibly infinite dimensional, and thus $\in H_2$. Let $\phi_i(s) = \frac{\sqrt{2\lambda}}{s+\lambda}(\frac{s-\lambda}{s+\lambda})^i$, $\lambda > 0$ denote the Laguerre functions [5,12,16]. It is known that $\phi_i(s)$ is a complete orthonormal basis of H_2. Let $\sum_{i=0}^{\infty} \beta_i \phi_i(s)$ be the Fourier-Laguerre expansion of $\Delta(s)$ [5,12,16]. It is well known that

$$(\beta_0, \beta_1, \ldots, \beta_n) = \arg\min_{\alpha_i} \int_{-\infty}^{\infty} |\Delta(j\omega) - \sum_{i=0}^{n} \alpha_i \phi_i(j\omega)|^2 d\omega \qquad (1)$$

for any n and moreover $\Delta(s) = \sum_{i=0}^{\infty} \beta_i \phi_i(s)$ in H_2 sense. Notice that the β_i's have a nice nesting property, i.e. each β_i is the projection of $\Delta(s)$ on $\phi_i(s)$ and is independent of all other β_i's due to the orthogonality of $\phi_i(s)$. For the quantification of the uncertainty bound, it is natural to consider $|\sum_{i=0}^{\infty} \beta_i \phi_i(j\omega)|$ as the nth order approximation of $|\Delta(j\omega)|$ in H_2 sense. However, it can be shown that $|\sum_{i=0}^{n} \beta_i \phi_i(j\omega)|$ is not an upper bound of $|\Delta(j\omega)|$. In fact, from the well known Bessel inequality, we have

$$\left\| \sum_{i=0}^{n} \beta_i \phi_i(j\omega) \right\|_2 \leq \|\Delta(j\omega)\|_2 \qquad \forall n$$

despite the fact that $\sum_{i=0}^{n} \beta_i \phi_i(j\omega) \rightarrow \Delta(j\omega)$, as $n \rightarrow \infty$ in H_2 sense. If the intended use of the uncertainty bound is for the worst case design, the above quantification method is certainly not very useful. We shall now propose a simple method based on the Fourier-Laguerre series which provides a truly upper bound on the uncertainty while also maintaining the convergence results and the nesting property.

Assume that $\Delta(s) \neq 0$, then the projection β_i of $\Delta(s)$ on some $\phi_i(s)$ is not zero. Without loss of generality, we assume

(A3) $\beta_0 \neq 0$.

This assumption is not a restriction. If $\Delta(s) \neq 0$, let $m \geq 0$ be the least integer such that $\beta_m \neq 0$. Then by redefining $\phi'_i(s) = \phi_{m+i}(s)$ and $\beta'_i = \beta_{m+i}$, $i = 0, 1, 2, \ldots$, it follows that $\beta'_0 \neq 0$.

Now consider

$$J_n = \int_{-\infty}^{\infty} |k_n \Delta(j\omega) - \sum_{i=0}^{n} \alpha_i \phi_i(j\omega)|^2 d\omega \tag{2}$$

with $\alpha_0 = 1$. Let

$$(k_n^*, \alpha_1^*, \ldots, \alpha_n^*) = \arg\min_{k_n, \alpha_1, \ldots, \alpha_n} J_n \tag{3}$$

and define

$$\gamma_n(jw) = \frac{1}{k_n^*} \sum_{i=0}^{n} \alpha_i^* \phi_i(jw) \quad \text{with} \quad \alpha_0^* = 1 \tag{4}$$

Then it can be shown that

Theorem 1. *Consider (2) with $k_n^*, \alpha_i^*, \beta_i$ and $\gamma_n(jw)$ defined as in (1),(3) and (4). Suppose $\beta_0 \neq 0$, we then have*

(1) *There exist $M, \rho > 0$ such that $M \geq |k_n^*| \geq |k_{n-1}^*| \geq \rho > 0$ for all n and*

$$\alpha_i^* = k_n^* \beta_i , \ i = 1, 2, \ldots, n \ \text{and} \ \gamma_n(jw) = \frac{\phi_0(jw)}{k_n^*} + \sum_{i=1}^{n} \beta_i \phi_i(jw) \tag{5}$$

where k_n^* satisfies the recursive formula

$$\frac{1}{k_n^*} = \frac{1}{k_{n-1}^*} - \frac{\beta_n^2}{\beta_0} \tag{6}$$

$$= \left(1 + \frac{\beta_n^2}{\beta_{n-1}^2}\right) \frac{1}{k_{n-1}^*} - \frac{\beta_n^2}{\beta_{n-1}^2} \frac{1}{k_{n-2}^*}, \ \ \text{if } \beta_{n-1} \neq 0$$

(2) $\|\gamma_n(jw)\|_2 \geq \|\Delta(jw)\|_2 \quad \forall n$
(3) *There exists $M_1 > 0$ such that*

$$\min J_{n+1} \leq \min J_n \leq \frac{M_1}{n} \qquad \forall n$$

(4) *There exists $M_2 > 0$ such that*

$$\|\gamma_{n+1}(jw) - \Delta(jw)\|_2^2 \leq \|\gamma_n(jw) - \Delta(jw)\|_2^2 \leq \frac{M_2}{n} \qquad \forall n$$

Proof. Part(1). Note that $\Delta(s) = \sum_{i=0}^{\infty} \beta_i \phi_i(s)$ in H_2 sense and $\phi_i(s)$'s are orthonormal, therefore

$$J_n = \left\| k_n \Delta - \sum_{i=0}^{n} \alpha_i \phi_i \right\|_2^2 = \left\| \sum_{i=0}^{n} (k_n \beta_i - \alpha_i) \phi_i + \sum_{i=n+1}^{\infty} k_n \beta_i \phi_i \right\|_2^2$$

$$= \sum_{i=0}^{n} (k_n \beta_i - \alpha_i)^2 + \sum_{i=n+1}^{\infty} (k_n \beta_i)^2$$

Now $\alpha_0 = 1$ implies

$$J_n = (k_n \beta_0 - 1)^2 + \sum_{i=1}^{n} (k_n \beta_i - \alpha_i)^2 + k_n^2 \sum_{i=n+1}^{\infty} \beta_i^2$$

Let

$$f(k) = (k \beta_0 - 1)^2 + k^2 \sum_{i=n+1}^{\infty} \beta_i^2$$

The optimal $k_n^*, \alpha_1^*, \ldots, \alpha_n^*$ must then satisfy

$$\alpha_i^* = k_n^* \beta_i \quad i = 1, 2, \ldots, n \quad \text{and} \quad k_n^* = \arg\min_k f(k)$$

Taking derivative of $f(k)$ with respect to k and setting it to zero yields

$$0 = \frac{df}{dk} |_{k=k_n^*} = (k_n^* \beta_0 - 1)\beta_0 + k_n^* \sum_{i=n+1}^{\infty} \beta_i^2$$

Thus

$$k_n^* = \frac{\beta_0}{\beta_0^2 + \sum_{i=n+1}^{\infty} \beta_i^2}.$$

Since by hypothesis, $\beta_0 \neq 0$, it follows that

$$\left| \frac{1}{\beta_0} \right| \geq |k_n^*| \geq |k_{n-1}^*| \geq \frac{|\beta_0|}{\|\Delta\|_2^2} > 0$$

$$\frac{1}{k_n^*} = \frac{1}{k_{n-1}^*} - \frac{\beta_n^2}{\beta_0} = \left(1 + \frac{\beta_n^2}{\beta_{n-1}^2}\right) \frac{1}{k_{n-1}^*} - \frac{\beta_n^2}{\beta_{n-1}^2} \frac{1}{k_{n-2}^*}$$

and therefore

$$\gamma_n(j\omega) = \frac{\phi_0(j\omega)}{k_n^*} + \sum_{i=1}^{n} \beta_i \phi_i(j\omega)$$

Part (2).

$$0 = k_n \frac{\partial J_n}{\partial k_n} = k_n \int \left\{ \Delta \left(k_n \Delta^* - \sum \alpha_i \phi_i^* \right) + \left(k_n \Delta - \sum \alpha_i \phi_i \right) \Delta^* \right\} d\omega$$

where the superscript * on Δ and ϕ_i indicates complex conjugate, and this implies

$$\int 2k_n^2|\Delta|^2 d\omega = \int \left(k_n^2 \left(\frac{1}{k_n} \sum \alpha_i \phi_i^* \right) \Delta + k_n^2 \left(\frac{1}{k_n} \sum \alpha_i \phi_i \right) \Delta^* \right) d\omega$$

when $(k_n, \alpha_1, \ldots, \alpha_n) = (k_n^*, \alpha_1^*, \ldots, \alpha_n^*)$. Now

$$0 \leq J_n = \int \left\{ k_n^2 |\Delta|^2 - k_n^2 \left(\frac{1}{k_n} \Delta \sum \alpha_i \phi_i^* \right) - k_n^2 \left(\frac{1}{k_n} \sum \alpha_i \phi_i \Delta^* \right) \right.$$
$$\left. + k_n^2 \left(\frac{1}{k_n^2} | \sum \alpha_i \phi_i|^2 \right) \right\} d\omega |_{\alpha_i = \alpha_i^*, k_n = k_n^*}$$

It follows that

$$\int_{-\infty}^{\infty} |\Delta(jw)|^2 dw \leq \int_{-\infty}^{\infty} |\gamma_n(jw)|^2 dw$$

Part (3). $\Delta \in H_2$, so $\|\Delta(jw) - \sum_{i=0}^{n} \beta_i \phi_i(jw)\|_2^2 \leq \frac{M}{n}$ for some $M > 0$ [12]. Since $\beta_0 \neq 0$, we have

$$\left\| \frac{\Delta}{\beta_0} - \phi_0 - \sum_{i=1}^{n} \frac{\beta_n}{\beta_0} \phi_i \right\|_2^2 \leq \frac{1}{n} \frac{M}{|\beta_0|^2}$$

Now $\quad \min J_{n+1} = \min \int \left| k_{n+1}\Delta - \phi_0 - \sum_{i=1}^{n+1} \alpha_i \phi_i \right|^2 d\omega$

$$\leq \min \int \left| k_n \Delta - \phi_0 - \sum_{i=1}^{n} \alpha_i \phi_i \right|^2 d\omega = \min J_n$$

$$\leq \min \int \left| \frac{\Delta}{\beta_0} - \phi_0 - \sum_{i=1}^{n} \alpha_i \phi_i \right|^2 d\omega \leq \frac{M_1}{n}$$

Part (4). From part (3) we have

$$(k_{n+1}^*)^2 \|\Delta - \gamma_{n+1}\|_2^2 \leq (k_n^*)^2 \|\Delta - \gamma_n\|_2^2 \leq \frac{M_1}{n}$$

and from (1) $\quad |k_{n+1}^*| \geq |k_n^*| \geq |k_0^*|$. Hence

$$\|\gamma_{n+1}(jw) - \Delta(jw)\|_2^2 \leq \|\gamma_n(jw) - \Delta(jw)\|_2^2 \leq \frac{1}{n} \frac{M_1}{(k_0^*)^2} \leq \frac{M_2}{n}$$

This completes the proof. $\qquad \qquad \square$

Remarks.

1. An upper bound $\gamma_n(jw)$ of $\Delta(jw)$ is obtained in H_2 sense. The emphasis is given to the extraction of the uncertainty bound, not to the modeling of uncertainty. Note that even for the first order approximation of uncertainty, n=1, we have $\|\gamma_1(jw)\|_2 \geq \|\Delta(jw)\|_2$, although $\gamma_1(jw)$ is unlikely to be a good estimate of $\Delta(jw)$.

2. The convergence result $\|\gamma_{n+1}(jw) - \Delta(jw)\|_2^2 \leq \|\gamma_n(jw) - \Delta(jw)\|_2^2 \leq \frac{M_2}{n}$ implies that as the order of the approximant increases, the bounds gets better and better, converging asymptotically to the actual bound.

3. $\gamma_n(jw)$ has a nice nesting property. Once $\gamma_{n-1}(jw)$ is known, only β_n is required to calculate $\gamma_n(jw)$ and note that β_n is independent of all the previous β_i's due to the orthogonality of $\phi_i(jw)$. We will discuss the issue of on line identification to obtain β_i in the next section.

2.2 On Line Identification of the Uncertainty Bound

Here, we shall present some on line algorithms to obtain the upper bound, $\gamma_n(jw)$, in H_2 sense. There are basically two approaches to achieve this. The first one is the one shot algorithm which solves $(k_n^*, \alpha_1^*, \ldots, \alpha_n^*)$ for every given order n of approximation. The other one exploits the orthogonality of ϕ_i, making the calculations recursive.

One-shot Technique. In the one shot technique, we will compute the entire vector $\theta^* = (k_n^*, \alpha_1^*, \ldots, \alpha_n^*)$, which is the solution of (2), for every n. More precisely, for each n, we compute all the $n+1$ components of the vector θ^*.

Let $\theta = (k_n, \alpha_1, \ldots, \alpha_n)^T$ and $Q(s) = (\Delta(s), -\phi_1(s), \ldots, -\phi_n(s))^T$. Then, since $\alpha_0 = 1$, we have

$$k_n \Delta(jw) - \sum_{i=0}^{n} \alpha_i \phi_i(jw) = k_n \Delta(jw) - \phi_0(jw) - \sum_{i=1}^{n} \alpha_i \phi_i(jw)$$

$$= Q^T(jw)\theta - \phi_0(jw)$$

Thus $\theta^* = (k_n^*, \alpha_1^*, \ldots, \alpha_n^*)$ is the solution of (2.2)

$$\arg\min_{k_n, \alpha_1, \ldots, \alpha_n} \int_{-\infty}^{\infty} |k_n \Delta(jw) - \sum_{i=0}^{n} \alpha_i \phi_i(jw)|^2 dw$$

if and only if

$$\theta^* = \arg\min_{\theta} J_n = \arg\min_{\theta} \int_{-\infty}^{\infty} |Q^T(jw)\theta - \phi_0(jw)|^2 dw$$

Differentiating $J_n(\theta)$ and setting it to zero i.e. $\frac{\partial J}{\partial \theta}|_{\theta=\theta^*} = 0$ yields

$$\int_{-\infty}^{\infty} Q(jw)Q^T(-jw)dw\,\theta^* = \int_{-\infty}^{\infty} \phi_0(jw)Q(-jw)dw$$

If $\int_{-\infty}^{\infty} Q(j\omega)Q^T(-j\omega)d\omega$ is nonsingular, which is related to persistent excitation condition, then the solution θ^* is unique and given by

$$\theta^* = \left\{ \int_{-\infty}^{\infty} Q(jw)Q^T(-jw)dw \right\}^{-1} \left\{ \int_{-\infty}^{\infty} \phi_0(jw)Q(-jw)dw \right\} \quad (7)$$

Although the above equation provides an exact expression for θ^* and consequently the bound $\gamma_n(jw)$, it involves the unknown uncertainty $\Delta(jw)$. To get around this difficulty, we shall use the adaptive identifier shown in Fig. 1. The identifier has two parts: the plant identifier and the uncertainty identifier. Note that the complexity of the estimate is restricted i.e. the order of the estimate is fixed at some value. At time t, the estimated model $\hat{g}_t$ is given by the plant identifier. The uncertainty identifier compares the outputs of the true system and the estimate and then produces an estimate of the uncertainty bound. The primary purpose of this paper is the quantification of uncertainty bound. The plant identifier here can be any standard one, as long as it converges. In Fig. 1, $y_t(t)$ and $y(t)$ are the outputs of $\hat{g}_t$ and the plant $g(s)$ respectively and $y_{\phi_i}(s) = \phi_i(s)u(s)$, $i = 0, 1, 2, \ldots, n$.

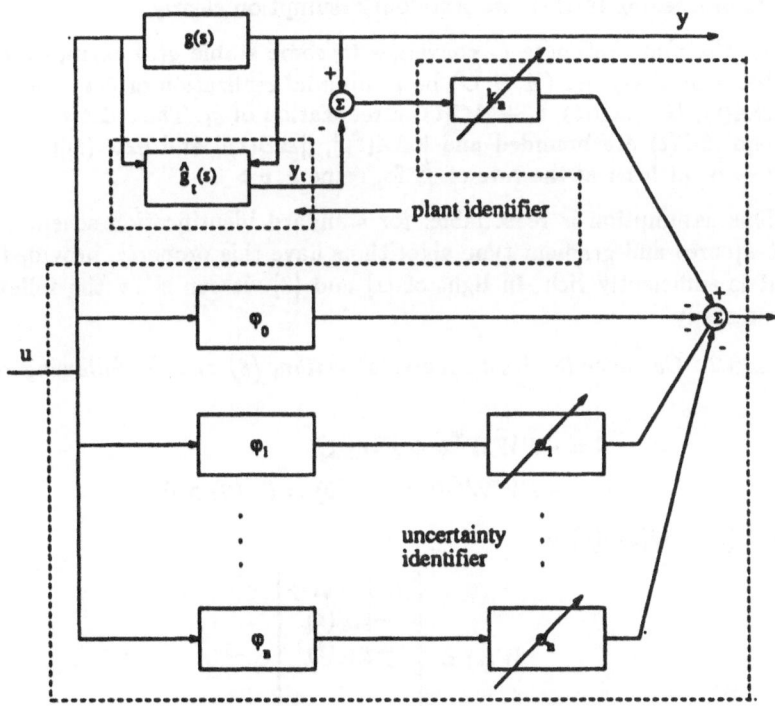

Fig. 1   Identifier

Let

$$\hat{W} = \begin{bmatrix} y(t) - y_t(t) \\ -y_{\phi_1}(t) \\ -y_{\phi_2}(t) \\ \vdots \\ -y_{\phi_n}(t) \end{bmatrix}.$$

Define the least squares update law as

$$\dot{\theta} = -P\hat{W}\hat{W}^T\theta + P\hat{W}y_{\phi_0}$$
$$\dot{P} = -P\hat{W}\hat{W}^T P \; ; \quad P(0) = P^T(0) > 0 \tag{8}$$

Notice that every signal in the update law is measurable. We will now show that $\theta(t)$ obtained from (8) will converge asymptotically to θ^* of equation (2) or equation (7) and consequently the uncertainty bound $\gamma_n(jw)$ of equation (4). We prove our claim in two steps. First, we show the fact that although the plant estimate $\hat{g}_t$ changes with time, it can be replaced by a fixed $\hat{g}(s)$ as long as $\hat{g}_t$ converges to some $\hat{g}(s)$ asymptotically.

In other words, all signals involving $\hat{g}_t$ can be replaced by those involving $\hat{g}(s)$ without affecting any of the convergence results of the uncertainty iden-tifier e.g. we can replace $y_t(t)$ by $\hat{y}(t)$ where $\hat{y}(t)$ is the output of $\hat{g}(s)$ with input $u(t)$ and $\tilde{y}(t) = y(t) - y_t(t)$ can be replaced by $y_\Delta(t) = y(t) - \hat{y}(t)$. Before proceeding further, we state our assumption clearly.

(A4) The plant estimate $\hat{g}_t$ converges to some stable $\hat{g}(s)$ asymptotically. More precisely, let (A, B, C) be a minimal realization of $\hat{g}(s)$ and $(A + \Delta A(t), B + \Delta B(t), C + \Delta C(t))$ a realization of $\hat{g}_t$. Then $\Delta A(t)$, $\Delta B(t)$, and $\Delta C(t)$ are bounded and $\|\Delta A(t)\|$, $\|\Delta B(t)\|$ and $\|\Delta C(t)\| \to 0$ as $t \to \infty$ at least at the rate of $\frac{k}{t^v}$ for some $k, v > 0$

This assumption is reasonable, for standard identification schemes e.g, least-squares and gradient type algorithms have this property, provided the input is sufficiently rich. In light of [1] and [2] we can show the following lemma easily

Lemma 2. *Consider the least squares algorithm (8) and the following equa-tion*

$$\dot{\hat{\theta}} = -PWW^T\hat{\theta} + PWy_{\phi_0}$$
$$\dot{P} = -PWW^T P \; ; \quad P(0) = P^T(0) > 0$$

with $W(s) = Q(s)u(s)$ *or*

$$W(t) = \begin{bmatrix} y(t) - \hat{y}(t) \\ -y_{\phi_1}(t) \\ -y_{\phi_2}(t) \\ \vdots \\ -y_{\phi_n}(t) \end{bmatrix}.$$

(Notice that except for the first entry, $W(t)$ and $\hat{W}(t)$ are identical. The difference in the first entry indicates the substitution of $\hat{g}_t$ by $\hat{g}$, i.e. the time varying estimate $\hat{g}_t$ is replaced by $\hat{g}(s)$ which is time invariant). Then provided W is persistently exciting, we have

$$\lim_{t \to \infty} (\theta(t) - \hat{\theta}(t)) = 0$$

The importance of the above fact is that in as far as the convergence results are concerned, we can either identify the plant estimate and the uncertainty bound simultaneously, or we may identify the plant first and once its estimate $\hat{g}(s)$ is obtained, we fix $\hat{g}_t$ in Fig. 1 by $\hat{g}(s)$ and then obtain the uncertainty bound. The two approaches provide the same θ and consequently the same bound $\gamma_n(jw)$. From Theorem 3.3 of [1], we now have

Theorem 3. *Consider the least squares algorithm (8). Suppose $W(t)$ is persistently exciting. Then*

$$\lim_{t \to \infty} \theta(t) = \left\{ \lim_{t \to \infty} \frac{1}{t} \int_0^t W W^T dt \right\}^{-1} \left\{ \lim_{t \to \infty} \frac{1}{t} \int_0^t W y_{\phi_0} dt \right\} = R_w^{-1} R_{w y_{\phi_0}}$$

If the spectral measure of the input $S_u(\omega)$ is a constant, then

$$\lim_{t \to \infty} \theta(t) = R_w^{-1} R_{w y_{\phi_0}} = \theta^*$$

where θ^* is given by (7).

Remarks.

1. An upper bound $\gamma_n(jw)$ is obtained by least squares algorithm such that $\|\gamma_n(jw)\|_2 \geq \|\Delta(jw)\|_2$ for each n.
2. The persistent excitation condition as required in Theorem 2 can be satisfied provided the input has at least $n+1$ spectral lines.
3. The least squares algorithm of (2.8) can be replaced by the gradient type algorithms. In that case, $\theta(t)$ does not converge to the actual value but converges to a neighborhood of $R_w^{-1} R_{w y_{\phi_0}}$ with radius $\eta(\epsilon)$ where ϵ is the gain of the algorithm and η is a class k function.
4. The convergence rate of θ depends on that of the plant estimate $\hat{g}_t$. Suppose $\hat{g}_t$ converges to $\hat{g}(s)$ exponentially and the gradient type algorithm is used to quantify the uncertainty, then $\theta(t)$ converges to a neighborhood of $R_w^{-1} R_{w y_{\phi_0}}$ exponentially.

Recursive Method. Another technique for on line uncertainty quantification which requires less computation is the recursive method. This method exploits the orthonormal properties of the Laguerre functions $\phi_i(s)$. In contrast to the one shot technique, here, for each n, we identify only the nth component of an n dimensional vector. This scalar value along with $\gamma_{n-1}(jw)$ can be used to obtain $\gamma_n(jw)$, for any n.

From Theorem 1, we have

$$\gamma_n(jw) = \frac{\phi_0(jw)}{k_n^*} + \sum_{i=1}^{n} \beta_i \phi_i(jw)$$

where k_n^* satisfies the recursive formula

$$\frac{1}{k_n^*} = \frac{1}{k_{n-1}^*} - \frac{\beta_n^2}{\beta_0} = \left(1 + \frac{\beta_n^2}{\beta_{n-1}^2}\right) \frac{1}{k_{n-1}^*} - \frac{\beta_n^2}{\beta_{n-1}^2} \frac{1}{k_{n-2}^*} \tag{9}$$

The orthonormality of the ϕ_i implies that

$$(\beta_0, \beta_1, \ldots, \beta_n) = \arg\min_{\alpha_i} \int_{-\infty}^{\infty} |\Delta(jw) - \sum_{i=0}^{n} \alpha_i \phi_i(jw)|^2 dw$$

if and only if for each i

$$\beta_i = \arg\min_{\alpha_i} \int_{-\infty}^{\infty} |\Delta(jw) - \alpha_i \phi_i(jw)|^2 dw \tag{10}$$

Thus from (10), it is clear that the β_i's can be obtained independent of each other. The recursive method may now be stated as follows:

Initial step: Apply the one shot algorithm for any fixed n. Determine the corresponding $k_n, \beta_1, \ldots, \beta_n$ and hence $\gamma_n(jw)$.

Step 1: Determine recursively the uncertainty bound $\gamma_{n+1}(jw)$ from $\gamma_n(jw)$, $n = 0, 1, 2 \ldots$ i.e. suppose $\gamma_n(jw)$ (or equivalently $k_n, \beta_1, \ldots, \beta_n$) is available. To obtain $\gamma_{n+1}(jw)$, it follows from (9) that one needs to obtain only the scalar β_{n+1} using (10).

We use the adaptive identifier shown in Fig. 2 to obtain β_i for any i. Let $W(s) = \phi_i(s)u(s)$ and $\tilde{y}(t) = y(t) - y_t(t)$. Define the least squares update law

$$\dot{\alpha} = -pW^2\alpha + pW\tilde{y}$$
$$\dot{p} = -(pW)^2 \quad ; \quad p(0) > 0 \tag{11}$$

where α is a scalar. Then

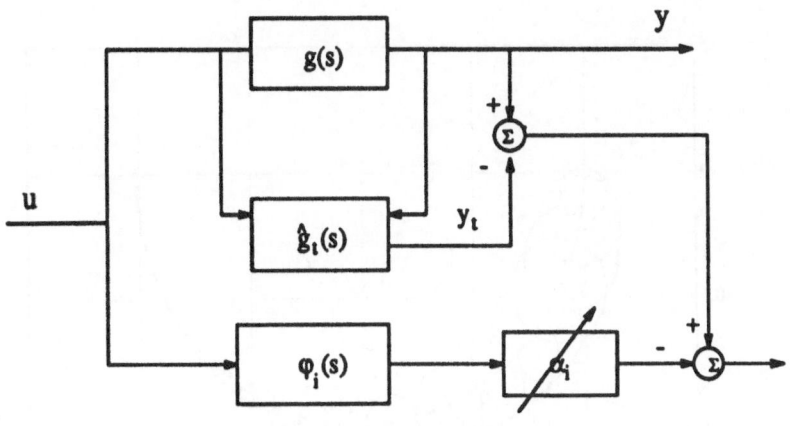

Fig. 2

Theorem 4. *Consider the least squares algorithm (11). Suppose $W(t)$ is persistently exciting. Then*

$$\lim_{t\to\infty} \alpha(t) = \left\{ \lim_{t\to\infty} \frac{1}{t} \int_0^t W^2 dt \right\}^{-1} \left\{ \lim_{t\to\infty} \frac{1}{t} \int_0^t W\tilde{y} dt \right\} = R_w^{-1} R_{w\tilde{y}}$$

If the spectral measure of the input $S_u(\omega)$ is a constant, then

$$\lim_{t\to\infty} \alpha(t) = R_w^{-1} R_{w\tilde{y}} = \beta_i$$

Proof. Similar to that of Theorem 2.   □

To illustrate the proposed methods we give two examples.

Example 1. Consider the true but unknown system

$$g(s) = \frac{2}{s+1} \frac{230}{s^2 + 30s + 229}$$

The estimate is chosen to be a first order transfer function. The identifier shown in Fig. 1 is used to obtain $\hat{g}(s)$ as well as the bound $\gamma_n(j\omega)$ in H_2 sense. For simulation, $\lambda = 3$ and $u(t) = 1 + \sum_{i=1}^{20} \sin(0.5*i*t)$. The estimate $\hat{g}_t$ converges to

$$\hat{g}(s) = \frac{1.756}{s + 0.8688}.$$

Fig.3 shows the polar plots of the true plant $g(j\omega)$ and estimate $\hat{g}(j\omega)$ plus the uncertainty bounds given by $|\gamma_6(j\omega)|$ respectively. The bounds shown in the plots are a set of circles centered on the estimate $\hat{g}(j\omega)$ with radius $|\gamma_6(j\omega)|$. For $n = 6$, $|\gamma_6(jw)|$ provides a very tight bound. The norm used for this example is H_2.

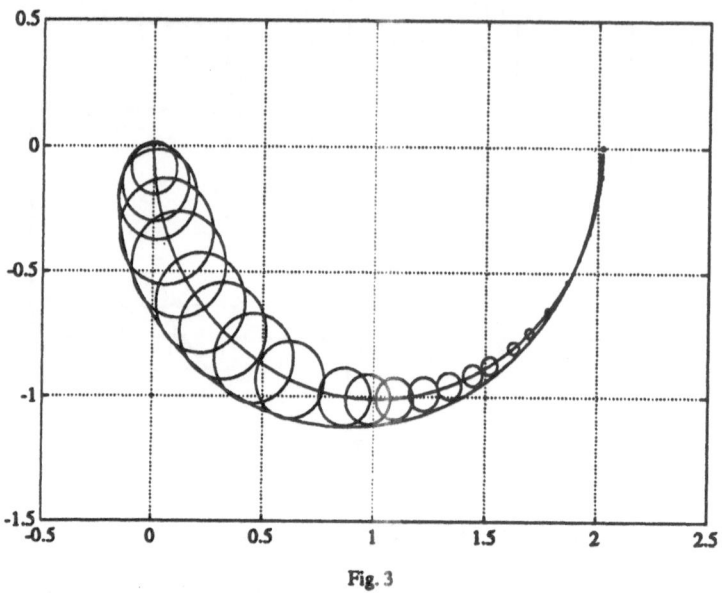

Fig. 3

2.3 Determination of the Order n of $r_n(j\omega)$

The results of previous sections provide a method for obtaining hard upper
bounds on the model uncertainty adaptively. But once the process has been
carried out for a fixed order n, the question remains as to how good the
quantification of the model uncertainty is. This is an universal issue for any
approximation problem and is referred to as the remainder theory. Note that
for a fixed order n, the identifier described in the previous section gives a nth
order approximation $\gamma_n(j\omega)$ on the unknown model uncertainty $\Delta(j\omega)$ such
that $\|\gamma_n(j\omega)\|_2 \geq \|\Delta(j\omega)\|_2$ and $\|\gamma_n(j\omega) - \Delta(j\omega)\|_2 \leq \frac{M_2}{n} \cdot \|\gamma_n(j\omega)\|_2$ is an
upper bound on $\|\Delta(j\omega)\|_2$ and converges to $\|\Delta(j\omega)\|_2$. But how close $\gamma_n(j\omega)$
is to $\Delta(j\omega)$ depends on the constant M_2 which in turn is a function of the
unknown uncertainty $\Delta(j\omega)$. Hence, without knowing anything about $\Delta(j\omega)$
it seems impossible to justify whether or not an nth order quantification
$\gamma_n(j\omega)$ is enough for the unknown $\Delta(j\omega)$.

   This problem is important in practice simply because in order to use
$\gamma_n(j\omega)$ as the upper bound of $\Delta(j\omega)$ for control design, it is necessary to
have some confidence on how good $\gamma_n(j\omega)$ is. Of course, $\gamma_n(j\omega)$ is an upper
bound on $\Delta(j\omega)$ for all n. If $\gamma_n(j\omega)$ is a loose bound, the control design could
be very conservative. A heuristic way to deal with this problem is to check
$\|\gamma_n(j\omega) - \gamma_{n-1}(j\omega)\|_2$. If it is less than some prescribed error, then $\gamma_n(j\omega)$
is considered to be close to $\Delta(j\omega)$. The difficulty with this approach is that
it may or may not work well. In this section, we propose an alternative. The

idea is to construct a hard lower bound $\gamma'_n(j\omega)$ on $\Delta(j\omega)$ such that

$$\|\gamma_n{}'(j\omega)\|_2 \leq \|\Delta(j\omega)\|_2 \quad \forall n$$

and

$$\|\gamma'_n(j\omega) - \Delta(j\omega)\|_2 \to 0 \quad \text{as} \quad n \to \infty \qquad (12)$$

Then since $\|\gamma'_n(j\omega)\|_2 \leq \|\Delta(j\omega)\|_2 \leq \|\gamma_n(j\omega)\|_2$ and $\|\gamma_n(j\omega)\| - \|\gamma'_n(j\omega)\| \to 0$, $\|\gamma_n(j\omega)\| - \|\gamma'_n(j\omega)\|$ can be used to justify how good $\gamma_n(j\omega)$ is. If $\|\gamma_n(j\omega)\| - \|\gamma'_n(j\omega)\|$ is small, then $\gamma_n(j\omega)$ is guaranteed to be a good one.

Generation of the lower bound $\gamma'_n(j\omega)$ is very easy. In fact, it is simply a by-product of $\gamma_n(j\omega)$. Suppose a recursive method as in Sect. 2.2.2 is used to generate $\gamma_n(j\omega)$ along with the coefficients $k_n^*, \beta_0, \ldots, \beta_n$. Define

$$\gamma'_n(j\omega) = \sum_{i=1}^{n} \beta_i \phi(j\omega)$$

Then by the Bessel inequality and the completeness of the Laguerre function $\phi_i(j\omega)$, we have (12).

3 Quantification of Uncertainty in H_∞ and Pointwise Sense

3.1 Theory for H_∞ and Pointwise Quantification

In this section, we present results on the quantification of uncertainty in H_∞ and pointwise sense using the concept of Fourier-Laguerre series. The Hardy space H_∞ consists of all bounded analytic functions on the open right half plane with the H_∞ norm being defined as

$$\|f\|_\infty = \sup_{\omega \in [0\infty]} |f(jw)|$$

for every $f \in H_\infty$.

It is assumed that the uncertainty $\Delta(s) = g(s) - \hat{g}(s)$ is a strictly proper, stable rational function of possibly infinite dimension and that the ϕ_i are the Laguerre functions as described earlier.

Consider the function $H(s)$ defined as

$$H(s) = \frac{\sqrt{2\lambda}}{s + \lambda} \Delta(s), \quad \lambda > 0 \qquad (13)$$

Let $\sum_{i=0}^{\infty} h_i \phi_i(s)$ denote the Fourier-Laguerre expansion of $H(s)$ i.e.

$$H(s) = \sum_{i=0}^{\infty} h_i \phi_i(s) \qquad (14)$$

where the equality is in H_2 sense. Define

$$r_n(s) = \sum_{i=0}^{n} h_i \left(\frac{s-\lambda}{s+\lambda}\right)^i = \sum_{i=0}^{n} h_i E_i(s) \tag{15}$$

where $E_i(s) = \left(\frac{s-\lambda}{s+\lambda}\right)^i$. We then have the following theorem from [5,12]

Theorem 5. *If $(s-\lambda)\frac{d\Delta}{ds} \in H_2$, then $r_n(s)$ converges in H_∞ sense to $\Delta(s)$ as $n \to \infty$, that is*

$$\|r_n(jw) - \Delta(jw)\|_\infty \le \xi(n) \to 0 \ \text{as} \ n \to \infty \tag{16}$$

where

$$0 \le \xi(n) \le \min\left\{\frac{K}{n}, \frac{1}{\sqrt{2\lambda n}}\left(\left\|(s-\lambda)\frac{d\Delta}{ds}\right\|_2^2 - \sum_{i=0}^{n}\left(\sqrt{2\lambda}\, ih_i\right)^2\right)^{1/2}\right\}$$

for some $K > 0$.

Notice that although the above theorem guarantees that $r_n(jw)$ approximates $\Delta(jw)$, it is not necessarily true that $r_n(jw)$ is an upper bound of $\Delta(jw)$ which is what we are looking for. To this end, define

$$\gamma_n(s) = \sum_{i=0}^{n} h_i E_i(s) \left(1 + \frac{\xi(n)}{|\sum_{i=0}^{n} h_i E_i(s)|}\right)$$

$$= \xi(n) \qquad \text{if} \ \sum_{i=0}^{n} h_i E_i(jw) = 0 \tag{17}$$

Then it follows that

Theorem 6. *Let $\gamma_n(s)$ be as defined in (17) where the coefficients h_i are the Laguerre coefficients of $H(s)$ as in (14). Then*

1. $\|\Delta(jw) - \gamma_n(jw)\|_\infty \le 2\xi(n) \to$ *as* $n \to \infty$
2. $|\gamma_n(jw)| \ge |\Delta(jw)| \ \forall n$ *and all* ω
3. $\|\gamma_n(jw)\|_\infty \ge \|\Delta(jw)\|_\infty \ \forall n$

Proof. Part(1).

$$\|\Delta(jw) - \gamma_n(jw)\|_\infty = \left\|\Delta(jw) - \sum_{i=0}^{n} h_i E_i(jw) - \frac{\sum_{i=0}^{n} h_i E_i(jw)}{|\sum_{i=0}^{n} h_i E_i(jw)|}\xi(n)\right\|_\infty$$

$$\le \left\|\Delta(jw) - \sum_{i=0}^{n} h_i E_i(jw)\right\|_\infty + \|\xi(n)\|_\infty$$

$$= \|\Delta(jw) - r_n(jw)\|_\infty + \xi(n) \le 2\xi(n)$$

Part (2). We have from (16) $|\Delta(jw) - r_n(jw)| \leq \xi(n)$, $\forall w, n$. This implies

$$|\Delta(jw)| \leq |\sum_{i=0}^{n} h_i E_i(jw)| + \xi(n)$$

$$= |\sum_{i=0}^{n} h_i E_i(jw)| \left(1 + \frac{\xi(n)}{|\sum_{i=0}^{n} h_i E_i(jw)|}\right) = |\gamma_n(jw)|$$

Part (3) Follows from part (2). This completes the proof. □

Remarks.

1. The above theorem not only guarantees that $\gamma_n(jw)$ approximates $\Delta(jw)$ as $n \to \infty$ but also that $|\gamma_n(jw)|$ is an upper bound of $|\Delta(jw)|$ in both H_∞ and pointwise sense. The emphasis is again given to the extraction of the upper bound of the uncertainty, not to the modeling of undermodeling. Note that even for $n = 1, |\gamma_n(jw)| \geq |\Delta(jw)|$.

2. The order selection of $\gamma_n(j\omega)$ can be done in a similar way as for H_2 uncertainty quantification described in Sect 2.3. Let

$$\gamma_n'(j\omega) = \sum_{i=0}^{n} h_i E_i(s) \left(1 + \xi(n)|\sum_{i=0}^{n} h_i E_i(s)|\right) \text{ if } 1 - \frac{\xi(n)}{|\sum_{i=0}^{n} h_i E_i(s)|} \leq 0$$

$$= 0 \qquad \text{if} \quad 1 - \frac{\xi(n)}{|\sum_{i=0}^{n} h_i E_i(s)|} < 0$$

It can be shown easily that $|\gamma_n'(j\omega)| \leq |\Delta(j\omega)|, \forall w$ and all n and $|\gamma_n'(j\omega) - \gamma_n(j\omega)| \to 0$ as $n \to \infty$. Therefore $|\gamma_n(j\omega)| - |\gamma_n'(j\omega)|$ can be used to justify how good the upper bound $|\gamma_n(j\omega)|$ is.

In quantifications of uncertainty bounds in H_∞ and pointwise sense, an estimate of the maximum value of $\|(s - \lambda)\frac{d\Delta}{ds}\|_2$ is required. While the issue of obtaining a bound on $\|(s - \lambda)\frac{d\Delta}{ds}\|_2$ will be discussed in Sect. 3.3, we first present methods to determine the Laguerre coefficients on line in the next section.

3.2 On Line Identification of Laguerre Coefficients h_i

We assume in this section that $\xi(n)$ is available. Then from (17), it is seen that the only unknowns are the constants h_i which are the Fourier-Laguerre coefficients of $H(s) = \frac{\sqrt{2\lambda}}{s+\lambda}\Delta(s)$. Let $\theta = (h_0, h_1, \ldots, h_n)^T$ and $Q(s) = (\phi_0, \phi_1, \ldots, \phi_n)^T$. Then $\theta^* = (h_0^*, h_1^*, \ldots, h_n^*)^T$ are the Fourier-Laguerre coefficients if and only if

$$\theta^* = \arg\min_\theta J = \arg\min_\theta \int_{-\infty}^{\infty} |H(jw) - \theta^T Q(jw)|^2 d\omega \qquad (18)$$

If $\int_{-\infty}^{\infty} Q(jw)Q^T(-jw)dw$ is nonsingular, which is related to persistent excitation condition, then similar to the solution of (2), the solution θ^* of (18) is unique and given by

$$\theta^* = \left\{ \int_{-\infty}^{\infty} Q(jw)Q^T(-jw)dw \right\}^{-1} \left\{ \int_{-\infty}^{\infty} H(jw)Q(-jw)dw \right\} \qquad (19)$$

As before we can obtain θ^* using the one shot and the recursive method.

One Shot Method. Consider the identifier in Fig. 4. Let $W(s) = Q(s)u(s)$ and $\tilde{y}(t)$ is the output of the filter $\frac{\sqrt{2\lambda}}{s+\lambda}$ with input $y(t) - y_i(t)$. Using the least squares update law

$$\dot{\theta} = -PWW^T\theta + PW\tilde{y}$$
$$\dot{P} = -PWW^TP ; \qquad P(0) = P^T(0) > 0 \qquad (20)$$

we have

Theorem 7. *Consider the least squares algorithm (20). Suppose $W(t)$ is persistently exciting. Then*

$$\lim_{t\to\infty} \theta(t) = \left\{ \lim_{t\to\infty} \frac{1}{t} \int_0^t WW^T dt \right\}^{-1} \left\{ \lim_{t\to\infty} \frac{1}{t} \int_0^t W\tilde{y}dt \right\} = R_w^{-1}R_{w\tilde{y}}$$

If the spectral measure of the input $S_u(\omega)$ is a constant, then

$$\lim_{t\to\infty} \theta(t) = R_w^{-1}R_{w\tilde{y}} = \theta^*$$

where θ^* is given by (19).

Recursive Technique. The Laguerre functions $\{\phi_i\}$ are orthonormal, therefore the Laguerre coefficients h_i's satisfy

$$h_i^* = \arg\min_\alpha \int_{-\infty}^{\infty} |H(jw) - \alpha\phi_i(jw)|^2 dw \qquad (21)$$

which can be obtained independent of each other and thus satisfy a nesting property. Using a technique similar to that discussed in Sect. 2.2.2, we can obtain the coefficients h_i^* and consequently $\gamma_n(jw)$. Note, however, here the initial step discussed in Sect. 2.2.2 is not necessary. More precisely, let $W(s) = \phi_i(s)u(s)$ and $\tilde{y}(t)$ be the output of the filter $\frac{\sqrt{2\lambda}}{s+\lambda}$ with the input $y(t) - y_i(t)$. Then

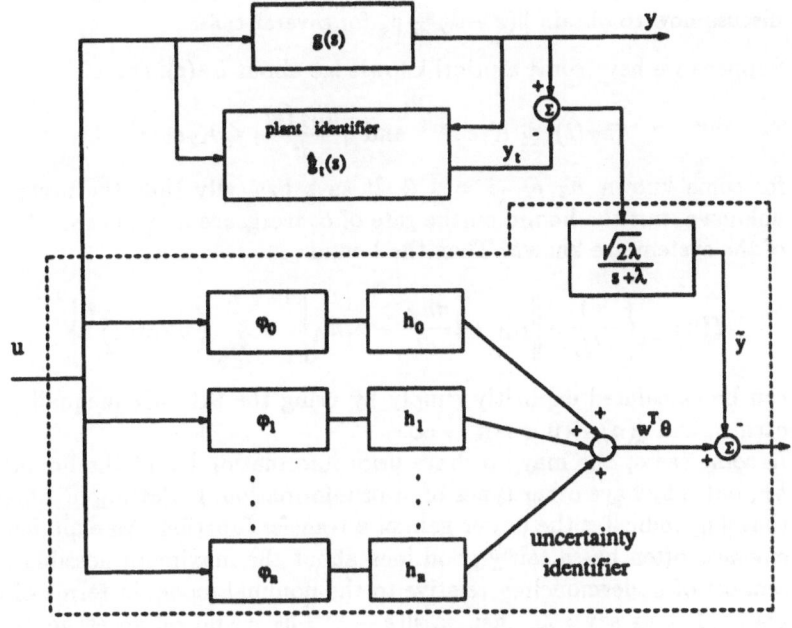

Fig. 4

Corollary 8. *Consider (11) with W and $\tilde{y}$ defined above. Then*

$$\lim_{t \to \infty} \alpha(t) = \left\{ \lim_{t \to \infty} \frac{1}{t} \int_0^t W^2 dt \right\}^{-1} \left\{ \lim_{t \to \infty} \frac{1}{t} \int_0^t W\tilde{y}dt \right\} = R_w^{-1} R_{w\tilde{y}}$$

If the spectral measure of the input $S_u(\omega)$ is a constant, then

$$\lim_{t \to \infty} \alpha(t) = R_w^{-1} R_{w\tilde{y}} = h_i$$

which is the solution of (21).

3.3 Prior Information

Unlike H_2 quantification, where essentially no prior information was required, H_∞ and pointwise quantification require an explicit bound on $\xi(n)$. Observe that $|\gamma_n(j\omega)| = |r_n(j\omega)| + \xi(n)$. Since $0 \le \xi(n) \to 0$ as $n \to \infty$, the contribution of the second term is significant for low order approximation and decreases gradually as the order n increases.

In the real world, the plant estimate (or nominal model) is known after plant identification and so is its impulse response $h_e(t)$. Let $h_\Delta(t)$ denote the impulse response of $\Delta(s)$ and $h_T(t)$ that of the true but unknown system. Then $h_\Delta(t) = h_T(t) - h_e(t)$. Notice that

$$\left\| (s - \lambda)\frac{d\Delta}{ds} \right\|_2 = \| h_\Delta + tdh_\Delta dt - \lambda th_\Delta \|_2.$$

We discuss how to obtain $\left\|(s - \lambda)\frac{d\Delta}{ds}\right\|_2$ for several cases.

1. Suppose we have some a priori knowledge about $h_T(t)$, say

$$|h_T(t)| \le K_1 e^{-\rho_1 t} \quad \text{and} \quad \left|\frac{dh_T(t)}{dt}\right| \le K_2 e^{-\rho_2 t}$$

for some known K_i, ρ_i, $i = 1, 2$. It says basically that the system is unknown, but the bound on the rate of convergence and the smoothness of the system are known. Then the bound

$$\xi(n) \le \left\{\frac{1}{\sqrt{2\lambda n}}\left\|h_\Delta + t\frac{dh_\Delta}{dt} - \lambda t h_\Delta\right\|_2^2 - \sum_{i=0}^n \left(\sqrt{2\lambda} i h_i\right)^2\right\}^{1/2}$$

can be calculated explicitly simply by using the Schwarz inequality for each n, and $\xi(n) \to 0$ as $n \to \infty$.

2. In some cases, one may not have prior information about the bound on h_T, but may have other types of prior information. For example, observe that $\|.\|_2$ indicates the power gain of a transfer function. An experienced engineer often has a fairly good idea about the maximum possible percentage of undermodeling relative to the nominal model in terms of the power. Let us say $x\%$, then $x\%\|(s - \lambda)\frac{d\hat{g}}{ds}\|_2$ would be an estimate for bound of $\|(s - \lambda)\frac{d\Delta}{ds}\|_2$, where $\hat{g}(s)$ is the plant estimate.

3. In the absence of any form of prior information we may still find a reasonable estimate of $\xi(n)$. The idea is to first apply the H_2 quantification methods of Sect. 2 to obtain an estimate $\gamma_n(jw)$ of $\Delta(jw)$ in H_2 sense. We then consider $\|(s - \lambda)\frac{d\gamma_n}{ds}\|_2$ as an estimate of $\|(s - \lambda)\frac{d\Delta}{ds}\|_2$. In this case, of course, the upper boundedness of the H_∞ quantification cannot be guaranteed. However, for a reasonably large n, $|\gamma_n(jw)|$ is a good approximation of $|\Delta(jw)|$ due to the convergence of $|\gamma_n(jw)|$

4. Another way to determine $\|(s - \lambda)\frac{d\Delta}{ds}\|_2$ in the absence of prior information is through the Fourier-Laguerre coefficients h_i of $H(s)$. From [5], it is known that

$$\left\|(s - \lambda)\frac{d\Delta}{ds}\right\|_2^2 = 2\lambda \sum_{i=0}^\infty i^2 h_i^2$$

with h_i's are the Fourier-Laguerre coefficients of $H(s)$ of (13) which can be obtained by on line least squares algorithm. This implies that the nth order approximation of $\|(s - \lambda)\frac{d\Delta}{ds}\|_2^2 \approx 2\lambda \sum_{i=0}^n i^2 h_i^2$ is available. Therefore

$$\xi(n) \le \min\left\{\frac{K}{n}, \frac{1}{\sqrt{2\lambda n}}\left(\left\|(s - \lambda)\frac{d\Delta}{ds}\right\|_2^2 - \sum_{i=0}^n (\sqrt{2\lambda} i h_i)^2\right)^{1/2}\right\}$$

$$\le \frac{1}{\sqrt{2\lambda n}}\left\|(s - \lambda)\frac{d\Delta}{ds}\right\|_2 \approx \frac{1}{\sqrt{n}}\left(\sum_{i=0}^n i^2 h_i^2\right)^{1/2}$$

which can be obtained on line for each n.

5. If more prior information is available, then the performance of the above schemes can be improved. For example, if we know the bounds not only on h_T and $\frac{dh_T(t)}{dt}$ but also on $\frac{d^2h_T(t)}{dt^2}$ then from [12], it can be shown that $\xi(n) \leq \frac{K}{n}$ for some known $K > 0$, where K depends on h_T, h'_T and h''_T and can be calculated explicitly. In turn, a bound on $\xi(n)$ can also be determined explicitly.

To illustrate the H_∞ quantification results, we consider the same examples as in Sect. 2.

Example 2. Consider the system $g(s) = \frac{2}{s+1} \frac{230}{s^2+30s+229}$ as in Example 1. With the same setting, $\lambda = 3$ and $u(t) = 1 + \sum_{i=1}^{20} \sin(0.5 * i * t)$, the plant estimate converges to

$$\hat{g}(s) = \frac{1.756}{s + 0.8688}.$$

Since $\left\| (s-3)\frac{d\Delta(s)}{ds} \right\|_2$ is unknown, apply the H_2 quantification first obtaining the estimate $\gamma_n(s), n = 0$. We approximate

$$\left\| (s-3)\frac{d\Delta(s)}{ds} \right\|_2$$

by

$$\left\| (s-3)\frac{d\gamma_0(s)}{ds} \right\|_2$$

and then calculate $\xi(n)$ for each n. Fig. 5 shows the results of H_∞ quantification. Solid line is the actual uncertainty bound $|\Delta(j\omega)| = |g(j\omega) - \hat{g}(j\omega)|$, and $|\gamma_5(j\omega)|$ and $|\gamma_10(j\omega)|$ denote the derived upper bound of the uncertainty for $n = 5$ and $n = 10$ respectively. We see that each $|\gamma_n(j\omega)|$ provides an upper bound on $|\Delta(j\omega)|$ and the bounds get tighter and tighter as n increases. When $n = 10$, an almost perfect bound is achieved.

4 Concluding Remarks

This paper has presented uncertainty bound quantification schemes when dealing with restricted complexity models. It was shown that the uncertainty approximants not only converge to the actual uncertainty asymptotically in H_2, H_∞ and pointwise sense but also form an upper bound for the uncertainty for any order of the approximant (i.e any n) in the respective topologies. Another nice feature of the methods proposed is the simplicity of the resulting algorithms. Simulations show that the proposed methods provide a realistic upper bound and moreover the bound description is consistent with the requirement of robust control designs.

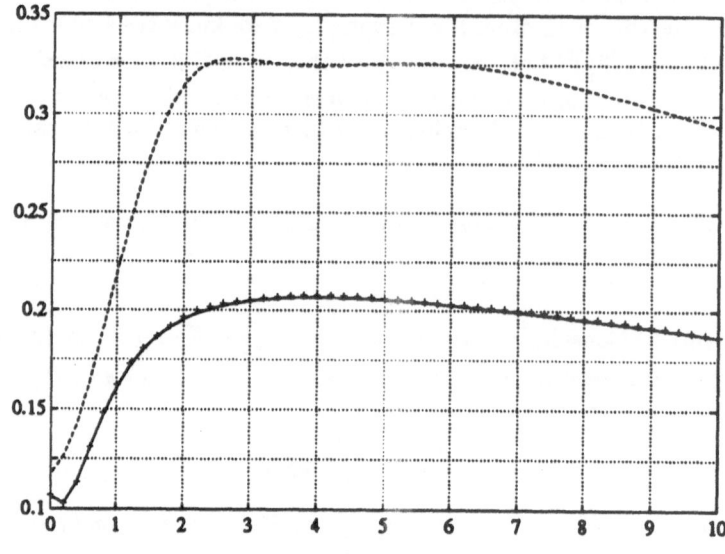

Fig. 5, lambda=3, -.-n=5 + n=10

A potentially important application of the theory presented here is in the area of robust adaptive or adaptive robust control, since it allows the combination of adaptive identification and robust control design.

The main purpose of this paper is quantification of uncertainty bound, the effects of noise has not been discussed. It is interesting to see how these methods would work in the presence of noise.

5 Acknowledgements

The authors wish to thank Professor Roberto Tempo of the CNR of Italy for an inspirational discussion which resulted in Sect. 2.3 of this paper.

References

1. E.W. Bai, Adaptive quantification of model uncertainties by rational approximation, IEEE Trans. on Automat. Contr., Vol. 36, pp. 441-453, 1991.
2. E.W. Bai and S. Raman, Adaptive quantification of model uncertainties for possibly unstable systems, Amer. Contr. Conf., pp. 1204-1209, 1991.
3. K. Glover, J. Lam and J.R. Partington, Rational approximation of a class of infinite dimensional systems, to be published.
4. G.C. Goodwin and M.E. Salgado, A stochastic embedding approach for quantifying uncertainty in the estimation of restricted complexity model, Int. J. ACSP, Vol.3, pp. 333-356, 1989.

5. G.Gu, P.P. Khargonekar and E.B. Lee, Approximation of infinite dimensional systems, IEEE Trans. on Automat. Contr., pp. 610-618, 1989.

6. G.Gu and P.P. Khargonekar, Linear and nonlinear algorithms for identification in H_∞ with error bounds, Proc. of ACC, pp. 64-69, 1991

7. A.J. Helmicki, C.A. Jacobson and C.N. Nett, Control oriented system identification: A worst case/deterministic approach in H_∞, IEEE Trans. on Automat. Contr., Vol.36, pp. 1163-1176, 1991.

8. R.L. Kosut, Adaptive control via parameter set estimation, Int. J. ACSP Vol.2, pp. 371-400, 1988.

9. M. Lau, R.L. Kosut and S. Boyd, Parameter set estimation of systems with uncertain nonparametric dynamics and disturbances, Contr. and Decision Conf., pp. 3162-3167, 1990.

10. L. Ljung, System identification: theory for the user, Prentice-Hall, Englewood Cliffs, NJ, 1987.

11. P.M. Mäkilä, Approximation of stable systems by Laguerre filters, Automatica, pp. 333-346, 1990.

12. P.M. Mäkilä, Laguerre series approximation of infinite dimensional systems, Automatica, pp. 985-995, 1990

13. P.J. Parker and R.R. Bitmead, Adaptive frequency response identification, Contr. and Decision Conf., pp. 348-353, 1987.

14. J.R. Partington, Robust identification in H_∞, J. Math & Appl. (to appear).

15. R.S. Smith and J.C. Doyle, Model invalidation: a connection between robust control and identification, Amer. Contr. Conf., pp.1435-1440, 1989

16. B. Wahlberg, System identification using Laguerre models, IEEE Trans. on Auto. Contr., Vol. 36, pp. 551-562, 1991

A Mixed Deterministic-Probabilistic Approach for Quantifying Uncertainty in Transfer Function Estimation

Douwe K. de Vries and Paul M.J. Van den Hof

Mechanical Engineering, Systems and Control Group,
Delft University of Technology, Mekelweg 2, 2628 CD Delft, The Netherlands.

Abstract:

A procedure is presented to obtain an estimate of the transfer function of a linear system together with an upper bound on the error, using only limited a priori information on the data generating process. By employing a periodic input signal, together with a non-parametric Emperical Transfer Function Estimate (ETFE) over each period, and by averaging over a number of estimates, the statistics of the resulting model asymptotically can be obtained from the data. The model error consists of two parts: a probabilistic part, due to the stochastic noise disturbance on the data, and a deterministic part, due to the bias in the estimate. The latter is explicitly bounded with a deterministic error bound, while the former asymptotically results from an F-distribution. For this analysis no assumptions are made on the distribution of the noise. A mixed deterministic-probabilistic error bound is achieved, clearly distinguishing the different sources of uncertainty.

1 Introduction

In the systems and control community there is a growing interest in merging the problems of system identification and (robust) control system design. This interest is based on the conviction that, in many situations, models obtained from process experiments will be used as a basis for control system design. On the other hand, in model-based robust control design, models and model uncertainties have to be available that are essentially provided by, or at least validated by, measurement data from the process.

Recently several approaches to the identification problem have been presented, considering the identification in view of the control design. By far the most attention has been paid to the construction of deterministic (hard) error bounds [6, 7, 8, 9]. In [1, 4, 5] identification procedures are presented that provide probabilistic (soft) error bounds.

In [6, 8] the a priori information consists of a finite number of corrupted samples of the transfer function of the system, and a weighted H_∞ bound on the measure of corruption. A model in H_∞ and a hard error bound that is valid on the whole unit circle is obtained. In [7, 9] hard model error bounds are constructed using an upper bound on the amplitude of the noise in the time domain. In [9] the hard error bound on the system's transfer function is obtained using a least squares FIR estimate and parameter set estimation techniques, whereas in [7] the bound is obtained by directly calculating the worst-case uncertainty from the time domain data using linear programming. In [2] a frequency domain estimation procedure based on the ETFE is proposed, by which a frequency dependent hard quantification of the model uncertainty can be obtained for a prespecified nominal model. The a priori information on the noise that is necessary (a hard bound on the DFT of the noise) however will be hard to obtain in practice.

In [4, 5] a stochastic embedding approach is used. The distribution of the error is assumed to be known, up to a number of free parameters. The free parameters of the distribution are estimated from the data, together with a least squares estimate of the system. The model error due to undermodelling is represented as a zero mean stochastic process. This results in a probabilistic description of the error in the least squares estimate. In [1] a periodic input signal is used, and an ETFE is made over each period of the input signal. The average over a number of these ETFE's provides the final estimate, and a probabilistic description of the error in this final estimate is presented. However, it is assumed that the noise is normally distributed, that the noise filter is known, and that the steady-state situation is reached before experimental data is taken.

In this paper we will use a stochastic description of the disturbances, based on the same and, in the authors' opinion, definitely sound arguments that are given in [4]. We will however consider the errors due to undermodelling, as deterministic. The input signal is also considered to be deterministic, because the input signal is known in the measurement interval. Hence, for the influence of the noise we will use a probabilistic description, whereas the errors due to undermodelling and unknown past inputs will be bounded with deterministic bounds. This constitutes the main deviation from existing methods: in the current literature on identification with error bounds either both the errors due to undermodelling and noise are considered as being deterministic, or they are both considered as being stochastic. As a result, the approach presented in this paper will yield error bounds that consist of elements with a probabilistic nature, and elements with a deterministic nature. We will call this kind of error bound a mixed deterministic-probabilistic error bound.

In this paper the ETFE is used to obtain a non-parametric frequency domain estimate $\hat{G}(e^{j\omega_k})$, which is only defined at a finite number of frequency points ω_k, and an error bound. The asymptotic distribution of this estimate is obtained through application of the central limit theorem, and by mutually comparing the information arising from different sections of the measured input-output data. This way of extracting the statistical properties of the noise is enabled by using a periodic input signal. Using a periodic input signal and distinguishing different sections of the data set can be thought of

as a repetition of similar experiments, which is a very appealing way to separate structural phenomena present in the data (i.e. the input-output system)
from random effects due to disturbances.

In comparison with previous work on hard error bounds, the probabilistic
setting used in this paper has the advantage that we do need only minor a
priori information. We do not need a hard error bound on the noise in the
time or frequency domain. Actually, through the repetition of experiments, a
corresponding probabilistic bound is estimated from the data. As opposed to
[4], the method proposed here has the property that the form of the distribution of the error is induced by the estimation procedure itself, so that we
do not have to choose it a priori. As parameters we have the frequency dependent variance of the noise, which is estimated from the data, and whose
estimation error is taken into account. As opposed to [1], the noise is not
assumed to be normally distributed, nor is the noise filter assumed to be
known, and the deviation from a steady-state situation is taken into account.

For brevity, all proofs are omitted; the reader is referred to [3].

2 Preliminaries

It is assumed that the plant, and the measurement data that is obtained from
this plant, allow a description

$$y(t) = G_o(q)u(t) + v(t) \tag{1}$$

with $y(t)$ the output signal, $u(t)$ a bounded deterministic input signal, $v(t)$
an additive output noise, q^{-1} the delay operator, and G_o the proper transfer
function of the system, being time-invariant and exponentially stable. The
transfer function can be written in its Laurent expansion around $z = \infty$, as

$$G_o(z) = \sum_{k=0}^{\infty} g_o(k)z^{-k} . \tag{2}$$

with $g_o(k)$ the impulse response of the plant. We will consider scalar (single
input, single output) systems. The output disturbance $v(t)$ is represented as

$$v(t) = H_o(q)e(t) \tag{3}$$

where $\{e(t)\}$ is a sequence of independent identically distributed random
variables with zero mean values, finite variance σ_e^2, and bounded moments of
order $2 + \delta$ for some $\delta > 0$, and where H_o is a proper transfer function that is
strictly stable. The noise filter $H_o(z)$ can be written in its Laurent expansion
around $z = \infty$, as

$$H_o(z) = 1 + \sum_{k=1}^{\infty} h_o(k)z^{-k} \tag{4}$$

Throughout the paper we will consider discrete time intervals for input and
output signals denoted by the integer intervals $T^N := [0, N-1]$, $T_{N_s}^N :=$

$[N_s, N + N_s - 1]$ with N and N_s appropriate integers. We will frequently use
a partioning of the time interval $T_{N_s}^N$ with $N = rN_o$ in r time-intervals of
length N_o, denoting $T_i := [(i - 1)N_o + N_s, iN_o + N_s - 1]$, $i = 1, .., r$.

With the subscript i we will indicate a variable that originates from the
i-th time interval T_i, e.g.

$$x_i(t) = x(t + (i - 1)N_o + N_s) \quad \text{where} \quad t \in T^{N_o} \tag{5}$$

$$X_i(e^{j\omega}) = \frac{1}{\sqrt{N_o}} \sum_{t=0}^{N_o-1} x_i(t)e^{-j\omega t} . \tag{6}$$

For a signal $x(t)$, defined on T^N, we will denote the N-point Discrete Fourier
Transform (DFT) by

$$X(e^{j\frac{2\pi k}{N}}) = \frac{1}{\sqrt{N}} \sum_{t=0}^{N-1} x(t)e^{-j\frac{2\pi k}{N}t} \quad \text{for} \quad k \in T^N . \tag{7}$$

Some specific sets of frequencies that arise in the DFT are denoted as

$$\Omega_N := \{\omega_k = \frac{2\pi k}{N} , k = 0, 1, ..., N - 1\} \tag{8}$$

$$\Omega_{N_o}^{u_i} := \{\omega_k \in \Omega_{N_o} \mid |U_i(e^{j\omega_k})| \neq 0\} . \tag{9}$$

Finally we will denote

$$\max_{t \in T^{N+N_s}} |u(t)| = \bar{u} .$$

Throughout this paper we will adopt a number of additional assumptions on
the system and the generated data.

Assumption 1. *We have as a priori information that*

i. *there exists a finite and known $\bar{u}^p \in \mathbb{R}$, such that $|u(t)| \leq \bar{u}^p$ for $t < 0$*
ii. *there exist finite and known M and ρ, with $M, \rho \in \mathbb{R}$, $\rho > 1$, such that
$|g_o(k)| \leq M\rho^{-k}$, for $k \in \mathbb{Z}_+$*

The a priori information on M and ρ need not be very tight in first instance,
as it can be improved using the measurement data. This will be discussed
later on. The a priori information on $\bar{u}^p$ is given by the actuator constraints.

3 The DFT of the Noise

In this section we will present a theorem dealing with the properties of the
DFT of the output disturbance. This result will be essential in the sequel of
this paper. The result is quite general, and it certainly is valuable in itself.

Theorem 2. *Consider $v_i(t)$ as defined in (3), (4), (5) and let $V_i(e^{j\omega_k})$ be the N_o-point DFT of $v_i(t)$ with $\omega_k \in \Omega_{N_o}$. Let*

$$
\check{V}_{N_o} = \begin{bmatrix} Re\{V_i(e^{j\omega_k})\} \\ Im\{V_i(e^{j\omega_k})\} \\ Re\{V_i(e^{j\omega_\ell})\} \\ Im\{V_i(e^{j\omega_\ell})\} \\ Re\{V_m(e^{j\omega_\ell})\} \\ Im\{V_m(e^{j\omega_\ell})\} \end{bmatrix} .
$$

Then, for $\omega_k, \omega_\ell \in \Omega_{N_o}$, $\omega_k \neq \omega_\ell$, $i, m = 1, ..., r$ and $i \neq m$, there holds

$$
\check{V}_{N_o} \in As\mathcal{N}(0, \Lambda)
$$

where Λ is a diagonal matrix with diagonal elements given by

$$
\begin{aligned}
var[Re\{V_i(e^{j\omega_k})\}] = var[Im\{V_i(e^{j\omega_k})\}] = \tfrac{1}{2}\sigma_e^2|H_o(e^{j\omega_k})|^2 \qquad &\omega_k \neq 0, \pi \\
var[Re\{V_i(e^{j\omega_k})\}] = \sigma_e^2|H_o(e^{j\omega_k})|^2 \qquad &\omega_k = 0, \pi \\
var[Im\{V_i(e^{j\omega_k})\}] = 0 \qquad &\omega_k = 0, \pi .
\end{aligned}
$$

The theorem states that the DFT of the noise is asymptotically normally distributed, with real and imaginary parts that are uncorrelated, and have equal variance for $\omega_k \neq 0, \pi$. Furthermore, asymptotically the DFT's of the noise for different frequencies are uncorrelated, and the DFT's of the noise over different intervals are uncorrelated. Note that uncorrelated jointly normally distributed random variables are independent.

4 Error Bound for Transfer Function Estimate

4.1 Introduction

In this section we will construct a nonparametric estimate $\hat{G}$ of the system's transfer function G_o, by averaging over a set of ETFE's. Note that the ETFE is only defined at a finite number of frequency points. We will establish an error bound $\alpha(\omega_k)$ such that

$$
|G_o(e^{j\omega_k}) - \hat{G}(e^{j\omega_k})| \leq \alpha(\omega_k) \tag{10}
$$

for a finite set of frequencies that will be specified later on. As mentioned in Sect. 1, we will use a probabilistic description of the noise, whereas both the error due to undermodelling and the input signal are considered as being deterministic. This results in an upper bound on the error which has both soft (probabilistic) and hard (deterministic) components, and consequently a statement as (10) can only be made within a prespecified probability.

In order to arrive at an error bound (10), we will pursue the following strategy. Experimental data is available over a time set of length N. This time set is composed of a first interval of length N_s, not used for identification, and consecutively r intervals of length N_o. We consider an input signal that is periodic with period N_o, such that in each of the r intervals the same input signal is applied. This repetition of experiments offers the opportunity to mutually compare the information arising from different intervals of the data, and consequently to formulate the statistics of the estimated transfer function. In other words: the noise contribution on the data is also identified on the basis of the experiments. As a result an error bound (10) can be specified without heavily relying on a priori knowledge of the noise.

Because of the periodicity of the input signal u it follows that

$$\Omega_{N_o}^{u_i} = \Omega_{N_o}^u \quad \text{for all} \quad i = 1, ..., r \ .$$

4.2 A Transfer Function Estimate

Define the following estimates

$$\hat{G}_i(e^{j\omega_k}) = \frac{Y_i(e^{j\omega_k})}{U_i(e^{j\omega_k})} \quad \text{for} \quad i = 1, 2, ..., r \quad \omega_k \in \Omega_{N_o}^u \tag{11}$$

$$\hat{G}(e^{j\omega_k}) = \frac{1}{r} \sum_{i=1}^{r} \hat{G}_i(e^{j\omega_k}) \ . \tag{12}$$

Note that we do not average over different frequencies, we only average over different estimates $\hat{G}_i(e^{j\omega_k})$ of $G_o(e^{j\omega_k})$ at the same frequency ω_k. Employing the system's equations, similar as in [2], it follows that

$$Y_i(e^{j\omega_k}) = G_o(e^{j\omega_k})U_i(e^{j\omega_k}) + R_i(e^{j\omega_k}) + V_i(e^{j\omega_k}) \tag{13}$$

with $R_i(e^{j\omega_k})$ a component which is due to unknown past inputs, i.e. input samples outside the time interval that is considered. In [2] it is shown that for $\omega_k \in \Omega_{N_o}$ this term is bounded by

$$|R_i(e^{j\omega_k})| \le \frac{\bar{u}^P + \bar{u}}{\sqrt{N_o}} \frac{M\rho(1 - \rho^{-N_o})}{(\rho - 1)^2} \rho^{-(i-1)N_o - N_o} \tag{14}$$

if $u(t)$ is periodic with period N_o for $t \in T^{N+N_o}$, and

$$|R_i(e^{j\omega_k})| \le \frac{\bar{u}^P + \bar{u}}{\sqrt{N_o}} \frac{M\rho(1 - \rho^{-N_o})}{(\rho - 1)^2} \tag{15}$$

if $u(t)$ is not periodic. Using (11) and (13) we can write

$$\hat{G}_i(e^{j\omega_k}) = G_o(e^{j\omega_k}) + S_i(e^{j\omega_k}) + \frac{V_i(e^{j\omega_k})}{U_i(e^{j\omega_k})} \tag{16}$$

with

$$S_i(e^{j\omega_k}) = \frac{R_i(e^{j\omega_k})}{U_i(e^{j\omega_k})} \tag{17}$$

the error due to the unknown past inputs for the i-th estimate $\hat{G}_i$. Because S_i only depends on the input and the system, it is a deterministic term. This yields the following result.

Proposition 3. *Consider the estimates $\hat{G}_i(e^{j\omega_k})$, $i = 1, ..., r$. For all $\omega_k \in \Omega^u_{N_o}$ there holds*

a. $\mathcal{E}\{[\} \hat{G}_i(e^{j\omega_k})] = G_o(e^{j\omega_k}) + S_i(e^{j\omega_k})$, *where $S_i(e^{j\omega_k})$ is given by (17) and is bounded using (14) or (15)*

b. $var[\hat{G}_i(e^{j\omega_k})] = \dfrac{var[V_i(e^{j\omega_k})]}{|U_i(e^{j\omega_k})|^2}$.

c. $\hat{G}_i(e^{j\omega_k})$ *asymptotically in N_o is normally distributed*

The above proposition states that the Emperical Transfer Function Estimate (ETFE) is asymptotically normally distributed, and asymptotically unbiased. However, the variance does not decrease with N_o, it is just the noise to signal ratio in the frequency domain. The averaging (12) is introduced in order to obtain an estimate with decreasing variance. However, $S_i(e^{j\omega_k})$ is unknown and varies with i, even if $U_i(e^{j\omega_k})$ is independent of i for $i = 1, ..., r$. Furthermore, $var[V_i(e^{j\omega_k})]$ is unknown, and $var[\hat{G}_i(e^{j\omega_k})]$ varies with i if $U_i(e^{j\omega_k})$ varies with i. Moreover, if the input is not periodic, the uncertainty due to the unknown past inputs $S_i(e^{j\omega_k})$ typically will dominate the error bound, see (15). Hence in general the estimate $\hat{G}(e^{j\omega_k})$ will be heavily biased, and it is not possible to obtain a satisfactory estimate of its variance.

In order to improve the above situation, we will use a periodic input signal. We will split up the analysis into two parts: first we will derive the properties of an intermediate variable, and next we will analyse the properties of the estimate (12). Define the intermediate variable $\tilde{G}_i(e^{j\omega_k})$ as

$$\tilde{G}_i(e^{j\omega_k}) = \hat{G}_i(e^{j\omega_k}) - S_i(e^{j\omega_k}) \tag{18}$$

$$= G_o(e^{j\omega_k}) + \frac{V_i(e^{j\omega_k})}{U_i(e^{j\omega_k})} . \tag{19}$$

Proposition 4. *Consider the intermediate variables $\tilde{G}_i(e^{j\omega_k})$, $\omega_k \in \Omega^u_{N_o}$. Let $u(t)$ be a periodic input signal with period N_o for all $t \in T^{N+N_o}$. Then asymptotically in N_o the random variables $\tilde{G}_i(e^{j\omega_k})$ and $\tilde{G}_\ell(e^{j\omega_k})$ are independent and identically distributed for all $i, \ell = 1, ..., r$, $i \neq \ell$.*

We denote the averaged intermediate variable, averaged over the different time intervals, as

$$\tilde{G}(e^{j\omega_k}) = \frac{1}{r} \sum_{i=1}^{r} \tilde{G}_i(e^{j\omega_k}) \quad \text{for} \quad \omega_k \in \Omega^u_{N_o} . \tag{20}$$

This averaged intermediate is going to be used in determining an error bound as in (10)

$$|G_o(e^{j\omega_k}) - \hat{G}(e^{j\omega_k})| \leq |G_o(e^{j\omega_k}) - \tilde{G}(e^{j\omega_k})| + |\tilde{G}(e^{j\omega_k}) - \hat{G}(e^{j\omega_k})| =$$

$$= |G_o(e^{j\omega_k}) - \tilde{G}(e^{j\omega_k})| + |S(e^{j\omega_k})| \tag{21}$$

with

$$S(e^{j\omega_k}) = \frac{1}{r} \sum_{i=1}^{r} S_i(e^{j\omega_k}) \ .$$

Considering the inequality (21), the first term on the right hand side is the variance contribution to the error due to the noise disturbance. The second term on the right hand side of (21) is a bias contribution, due to unknown past inputs. This deterministic term can be bounded by using (14). For a periodic input signal this term can be made small by choosing N_s. A bound on the first (stochastic) term has to be determined on the basis of its distribution. For the distribution of $\tilde{G}(e^{j\omega_k})$ we have the following results.

Proposition 5. *Consider the situation as in Proposition 4. Then for all* $\omega_k \in \Omega_{N_o}^u$ *there holds*

a. $\mathcal{E}\{[\} \tilde{G}(e^{j\omega_k})] = G_o(e^{j\omega_k})$.

b. *Asymptotically in* N_0 $var[\tilde{G}(e^{j\omega_k})] = \frac{1}{r}var[\tilde{G}_i(e^{j\omega_k})] = \frac{var[V_i(e^{j\omega_k})]}{r|U_i(e^{j\omega_k})|^2}$, *independent of* i

c. $\sqrt{r}\ \tilde{G}(e^{j\omega_k})$ *asymptotically in* N *is normally distributed*

As a result the asymptotic distribution of the estimate $\tilde{G}(e^{j\omega_k})$ is specified, although its variance still remains to be unknown. In the next steps we will quantify this variance on the basis of measurement data. To this end we use the following two estimates

$$\hat{\sigma}_r^2(\hat{G}(e^{j\omega_k})) = \frac{1}{r(r-1)} \sum_{\ell=1}^{r} |\hat{G}(e^{j\omega_k}) - \hat{G}_\ell(e^{j\omega_k})|^2 \tag{22}$$

$$\hat{\sigma}_r^2(\tilde{G}(e^{j\omega_k})) = \frac{1}{r(r-1)} \sum_{\ell=1}^{r} |\tilde{G}(e^{j\omega_k}) - \tilde{G}_\ell(e^{j\omega_k})|^2 \ . \tag{23}$$

Note that the first one of these estimates indeed can be calculated from data. However the second one is not available.

Proposition 6. *Consider the situation as in Proposition 4. Then the estimate* $\hat{\sigma}_r^2(\tilde{G}(e^{j\omega_k}))$ *is a consistent estimate of* $var[\tilde{G}(e^{j\omega_k})]$. *Asymptotically in* N_o *the variance of the estimate* $\hat{\sigma}_r^2(\tilde{G}(e^{j\omega_k}))$ *decays as* $\frac{1}{r-1}$.

Although this estimate is not available from data, we can bound the difference between the two estimates (22), (23).

Lemma 7. *Consider the estimates* $\hat{\sigma}_r^2(\hat{G}(e^{j\omega_k}))$, $\hat{\sigma}_r^2(\tilde{G}(e^{j\omega_k}))$ *as defined in* (22), (23). *Let* $u(t)$ *be a periodic input signal with period* N_o *for all* $t \in T^{N+N_s}$. *Then*

$$|\hat{\sigma}_r^2(\tilde{G}(e^{j\omega_k})) - \hat{\sigma}_r^2(\hat{G}(e^{j\omega_k}))| \le \epsilon(\omega_k) \tag{24}$$

with

$$\epsilon(\omega_k) = \frac{1}{r(r-1)} \sum_{i=1}^{r} (2|A_i(e^{j\omega_k})||B_i(e^{j\omega_k})| + |B_i(e^{j\omega_k})|^2) \qquad (25)$$

and

$$|A_i(e^{j\omega_k})| = |\hat{G}(e^{j\omega_k}) - \hat{G}_i(e^{j\omega_k})|$$

$$|B_i(e^{j\omega_k})| = \frac{1}{r} \sum_{m=1}^{r} |S_m(e^{j\omega_k})| + \frac{r-2}{r}|S_i(e^{j\omega_k})|$$

$$|S_i(e^{j\omega_k})| \le \frac{1}{\sqrt{N_o}} \frac{\bar{u}^P + \bar{u}}{|U_i(e^{j\omega_k})|} \frac{M\rho(1-\rho^{-N_o})}{(\rho-1)^2} \rho^{-(i-1)N_o-N_o} \;\;.$$

Clearly the difference between $\hat{\sigma}_r^2(\tilde{G}(e^{j\omega_k}))$ and $\hat{\sigma}_r^2(\hat{G}(e^{j\omega_k}))$ is due to the $S_i(e^{j\omega_k})$, i.e. the influence of the unknown past input signals. The difference is small if the $S_i(e^{j\omega_k})$ are small, which for a periodic input signal can be obtained by choosing N_s.

Using only the unknown intermediate transfer function $\tilde{G}(e^{j\omega_k})$, and its estimated variance $\hat{\sigma}_r^2(\tilde{G}(e^{j\omega_k}))$, an error bound with respect to the system's transfer function can be calculated asymptotically in N_o. Due to the fact that the intermediate variables $\tilde{G}_i(e^{j\omega_k})$ are independent and identically distributed, see Proposition 4, this results in an F distribution for the error, as formulated in the following proposition.

Lemma 8. *Consider the intermediate transfer function* $\tilde{G}(e^{j\omega_k})$, *(20), (19), and the estimate of its variance (23). Let the input signal be periodic with period* N_o *for* $t \in T^{N+N_s}$, *and let* $r > 1$. *Then as* $N_o \to \infty$

$$\frac{|G_o(e^{j\omega_k}) - \tilde{G}(e^{j\omega_k})|^2}{\hat{\sigma}_r^2(\tilde{G}(e^{j\omega_k}))} \to \begin{cases} F(2, 2(r-1)) & \text{for } \omega_k \ne 0, \pi \\ F(1, r-1) & \text{for } \omega_k = 0, \pi \end{cases}$$

for all $\omega_k \in \{\omega \in \Omega_{N_o}^u \mid \hat{\sigma}_r^2(\tilde{G}(e^{j\omega})) > 0\}$, *where* $F(n, d)$ *denotes the* F *distribution with* n *degrees of freedom in the numerator and* d *degrees of freedom in the denominator.*

Note that no assumptions are made on the distribution of the noise, and that the uncertainty in the estimated variance is taken into account by the F-distribution.

Combining Lemma's 7 and 8 and (21) leads to an error bound that can be calculated on the basis of data, in terms of a confidence interval. In formulating this confidence interval, we will adopt the following notation

$$F_\alpha(m, n) = \{P[x \le \alpha], \; x \in F(m, n)\}$$

which means that $F_\alpha(m, n)$ is the probability that $x \in F(m, n)$ is smaller than α.

Theorem 9. *Consider the transfer function* $\hat{G}(e^{j\omega_k})$, *(12), (11), and the estimate of its variance (22). Let the input signal be periodic with period N_o for $t \in T^{N+N_s}$, and let $r > 1$. Asymptotically in N_o there holds for all $\omega_k \in \{\omega \in \Omega_{N_o}^u \mid \hat{\sigma}_r^2(\hat{G}(e^{j\omega})) > \epsilon(\omega)\}$*

$$|G_o(e^{j\omega_k}) - \hat{G}(e^{j\omega_k})| \le \gamma_\alpha(\omega_k) + |S(e^{j\omega_k})| \quad w.p. \ge \begin{cases} F_\alpha(2, 2r-2) & \omega_k \ne 0, \pi \\ F_\alpha(1, r-1) & \omega_k = 0, \pi \end{cases}$$

where

$$\gamma_\alpha(\omega_k) = \sqrt{\alpha} \left(\hat{\sigma}_r^2(\hat{G}(e^{j\omega_k})) + \epsilon(\omega_k) \right)^{\frac{1}{2}}$$

$$|S(e^{j\omega_k})| \le \frac{1}{r\sqrt{N_o}} \frac{\bar{u}^P + \bar{u}}{|U_i(e^{j\omega_k})|} \frac{M\rho(1-\rho^{-N})}{(\rho-1)^2} \rho^{-N_s}$$

and $\epsilon(\omega_k)$ is given by (25).

The deterministic error terms are $\epsilon(\omega_k)$ and $S(e^{j\omega_k})$, where $\epsilon(\omega_k)$ is a function of the $S_i(e^{j\omega_k})$. These terms are due to the unknown past inputs, and typically will be small in comparison with the error due to the noise $\sqrt{\alpha}\,\hat{\sigma}_r(\hat{G}(e^{j\omega_k}))$. This is due to the fact that $S(e^{j\omega_k})$ and $S_i(e^{j\omega_k})$ have exponential convergence to zero with N_s. For the error due to the noise we have that, asymptotically in N_o, the variance decays as $1/r$ and the variance of the estimated variance decays as $1/(r-1)$.

Note that the above estimate is only defined at the finite number of frequency points $\Omega_{N_o}^u$. Theorem 9 provides the a priori information needed by [6, 8] except for the fact that the error bound provided is soft instead of hard.

5 Additional Results

Using a similar procedure as in Sect. 4, together with Theorem 2, we have obtained a number of additional results, see [3].

- By estimating a Finite Impulse Response (FIR) model on each $\hat{G}_i$ a set of FIR models can be obtained. Averaging over this set results in a FIR model with an error bound on the estimated parameters. Hence, an estimate with an error bound of the impulse response of the system is obtained, which can be used to improve the prior information M and ρ.
- An error bound for the transfer function of an estimated FIR model can be obtained. This results in an error bound that is valid on the whole unit circle, whereas the error bound of Theorem 9 is only defined at a finite number of frequency points.
- Estimates with an error bound of the response of the system to arbitrary input signals can be obtained. This is useful for validation, simulation and fault detection purposes.

6 Example

To illustrate our results a simulation was made of a fifth order system

$$G_o(z) = \frac{0.7027 - 0.8926z^{-1} + 0.24z^{-2} + 0.5243z^{-3} - 0.9023z^{-4} + 0.4z^{-5}}{1 - 2.4741z^{-1} + 2.8913z^{-2} - 1.9813z^{-3} + 0.8337z^{-4} - 0.1813z^{-5}}$$

whose impulse response $g_o(k)$ satisfies a bound given by $M_o = 2$ and $\rho_o = 1.23$. There was 10 percent (in amplitude) *uniformly* distributed colored noise (highpass filtered white noise) on the output.

As a priori information on the impulse response we choose $M = 3$ and $\rho = 1.2$. A periodic input signal was applied to the system. The input signal was chosen to obey $\bar{u}^p = 2$ and $\bar{u} = 1$. We used 1074 points with $N = 1024$, $N_o = 128$ and $N_s = 50$. The magnitude of the DFT of the input signal over one period, $|U_i(e^{j\omega_k})|$, is given in Fig. 1. Note that the frequency points where $|U_i(e^{j\omega_k})| > 0$ are not equidistant. The magnitude and frequency grid of the input express that we are especially interested in the behaviour of the system around $\omega = 0.88$ rad/s, and that we do not expect the system behaviour to change rapidly with frequency for the higher frequencies.

In Fig. 2 a Nyquist plot of the estimate $\hat{G}(e^{j\omega_k})$ of Theorem 9 is given, together with the estimated error bound and the true system. The probability level for the error bound is 99 %. Note that a very good estimate is obtained for those frequencies where $|U_i(e^{j\omega_k})|$, the magnitude of the DFT of the input signal, was chosen to be large. Note also that the error bound is tight, i.e. the actual error can be close to the upper bound.

7 Conclusions

In this paper a procedure is presented to obtain an estimate, together with an error bound, of the transfer function of a system, using only minor a priori information. The basis of our results is the derivation of the asymptotic distribution of a Discrete Fourier Transform (DFT) of a filtered sequence of independent random variables, the separation of the error in a deterministic and a probabilistic part, and the use of a periodic input signal.

By employing a periodic input signal a repetition of experiments is obtained. This repetition offers the possibility to mutually compare the information arising from different intervals of the data, and consequently to formulate the statistics of the estimated transfer function. More specifically, a non-parametric Empirical Transfer Function Estimate (ETFE) is made over each period of the input signal. Due to the periodicity of the input signal these estimates are approximately independent and identically distributed. Averaging over the estimates, which provides the final estimate of the system, now results in a fast decrease of the variance of the final estimate with the number of averages. Moreover, the final estimate is almost unbiased and its variance can be estimated consistently. The error in this final estimate can be

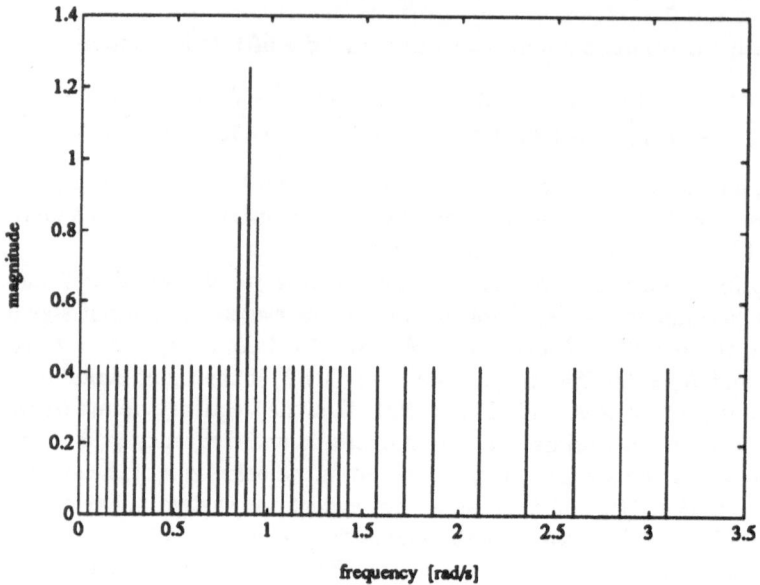

Fig. 1. Magnitude of the DFT over one period of the input signal, $|U_i(e^{j\omega_k})|$.

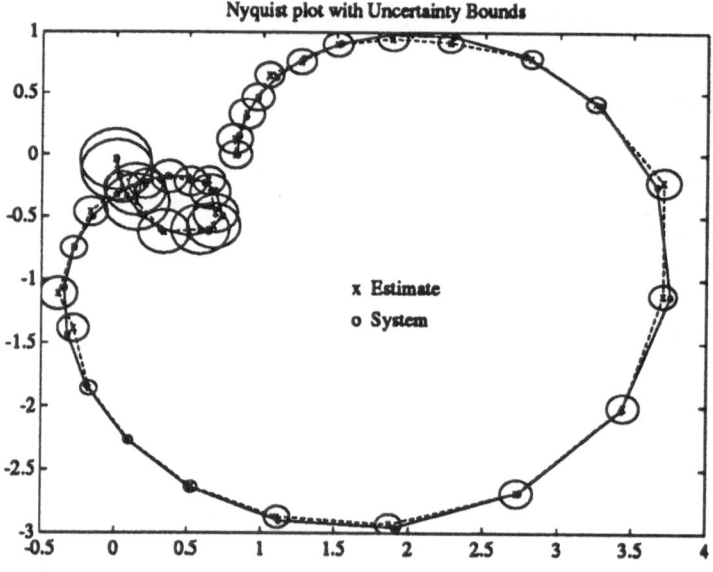

Fig. 2. Estimate $\hat{G}(e^{j\omega_k})$ with error bound, and true system $G_o(e^{j\omega_k})$ for $\omega_k \in \Omega_{N_o}^u$.

separated into two parts: a probabilistic part, due to the noise disturbance on the data, and a deterministic part, due to the bias in the estimate. The latter is explicitly bounded with a deterministic error bound. The former asymptotically has an F-distribution, so that a confidence interval can be specified. This results in a mixed deterministic-probabilistic error bound, which clearly distinguishes the different sources of uncertainty. No assumptions are made on the distribution of the noise, and the uncertainty in the estimated variance is taken into account by the F-distribution.

References

1. Bayard, D.S. (1992). Statistical plant set estimation using Schroeder-phased multisinusoidal input design. *Proc. American Control Conf.*, pp. 2988–2995.
2. De Vries, D.K. and P.M.J. Van den Hof (1992). Quantification of model uncertainty from data: input design, interpolation, and connection with robust control design specifications. *Proc. American Control Conf.*, pp. 3170–3175.
3. De Vries, D.K. and P.M.J. Van den Hof (1992). Quantification of uncertainty in transfer function estimation. *Report N-410, Mechanical Engineering Systems and Control Group, Delft Univ. Technology, The Netherlands.* Submitted to *Automatica*.
4. Goodwin, G.C., M. Gevers and B. Ninness (1992). Quantifying the error in estimated transfer functions with application to model order selection. *IEEE Trans. Automatic Contr.*, AC–37, pp. 913–928.
5. Goodwin, G.C. and M.E. Salgado (1989). Quantification of uncertainty in estimation using an embedding principle. *Proc. American Control Conf.*, pp. 1416–1421.
6. Gu, G. and P.P. Khargonekar (1992). A class of algorithms for identification in H_∞. *Automatica*, vol. 28, pp. 299–312.
7. Hakvoort, R.G. and P.M.J. Van den Hof (1992). Identification of model error bounds in ℓ_1- and H_∞-norm. *Proc. Workshop on the Modelling of Uncertainty in Control Systems, Univ. of California, Santa Barbara, June 18-20.* Springer-Verlag.
8. Helmicki, A.J., C.A. Jacobson and C.N. Nett (1990). Identification in H_∞: a robustly convergent, nonlinear algorithm. *Proc. American Control Conf.*, pp. 386–391.
9. Wahlberg, B. and L. Ljung (1991). On estimation of transfer function error bounds. *Proc. European Control Conf.*, pp. 1378–1383.

Estimation for Robust Control

Brett M. Ninness* and Graham C. Goodwin

Department of Electrical Engineering, University of Newcastle, Callaghan 2308, AUSTRALIA

1 Introduction

Until recently the work of those in the identification community and the work of those in the control community progressed relatively independently. Control theorists assumed knowledge of plant information in a certain format and identification theorists provided plant information in a format they assumed would be useful. The discovery that these two formats are not compatible for the purposes of robust control design is recent and has resulted in considerable research interest. This sudden research interest has led to a number of approaches to the problem. Almost universally these approaches seek to retain existing robust control theory and develop new estimation theory. This involves discarding stochastic analysis in favour of deterministic bounded disturbance models and then considering the worst possible estimation error. This obviously has the flavour of H_∞ design methods and indeed a wide body of research [19, 32, 37, 17, 18] has arisen using observed frequency response data to provide models in H_∞ with $\| \cdot \|_\infty$ norm error bounds that apply with probability one (wp1). Such models can obviously be bolted directly to existing H_∞ control design theory.

Another line of research [50, 42, 41, 43, 40] provides more explicit information; it provides frequency dependent bounds that apply wp1. These methods generally, but not always, use ideas from so-called 'set estimation' or 'bounded error' estimation theory that was developed in the early 1980's. Unfortunately, there is no robust controller synthesis theory available to take advantage of this more explicit information. One approach might be to use H_∞ synthesis methods with weighting functions derived from the explicit frequency dependent bounds available. This approach would discard the phase error information available.

---

* The author is grateful to the Centre for Industrial Control Science at the University of Newcastle, Australia and the Australian Telecommunications and Electronics Research Board for Funding Assistance.

The final line of research [34, 16, 15, 27, 30, 29, 24, 13], which is greatly overshadowed (in terms of current interest) by the other two, seeks to provide soft bounds. That is, to provide bounds that apply with probability $\alpha \leq 1$. This line of research aims not at finding new 'deterministic' estimation methods, but at preserving existing stochastic estimation theory while still accounting for the deterministic effects of undermodelling. Any control synthesis theory compatible with these sorts of models could, of course, only guarantee performance with probability α.

In this paper we discuss this latter line of research, and compare and contrast it to the former 'hard bounding' schools of thought. For this purpose we assume that the required model description is a frequency domain one so that we consider only linear time invariant plants. Furthermore, we assume that the description is to be obtained from some observed noise-corrupted data of short duration. We do not consider the noise free case or long data record cases. We believe these problems can be solved with existing theory by fitting a high order, high fidelity model to the data and then carrying out some sort of model order reduction if this is deemed desirable. The interest arises in short duration noise-corrupted data where undermodelling is introduced due to a bias versus variance tradeoff. Specifically, for given data, as the model order is increased the fitted model infidelity due to undermodelling (bias error) decreases but the infidelity due to noise (variance error) increases [25, 48]. The net result is that for short, noisy data records it is optimal to fit a lower order model than the true system to the data. The difficulty addressed by this paper is the question of quantifying the bias error introduced by this resultant undermodelling. That is, we seek to identify a model that consists not only of a nominal plant, but also error bounds around this nominal value that are constructed to take into account both stochastic disturbances and undermodelling.

The question of quantifying model errors has been widely addressed in existing literature using stochastic estimation theory, but only when these errors are introduced by noise alone. For example, suppose we choose a model structure that allows us to describe the observed data via a linear regression

$$Y = \Phi\theta + V \tag{1}$$

where Y and V are vectors of observed data and noise disturbances, Φ is a matrix of known regressors and θ is a vector of parameters. In this case an estimate $\hat{\theta}$ of θ may be obtained from the data by minimising a least squares criterion. The solution for θ is then well known as:

$$\hat{\theta} = \left(\Phi^T\Phi\right)^{-1}\Phi^T Y \tag{2}$$

Errors in this estimate are introduced by V. Say we know the statistics of V, for example we might assume[1]:

$$\mathcal{E}\{V\} = 0 \tag{3}$$

---

[1] $\mathcal{E}\{\cdot\}$ denotes expectation with respect to the random disturbance V

$$\mathcal{E}\left\{VV^T\right\} = \sigma_\nu^2 I \tag{4}$$

then provided Φ and V are independent we have

$$\mathcal{E}\left\{(\hat{\theta} - \theta_0)(\hat{\theta} - \theta_0)^T\right\} = \sigma_\nu^2 \left(\Phi^T\Phi\right)^{-1} \tag{5}$$

and this may be used (perhaps with scaling depending on conservatism) to provide quantification of the error $\hat{\theta} - \theta_0$. Note however that we are required to know the value σ_ν^2 which parameterises the *class* of disturbance sequences of which the one we observe (V) is a member. Usually σ_ν^2 is not known. However, it can also be estimated from the data. For example, a well known unbiased and consistent estimator is:

$$\hat{\sigma}_\nu^2 = \frac{\varepsilon^T\varepsilon}{N - p} \tag{6}$$

$$\varepsilon = \left[I - \Phi(\Phi^T\Phi)^{-1}\Phi^T\right]Y \tag{7}$$

where $N = \dim Y$ and $p = \dim\theta$. This sort of stochastic approach has been well tried and proven over the years. It relies on a key idea; for unknown disturbances, embed them in a *class* of signals that are parameterised by their *on average* properties. It then becomes feasible to estimate the class description parameterization; ie. σ_ν^2. This is a natural but ingenious progression when the signals themselves cannot be estimated and, in large part, is responsible for the practical success of stochastic estimation theory over the years. Motivated by this success we suggest the same approach of using class descriptors for undermodelling as well. As we shall prove, this allows parameters describing classes of undermodellings to be estimated from the data even when the particular undermodelling realisation itself cannot be estimated. As in the purely stochastic case, this leads to highly practical methods of providing models suitable for robust control since the prior knowledge assumed is relaxed to the point of being realistic.

We believe this property of estimability of embedding parameters to be a major advance. It is only feasible because we have chosen to use the same measurement theoretic description for undermodelling as is usually used for noise. That is, ensemble averages of undermodellings can be linked to time domain realisation averages available to us. This allows us to estimate the undermodelling ensemble averages from the data. In turn this allows us to specify error bounds with respect to the ensemble of possible undermodelling realisations. In this way estimation errors due to noise and undermodelling are handled in a consistent manner. Furthermore, the traditional and highly successful stochastic framework for estimation theory is retained.

This sort of result is not possible with the usual hard bounding approaches to undermodelling error quantification [50, 43, 40, 38, 31, 22, 19, 18]. To be sure, these approaches use the same framework; they consider classes of undermodellings and then provide estimation error bounds with respect to these classes. However, the classes of undermodellings are defined implicitly by bounding the effects of the undermodelling for bounded input. This makes estimation of a parameterisation for the class from the data impossible. We define the class of undermodellings explicitly. This allows us to use the effects of the undermodelling to estimate a parameterisation for the explicitly defined class.

2 Problem Formulation

As already described, we are interested in estimating the frequency response of LTI systems from observed noise corrupted data:

$$y_k = G_T(q^{-1})u_k + H(q^{-1})e_k \tag{8}$$

Here $\{u_k\}$ is a known input sequence, $\{e_k\}$ is an iid noise sequence and $G_T(q^{-1})$, $H(q^{-1})$ are stable rational transfer functions in the backward shift operator q^{-1}. We assume open loop conditions and decompose (8) as

$$y_k = G(q^{-1}, \theta_0)u_k + G_\Delta(q^{-1})u_k + \nu_k \tag{9}$$

Here $G(q^{-1}, \theta)$ is a transfer function of lower complexity than $G_T(q^{-1})$ and is parameterized by the vector $\theta \in \mathbb{R}^p$ where the dimension p is chosen to optimize the bias/variance tradeoff discussed in the previous section. For particular $\theta = \theta_0$ the unmodelled dynamics in $G_T(q^{-1})$ are described by $G_\Delta(q^{-1})$. Finally we have written $\nu_k = H(q^{-1})e_k$. The problem is to form an estimate $\hat\theta$ from the observed sequences $\{y_k\}$ and $\{u_k\}$ and to then place an error bound around the estimated frequency response $G(e^{-j2\pi f}, \hat\theta)$ that captures the true frequency response $G_T(e^{-j2\pi f})$. Here f is cyclic frequency normalised with respect to the sampling frequency.

The mainstream deterministic approaches to this problem exemplified in [50, 43, 40, 38, 31, 22, 19, 18] recognise that undermodelling is a deterministic phenomenon and so abandon a stochastic framework altogether. For example, in [50] it is assumed that $G(q^{-1}, \theta)u_k$ can be cast in a linear regressor form

$$y_k = \phi_k^T \theta + \nu_k \tag{10}$$

where

$$\nu_k = G_\Delta(q^{-1})u_k + H(q^{-1})e_k \tag{11}$$

Prior assumptions on magnitude and smoothness of $G_\Delta(e^{-j2\pi f})$ together with norm assumptions on $\{u_k\}$ and $\{e_k\}$ are translated into hard bounds of the form

$$|\nu_k| \le \delta \in \mathbb{R}^+ \tag{12}$$

so that a set for $\theta \in \mathbb{R}^p$ that is consistent with both assumptions and data can be formed:

$$\theta \in \{\beta \in \mathbb{R}^p : |y_k - \phi_k^T \beta| \le \delta\} \triangleq \Theta \tag{13}$$

Obviously this set takes into account the variance effects of $\{e_k\}$ and the bias effects of $G_\Delta(q^{-1})$ and so allows uncertainty sets for $G_T(e^{-j2\pi f})$ to be formed [50, 40]. The calculation of the feasible parameter set Θ is generally achieved by drawing on existing work in set estimation theory [14, 23, 35] or by exploiting algebraic properties of the model description [43, 40] directly.

Other work that examines the non-falsified data set theme in an asymptotic sense has been presented under the title of information based complexity or optimal algorithm theory [20, 12, 47]. Finally there is substantial literature on a hard bounding theme assuming a starting point of frequency domain data [36, 31, 19, 37, 17, 18].

We have three serious objections to these non-stochastic approaches. Firstly, it results in an abrupt abandonment of an existing highly successful approach to estimation as soon as an arbitrarily small amount of undermodelling is introduced; even when $\{\nu_k\}$ is in fact a realisation of a stochastic process.

Secondly, and more seriously than this aesthetic concern, these hard bound methods are very sensitive to the prior assumptions on $G_\Delta(q^{-1})$ and hence they almost always result in algorithms that are not useful in practice. Specifically, the exact bound δ in (12) will never be known. Instead the user must intelligently guess a bound μ based on prior knowledge. Now if this bound μ is less than the true bound δ then the calculation of Θ will be overly optimistic. It will be possible for Θ to collapse to an incorrect singleton. On the other hand, if μ is set conservatively so that $\mu > \delta$ then the size of Θ can be underbounded by a constant not depending on the amount of data observed. This is made clear in the following result:

Lemma 1. *Define d as the diameter of the largest ball that can be placed inside Θ. If energy in the regressors (columns of Φ) is bounded as $\|\phi_k\|_2 \leq \sigma_\phi$ and $\mu > \delta$ then we have the underbound:*

$$d > \frac{2(\mu - \delta)}{\sigma_\phi} > 0 \qquad (14)$$

Proof. See Appendix A.

That is, we cannot estimate θ_0 to arbitrary accuracy as more data is observed. In general then the volume of Θ is *polynomially* sensitive[2] to the assumption μ. Furthermore, μ is necessarily only able to be set vaguely. This inherent sensitivity to assumptions contradicts the original ambit of robust design which is to achieve insensitivity to prior information. We contrast this to stochastic estimation theory that is not nearly so sensitive to knowledge of the statistics of the noise. Indeed, it can be shown [6] from a Bayesian argument that in many cases hard bounding is subsumed by stochastic theory in a way that implicitly assumes an independent disturbance sequence. For a more complete discussion of stochastic vs set estimation theory see [6].

---

[2] the order of the polynomial $= p =$ number of parameters

Our final criticism involves the format of prior knowledge in these mainstream deterministic approaches. We believe it leads to overly conservative error bounds. For example, in the 'Estimation in H_∞' work [36, 19, 18] one is required to assume knowledge of two parameters ρ and M such that the true system transfer function evaluated at z^{-1}, $G_T(z)$, is analytic on $\mathbb{D}_\rho = \{z \in \mathbb{C} : |z| < \rho\}$, $\rho > 1$ and is bounded in magnitude by M on $\mathbb{D}_\rho$. This information is used to bound the estimation error term due to undermodelling by finding a bound on the impulse response $\{g_n\}$ of the true system. This latter bound is extremely cautious since it is derived[3] as follows. Since $G_T(z)$ is analytic on $\mathbb{D}_r$ for any $|r| < \rho$ then we have by application of Parseval's theorem

$$|g_n|^2 r^{2n} \le \sum_{k=0}^{\infty} |g_k|^2 r^{2k} = \frac{1}{2\pi} \int_{-\pi}^{\pi} |G_T(re^{j\varphi})|^2 d\varphi \le M^2 \qquad (15)$$

and this leads to

$$|g_n| \le \frac{M}{\rho^n} \qquad (16)$$

However, the first inequality in (15) is obviously grossly conservative since we have underbounded an infinite sum of positive terms by just *one* term in the sum ! Moreover, there is conservatism in bounding the integral by M^2 so that any bounds derived from (16) will be highly conservative. More details on this theme and an examination of stochastic and deterministic approaches to estimation in H_∞ can be found in [4].

These criticisms of the mainstream approach to estimation in the face of undermodelling lead us to a re-examination of the existing successful stochastic problem formulation to see how it can be extended to the case of undermodelling. Specifically, we note that the main ingredient in stochastic theory is to inject extra structure into the description of a disturbance signal ν_k. This is done by assuming that a disturbance is a function not only of time k, but also of the state of nature. This gives a description

$$\nu_k(\omega) \qquad \omega \in \{\Omega, \mathcal{F}, \mathbb{P}\} \qquad (17)$$

where now ω on a probability space $\{\Omega, \mathcal{F}, \mathbb{P}\}$ allows us to index disturbances over all possible states of nature in the same way that k allows us to index them over time. That is, an event ω from nature Ω arises to result in a particular disturbance realisation via the random variable mapping $\nu_k(\omega)$. The likelihood of whole sets of events $A \in \mathcal{F} \subseteq \Omega$ is determined as $\mathbb{P}(A) = \int_X \nu(x) f_\nu(x) dx$ where $\nu_k(A) \mapsto X \in \mathbb{R}$ and f_ν is the probability density function of ν. Results from probability theory then allow us to relate the values of averages of $\nu_k(\omega)$ over the two domains Ω and time.

Our contribution is to suggest that we also need to add extra structure into the description of $G_\Delta(q^{-1})$. Considering the remarkable success and

---

[3] In the literature (16) is derived by quoting Cauchy's estimate. Investigation [51] gives the derivation of Cauchy's estimate as (15)

precision of probability theory we elect to use the same description as used in (17):

$$G_\Delta(q^{-1}, \omega) \qquad \omega \in \{\Omega, \mathcal{F}_\Delta, \mathbb{P}_\Delta\} \tag{18}$$

That is, we add the extra structure into our problem formulation of allowing $G_\Delta(q^{-1}, \omega)$ to be a function on a measure space $\{\Omega, \mathcal{F}_\Delta, \mathbb{P}_\Delta\}$. In this way ω allows us to index over possible classes of undermodellings and the choice of the measure $\mathbb{P}_\Delta$ allows us to concentrate on some classes of undermodellings more than others.

Readers familiar with probability theory may note that with the normalisation $\mathbb{P}_\Delta(\Omega) = 1$ this problem description amounts to describing the undermodelling as a random variable. It is essential to note however that the description (18) does not remove the deterministic nature of the effects of $G_\Delta(q^{-1}, \omega)$. On the data set we observe we assume ω to be fixed so that G_Δ is a *fixed* stable transfer function. The utility of injecting the description (18) is to simply allow us to consider classes of undermodellings by drawing on the existing work of probability theory without discarding the known effects of undermodelling.

It remains to choose $\mathbb{P}_\Delta$. A choice we have found successful in both simulation and experiment is to first represent $G_\Delta(q^{-1}, \omega)$ by its IIR description $\{\eta_k(\omega)\}$:

$$G_\Delta(q^{-1}, \omega) = \sum_{k=1}^{\infty} \eta_k(\omega) q^{-k} \tag{19}$$

and to set

$$\int_\Omega \eta_k(\omega) d\mathbb{P}_\Delta(\omega) = 0 \tag{20}$$

$$\int_\Omega \eta_k(\omega)\eta_j(\omega) d\mathbb{P}_\Delta(\omega) = \begin{cases} \alpha\lambda^k & ; k = j \\ 0 & ; k \neq j \end{cases} \tag{21}$$

In this way, two parameters α, λ are required to specify $\mathbb{P}_\Delta$. They may be regarded as magnitude and Lipschitz smoothness constraints since (19), (20),(21) are equivalent to the assumptions [15, 16]:

$$\int_\Omega \left| G_\Delta(e^{-j2\pi f}, \omega) \right|^2 d\mathbb{P}_\Delta(\omega) = \frac{\alpha\lambda}{1 - \lambda} \tag{22}$$

$$\int_\Omega \left| G_\Delta(e^{-j2\pi f_1}, \omega) - G_\Delta(e^{-j2\pi f_2}, \omega) \right|^2 d\mathbb{P}_\Delta(\omega) \leq \frac{8\pi^2\alpha\lambda|f_1 - f_2|^2}{|1 - \lambda|^3} \tag{23}$$

Note that (22) and (23) are precisely the sort of assumptions on G_Δ used in much of the afore-mentioned hard bounding work with a crucial difference. In hard bounding work the assumptions are to hold over a class of undermodellings without prejudice for one particular undermodelling over another. In our work the choice of $\mathbb{P}_\Delta$ allows us to add extra structure to the assumptions so that some undermodellings are considered more important (or likely) than others. We assert that this is more realistic in practice. The description

(18) chosen by our approach avoids the sensitivity to prior assumptions that is inherent in hard bounding approaches. This arises because assumptions are not made on specific realisations of undermodellings. Assumptions are made on averages in a class (22),(23). Of course the hard bounding approach is subsumed by the description (18) since $\mathbb{P}_\Delta$ can be chosen to have compact support.

More complicated descriptions that (20),(21) may be made if more prior knowledge can be injected. For example, the impulse response definition

$$\int_\Omega \eta_k(\omega)d\mathbb{P}_\Delta(\omega) = 0 \tag{24}$$

$$\int_\Omega \eta_m(\omega)\eta_n(\omega)d\mathbb{P}_\Delta(\omega) = \begin{cases} \dfrac{\alpha\lambda}{(\rho^2-\lambda)}\left[\dfrac{\rho^{n+m}}{\rho^2} - \dfrac{\rho^n}{\lambda}\left(\dfrac{\lambda}{\rho}\right)^m\right] & ; m \le n \\[3mm] \dfrac{\alpha\lambda}{(\rho^2-\lambda)}\left[\dfrac{\rho^{n+m}}{\rho^2} - \dfrac{\rho^m}{\lambda}\left(\dfrac{\lambda}{\rho}\right)^n\right] & ; m > n \end{cases} \tag{25}$$

corresponds to [5]

$$\int_\Omega \left|G_\Delta(e^{-j2\pi f},\omega)\right|^2 d\mathbb{P}_\Delta(\omega) = \left(\dfrac{\alpha\lambda}{1-\lambda}\right)\dfrac{1}{|1-\rho e^{-j2\pi f}|^2} \tag{26}$$

which specifies more about the high frequency behavior of G_Δ than (22) at the expense of having to supply a value for the extra parameter ρ.

3 Undermodelling Error Quantification

We now show how the description (20),(21) can be used to provide error bounds on estimated frequency responses that account for both noise and undermodelling. We begin by putting (9) into a linear regressor form:

$$y_k = \phi_k^T \theta_0 + \psi_k^T \eta(\omega) + \nu_k(\omega) \tag{27}$$

where

$$\phi_k^T = [B_1(q^{-1})u_k, \cdots, B_p(q^{-1})u_k] \tag{28}$$

$$\psi_k^T = [u_{k-1}, \cdots, u_{k-L}] \tag{29}$$

$$\eta^T(\omega) = [\eta_1(\omega), \cdots, \eta_L(\omega)] \tag{30}$$

Here $B_1(q^{-1}), \cdots, B_p(q^{-1})$ are fixed stable transfer functions forming a basis for approximating $G_T(q^{-1})$. Examples are Laguerre polynomials in q^{-1} studied by Wahlberg and Mäkila [7, 39] and Kautz polynomials in q^{-1} studied by Wahlberg [49]. Notice that by the definition of ψ_k and $\eta(\omega)$ we have truncated the infinite series (19) at L terms. We are justified in this by the assumption of G_Δ being stable. The choice of L is further studied in [3].

Given N point observations of the sequences $\{y_k\}, \{u_k\}$ they may be represented in vector form through (27) as:

$$Y = \Phi\theta_0 + \Psi\eta(\omega) + V(\omega) \tag{31}$$

where

$$Y^T = [y_1, \cdots, y_N] \tag{32}$$
$$\Phi^T = [\phi_1, \cdots, \phi_N] \tag{33}$$
$$\Psi^T = [\psi_1, \cdots, \psi_N] \tag{34}$$
$$V(\omega) = [\nu_1(\omega), \cdots, \nu_N(\omega)] \tag{35}$$

The least squares estimate $\hat{\theta}$ of θ_0 is then given by

$$\hat{\theta} = (\Phi^T\Phi)^{-1}\Phi^TY \triangleq PY \tag{36}$$
$$= \theta_0 + P\Psi\eta(\omega) + PV(\omega) \tag{37}$$

If we assume $\{\nu_k(\omega)\}$ and $\{\eta_k(\omega)\}$ to be independent and have zero ensemble mean then we immediately have

$$\mathcal{E}_\theta\left\{\hat{\theta} - \theta_0\right\} = 0 \tag{38}$$

$$\mathcal{E}_\theta\left\{(\hat{\theta} - \theta_0)(\hat{\theta} - \theta_0)^T\right\} = P(\alpha\Psi C_\lambda\Psi^T + C_\nu)P^T \tag{39}$$

where

$$C_\lambda = \operatorname*{diag}_{1 \leq k \leq L}\{\lambda^k\} \qquad C_\nu = \mathcal{E}_\nu\{VV^T\} \tag{40}$$

Note that we are using the shorthand expectation operator notation

$$\mathcal{E}_x\{\cdot\} = \int_\Omega \cdot d\mathbb{P}_x \tag{41}$$

The first and second order properties of (38),(39) give some indication of how the undermodelling and noise influence the low order estimate $\hat{\theta}$. What we require for robust control system design is not this sort of information, but rather frequency domain information. This can be provided by noting that for some $\omega \in \Omega$ the frequency response of the observed system is

$$G_T(e^{-j2\pi f}) = G(e^{-j2\pi f}, \theta_0) + G_\Delta(e^{-j2\pi f}, \omega) \tag{42}$$
$$= \Lambda(f)\theta_0 + \Pi(f)\eta(\omega) \tag{43}$$

where

$$\Lambda(f) = [B_1(e^{-j2\pi f}), \cdots, B_p(e^{-j2\pi f})] \tag{44}$$
$$\Pi(f) = [e^{-j2\pi f}, e^{-j4\pi f}, \cdots, e^{-jL2\pi f}] \tag{45}$$

Therefore the frequency response estimation error is

$$G_T(e^{-j2\pi f}) - G(e^{-j2\pi f}, \hat{\theta}(\omega)) = \Lambda(f)(\theta_0 - \hat{\theta}(\omega)) + \Pi(f)\eta(\omega) \tag{46}$$

In order to extract both magnitude and phase error information we follow
[50] and rewrite (46) as

$$\tilde{g}(f) = \Gamma(f)(\rho_0 - \hat{\rho}(\omega)) \tag{47}$$

where

$$\tilde{g}(f) = \begin{bmatrix} \mathrm{Re}\left\{ G_T(e^{-j2\pi f}) - G(e^{-j2\pi f}, \hat{\theta}(\omega)) \right\} \\ \mathrm{Im}\left\{ G_T(e^{-j2\pi f}) - G(e^{-j2\pi f}, \hat{\theta}(\omega)) \right\} \end{bmatrix} \tag{48}$$

$$\Gamma(f) = \begin{bmatrix} \mathrm{Re}\left\{\Lambda(f)\right\} \ \mathrm{Re}\left\{\Pi(f)\right\} \\ \mathrm{Im}\left\{\Lambda(f)\right\} \ \mathrm{Im}\left\{\Pi(f)\right\} \end{bmatrix} \tag{49}$$

$$\rho_o^T = \begin{bmatrix} \theta_0^T, 0 \end{bmatrix} \tag{50}$$

$$\hat{\rho}(\omega)^T = \begin{bmatrix} \hat{\theta}(\omega)^T, \eta(\omega)^T \end{bmatrix} \tag{51}$$

Now (47) gives a description of the error at frequency f for a specific undermodelling indexed by a specific $\omega \in \Omega$. We find the average error over the
class of all possible undermodellings (and noise realisations) indexed by ω to
be given by

$$\mathcal{E}_{\tilde{g}}\left\{ \tilde{g}(f) \right\} = 0 \tag{52}$$

$$\mathcal{E}_{\tilde{g}}\left\{ \tilde{g}(f)\tilde{g}(f)^T \right\} = \Gamma(f)\Upsilon\Gamma(f)^T \tag{53}$$

where

$$\Upsilon = \begin{bmatrix} P(\alpha\Psi C_\lambda \Psi^T + C_\nu)P^T & -\alpha P\Psi C_\lambda \\ -\alpha C_\lambda \Psi^T P^T & \alpha C_\lambda \end{bmatrix} \tag{54}$$

and (52),(53) follow from the same assumptions on $\eta(\omega)$ and $V(\omega)$ that gave
(38),(39). The expressions (52),(53) allow us to specify confidence regions for
the frequency response error. For example, if the measures $\mathbb{P}_\nu$ and $\mathbb{P}_\Delta$ are
chosen so as to correspond to Gaussian shaped density functions then

$$\tilde{g}(f)^T \left[\Gamma(f)\Upsilon\Gamma(f)^T \right]^{-1} \tilde{g}(f) \tag{55}$$

will have a χ_2^2 shaped density function. Consequently, we can draw ellipses of
values of $\tilde{g}(f)$ where (55) is a constant chosen to make the integral of the χ_2^2
density function up to this constant close to one. The interior of the ellipse
represents possible magnitude and phase responses for $G_T(e^{-j2\pi f})$ that are
valid save for a few unimportant[4] possibilities for $G_\Delta(e^{-j2\pi f})$. An example
of how this information appears when superimposed on true and estimated
Nyquist plots is shown in Fig. 1.

---

[4] unimportant with respect to the measure $\mathbb{P}_\Delta$

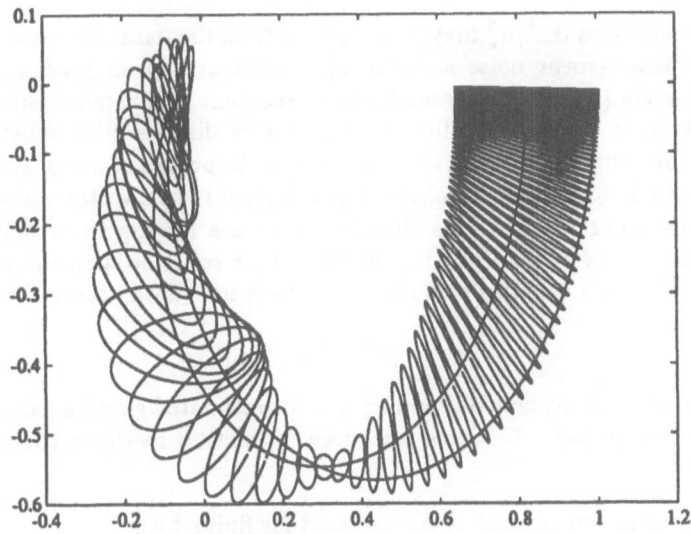

Fig. 1. Stochastic Embedding overbounds together with true and estimated Nyquist Plots. The ellipsoidal error bounds are centred on the estimate

4 Specification of Undermodelling Class Descriptors α,λ

A key question now, of course, is how do we obtain the descriptors α,λ and σ_ν^2 ? In hard bounding work the equivalent descriptors are assumed to be obtained from prior knowledge. This is theoretically possible but seems to imply practical difficulties which are, as yet, unresolved. The major contribution of our work is that we have found that the descriptors α,λ and σ_ν^2 may be estimated from the available data.

Intuitively, the reason this can be done is that the formulations (21),(22) explicitly state assumptions over weighted averages of undermodellings in a class[5]. The whole point of stochastic estimation theory is to relate such ensemble averages to time averages of observed realisations. By examining these relations, parameters such as α,λ can be estimated from observed data. Our formulation for undermodelling, while still describing its effect deterministically thus allows us to use well established theory of random processes to arrive at an approach requiring a minimum of prior knowledge. This leads to a method which is practical rather than theoretical.

---

[5] Hard bounding work implicitly is a statement over averages of undermodellings with $\mathbb{P}_\Delta = \text{constant}$.

5 Estimation of Undermodelling and Noise Description

Here we show how $\alpha, \lambda, \sigma_\nu^2$ may be estimated from the data. This is analogous to how measurement noise variance σ_ν^2 is estimated from prediction error residuals, ε, via (6),(7). Motivated by this precedent we examine estimates of α and λ based upon ε. We first note that ε of dimension N must have a singular distribution of rank $N - p$ since Y is projected onto a space of dimension p in forming $\hat{\theta}$. To obtain a new full rank data vector we represent ε in a new co-ordinate system that forms a basis for the space orthogonal to the columns of Φ. Let R be any matrix whose columns span the subspace orthogonal to the columns of Φ. Then define a new data vector $Z \in \mathbb{R}^{N-p}$ as follows:

$$Z \triangleq R^T \varepsilon \qquad (56)$$

We will now examine how α, λ and σ_ν^2 may be estimated simultaneously from the full rank process Z. There are three estimation methods that we will consider:

1. Estimators constructed to be unbiased for finite data.
2. Estimators constructed to be unbiased for finite data and also locally minimum variance.
3. Maximum Likelihood Estimators.

The proofs of results about these estimators will be omitted in this paper due to space restrictions. Readers interested in the details can find them in [3].

5.1 Unbiased Estimators

We begin by examining unbiased estimation. This is only possible when λ is known and α, σ_ν^2 are the only quantities to be estimated. We formulate these estimators by combining (59) and (56) to obtain (by construction of R)

$$Z = R^T(Y - \hat{Y}) = R^T Y \qquad (57)$$
$$= R^T \Psi \eta(\omega) + R^T \Xi \mu(\omega) + R^T V(\omega) \qquad (58)$$

Note that we have assumed a linear regression form model

$$Y = \Phi\theta_0 + \Psi\eta(\omega) + \Xi\mu(\omega) + V(\omega) \qquad (59)$$
$$\Xi^T = [\gamma_1, \cdots, \gamma_N] \qquad (60)$$
$$\gamma_k^T = [u_{k-L}, u_{k-L-1}, \cdots] \qquad (61)$$
$$\mu(\omega) = [\eta_{L+1}(\omega), \eta_{L+2}(\omega), \cdots] \qquad (62)$$

This is the same as the formulation (31) which appears in [15] save for the infinite matrix Ξ and infinite vector $\mu(\omega)$. These infinite terms represent the tail of $G_\Delta(q^{-1}, \omega)u_k = \sum_{j=1}^{\infty} \eta_j(\omega)u_{k-j}$ that was neglected in (31) and

[15]. We include the terms now in order to formulate an asymptotic analysis Proceeding by using the undermodelling description (20),(21) we have:

$$\mathcal{E}_z\{Z\} = 0 \tag{63}$$

$$\mathcal{E}_z\{ZZ^T\} = \alpha Q + \alpha Q' + \sigma_\nu^2 R^T R \triangleq \Sigma \tag{64}$$

where

$$Q \triangleq R^T \Psi C_\lambda \Psi^T R \tag{65}$$

$$Q' \triangleq \lambda^L R^T \Xi C_\mu \Xi^T R \tag{66}$$

$$C_\mu \triangleq \operatorname*{diag}_{1 \le k \le \infty} \{\lambda^k\} \tag{67}$$

The approach to estimating α, σ_ν^2 is to now select two quadratic forms of the data vector Z, set each quadratic form to its expected value, and solve the resultant set of simultaneous equations for α and σ_ν^2. By construction this gives unbiased estimators. This form of estimation is known in the statistics literature as Henderson's method [45, 11]. To be explicit, take arbitrary symmetric matrices $A, B \in \mathbb{R}^{(N-p) \times (N-p)}, A \ne B$. It immediately follows by the definition of Σ:

$$\mathcal{E}_z\{Z^T A Z\} = \alpha \operatorname{Tr}\{AQ\} + \sigma_\nu^2 \operatorname{Tr}\{AR^T R\} \tag{68}$$

$$\mathcal{E}_z\{Z^T B Z\} = \alpha \operatorname{Tr}\{BQ\} + \sigma_\nu^2 \operatorname{Tr}\{BR^T R\} \tag{69}$$

Note for simplicity we have assumed L is chosen large enough that $Q' \approx 0$. Solving (68) and (69) for α and σ_ν^2 with the expected value operators removed gives the estimators

$$\hat{\alpha} = \frac{(Z^T B Z) \operatorname{Tr}\{AR^T R\} - (Z^T A Z) \operatorname{Tr}\{BR^T R\}}{\operatorname{Tr}\{AR^T R\} \operatorname{Tr}\{BQ\} - \operatorname{Tr}\{BR^T R\} \operatorname{Tr}\{AQ\}} \tag{70}$$

$$\hat{\sigma}_\nu^2 = \frac{(Z^T A Z) \operatorname{Tr}\{BQ\} - (Z^T B Z) \operatorname{Tr}\{AQ\}}{\operatorname{Tr}\{AR^T R\} \operatorname{Tr}\{BQ\} - \operatorname{Tr}\{BR^T R\} \operatorname{Tr}\{AQ\}} \tag{71}$$

There are many possible choices for A and B. All choices give unbiased estimators by construction. Not all give consistent estimators [3]. The intuitive road taken in the selection of appropriate A and B is to first remove the influence of θ and $\eta(\omega)$ from the data by projecting it out:

$$\hat{\sigma}_\nu^2 = \frac{Y^T P_\Gamma Y}{N - p - L} \tag{72}$$

$$P_\Gamma = I - \Gamma (\Gamma^T \Gamma)^{-1} \Gamma^T \tag{73}$$

$$\Gamma^T = [\Phi^T, \Psi^T] \tag{74}$$

We then construct the estimator for α by first attempting to 'pre-whiten' the component of Z due to $\eta(\omega)$ by multiplication with $Q^\dagger$ where $\dagger$ denotes pseudo-inverse. This leads to the estimator

$$\hat{\alpha} = \frac{Z^T Q^\dagger Z - \hat{\sigma}_\nu^2 \, \mathrm{Tr}\left\{Q^\dagger R^T R\right\}}{L} \qquad (75)$$

These estimators constructed intuitively are shown in [3] to correspond to choices

$$A = (P_\Phi R)^\dagger P_\Gamma (R^T P_\Phi)^\dagger \qquad B = Q^\dagger \qquad (76)$$

and hence the estimators are unbiased. To show they are consistent is very difficult unless we introduce the following assumptions[6]:

1. The measure $\mathbb{P}_\Delta$ is chosen to induce a Gaussian density function.
2. The sequence $\{z_k\}$ is an uncorrelated sequence. This can be generally achieved by using the Gram-Schmidt procedure to form the columns of Φ, Ψ and Ξ as an orthogonal set. We assume the columns are normalised at this time to $\|\Phi\| = \|\Psi\| = \|\Xi\| = N\sigma_u^2$ where σ_u^2 is some constant representing the power in the input signal $\{u_k\}$.

Under these assumptions the obvious choice for R is Ψ which gives $Q' = 0$ by orthogonality. Furthermore, we get the following consistency result:

Theorem 2. *For the case of orthogonal regressors, if we constrain the rate of growth of L as a function of N by:*

$$L(N) = \frac{\ln N}{\rho |\ln \lambda|} \qquad \rho > 1 \qquad (77)$$

then

$$\mathrm{Var}\left\{\sqrt{\frac{\ln N}{|\ln \lambda|}} \frac{\hat{\alpha}}{\alpha}\right\} = 2\rho + \mathcal{O}\left(N^{\frac{1}{\rho}-1} \ln N\right) \qquad \text{as } N \to \infty \qquad (78)$$

$$\mathrm{Var}\left\{\sqrt{N} \frac{\hat{\sigma}_\nu^2}{\sigma_\nu^2}\right\} \longrightarrow 2 \qquad \text{as } N \to \infty \qquad (79)$$

and the result for $\mathrm{Var}\left\{\hat{\sigma}_\nu^2\right\}$ *holds for the non-orthogonal regressor case.*

Proof. See [3]

The unbiased estimates (72),(74) are thus consistent. In fact they are asymptotically efficient:

---

[6] These assumptions are only introduced to make analysis tractable. They are not required to actually use the estimators

Theorem 3. *For the orthogonal regressor case, if we define*

$$\hat{\xi} = \left[\sqrt{\frac{\ln N}{\ln |\lambda|}} \frac{\hat{\alpha}}{\alpha}, \sqrt{L(N)} \frac{\hat{\sigma}_\nu^2}{\sigma_\nu^2} \right] \tag{80}$$

where $L(N) > \frac{\ln N}{|\ln \lambda|}$ $\forall N$ and $\lim_{N \to \infty} \frac{\ln N}{|\ln \lambda| L(N)} = \kappa$, then the asymptotic Cramér-Rao lower bound on the variance of unbiased estimates of ξ based upon the data $\{z_1, \cdots, z_{L(N)}\}$ is:

$$Cov\{\hat{\xi}\} \geq 2 \begin{bmatrix} 1 & 0 \\ 0 & \frac{1}{1-\kappa} \end{bmatrix} \quad as \ N \to \infty \tag{81}$$

Proof. See [3].

5.2 Locally Minimum Variance Estimators

In [9] C.R. Rao formulated a method for choosing A and B matrices to produce an estimator of the variance components. His method applied to our estimation problem involves estimating a linear functional of α and σ_ν^2

$$f(\alpha, \sigma_\nu^2) = f_1 \alpha + f_2 \sigma_\nu^2 \quad f_1, f_2 \in \mathbb{R} \tag{82}$$

by a quadratic form

$$\hat{f} = Y^T A Y \tag{83}$$

for some symmetric matrix $A \in \mathbb{R}^{N \times N}$. We would like to make the estimator invariant with respect to θ. This can be satisfied by

$$A\Phi = 0 \tag{84}$$

Furthermore, given (84) and assuming L large enough that $Q' \approx 0$

$$\mathcal{E}_y \{Y^T A Y\} = \text{Tr} \{A (\alpha \Psi C_\lambda \Psi^T + \sigma_\nu^2 I)\} \tag{85}$$

so for unbiasedness we need

$$\text{Tr} \{A\Psi C_\lambda \Psi^T\} = f_1 \quad \text{Tr} \{A\} = f_2 \tag{86}$$

Finally, if we write

$$\Psi \eta(\omega) + V(\omega) = \Psi C_\lambda^{1/2} \epsilon(\omega) + V(\omega) \tag{87}$$

where

$$\mathcal{E}_\epsilon \{\epsilon(\omega)\epsilon(\omega)^T\} = \alpha I \tag{88}$$

then the natural estimator for f would be:

$$\hat{f}(\omega) = \xi^T(\omega) \Delta \xi(\omega) \tag{89}$$

where

$$\xi^T(\omega) = [\epsilon^T(\omega), V^T(\omega)] \tag{90}$$

$$\Delta = \begin{bmatrix} \frac{f_1}{L} I & 0 \\ 0 & \frac{f_2}{N} I \end{bmatrix} \tag{91}$$

Unfortunately $\xi(\omega)$ is not available to us, but we can make the estimator (83) close to the ideal one (89) by minimising

$$\|U^T A U - \Delta\| \quad U \triangleq \left[\Psi C_\lambda^{1/2}, I\right] \tag{92}$$

In [9, 10] Rao shows that minimising (92) subject to (84) and (86) is equivalent to minimizing

$$\text{Tr}\left\{AUU^T AUU^T\right\} \tag{93}$$

subject to (84) and (86) and the solution is given by:

$$A = \Lambda^\dagger \left(\mu_1 \Psi C_\lambda \Psi^T + \mu_2 I\right) \Lambda^\dagger \tag{94}$$

where μ_1 and μ_2 are given by:

$$\mu_1 = \frac{\kappa_4 f_1 - \kappa_2 f_2}{\kappa_1 \kappa_4 - \kappa_2 \kappa_3} \quad \mu_2 = \frac{\kappa_1 f_2 - \kappa_3 f_1}{\kappa_1 \kappa_4 - \kappa_2 \kappa_3} \tag{95}$$

with

$$\kappa_1 \triangleq \text{Tr}\left\{\Lambda^\dagger \Psi C_\lambda \Psi^T \Lambda^\dagger \Psi C_\lambda \Psi^T\right\} \quad \kappa_2 \triangleq \text{Tr}\left\{\Lambda^\dagger \Psi C_\lambda \Psi^T \Lambda^\dagger\right\} \tag{96}$$

$$\kappa_3 \triangleq \text{Tr}\left\{\Lambda^\dagger \Lambda^\dagger \Psi C_\lambda \Psi^T\right\} \quad \kappa_4 \triangleq \text{Tr}\left\{\Lambda^\dagger \Lambda^\dagger\right\} \tag{97}$$

and

$$\Lambda \triangleq P_\phi \left(\Psi C_\lambda \Psi^T + I\right) P_\phi \tag{98}$$

$$P_\phi \triangleq I - \Phi \left(\Phi^T \Phi\right)^{-1} \Phi^T \tag{99}$$

If we want to estimate α we set $f_1 = 1, f_2 = 0$ and if we want to estimate σ_ν^2 we set $f_1 = 0, f_2 = 1$. Note also that when $\mathbb{P}_\Delta$ induces a Gaussian density we have [3]

$$\text{Var}\left\{Y^T AY\right\} = 2 \text{Tr}\left\{AUSU^T AUS^T\right\} \tag{100}$$

where

$$S \triangleq \begin{bmatrix} \alpha I & 0 \\ 0 & \sigma_\nu^2 I \end{bmatrix} \tag{101}$$

so that the estimator (83) is the minimum variance unbiased estimator for α and σ_ν^2 when they are equal and. by continuity, approximately minimum variance for $\alpha \approx \sigma_\nu^2$.

5.3 Maximum Likelihood Estimators

Finally we examine the method of maximum likelihood for estimating the complete triplet $\alpha, \lambda, \sigma_\nu^2$. To use this method we have to specify $\mathbb{P}_\Delta$ which we again do by assuming it induces a Gaussian density function. If we define β as

$$\beta^T \triangleq [\alpha, \lambda, \sigma_\nu^2] \qquad (102)$$

then the negative log-likelihood function for the data conditional upon β is:

$$\ell(Z \mid \beta) = \ln \det \Sigma(\beta) + Z^T \Sigma(\beta)^{-1} Z \qquad (103)$$

where again in the formulation of the estimator (but not in the subsequent asymptotic analysis) we assume L is large enough for $Q' \approx 0$ so

$$\Sigma(\beta) = \alpha R^T \Psi C_\lambda \Psi^T R + \sigma_\nu^2 R^T R \qquad (104)$$

We define the maximum likelihood estimate $\hat{\beta}_N$ of β based on the $N - p$ data points in Z as

$$\hat{\beta}_N = \arg \min_{\beta \in \mathcal{D}} \{\ell(Z \mid \beta)\} \qquad (105)$$

where $\mathcal{D} \subset \mathbb{R}^3$ is some compact connected set such that $\Sigma(\beta) > 0 \ \forall \beta \in \mathcal{D}$. The definition of $\Sigma(\beta)$ given in (104), the continuity of $\ln(\cdot)$ and the assumption of $\Sigma(\beta) > 0$ imply that $\ell(Z \mid \beta)$ is continuous on the compact domain $\mathcal{D}$ and hence is bounded and possesses a minimum on $\mathcal{D}$. If this minimum is in the interior of $\mathcal{D}$ then it is characterised as the solution of

$$\frac{\partial \ell(Z \mid \beta)}{\partial \beta} = 0 \qquad (106)$$

If λ is known and only α, σ_ν^2 are to be estimated then this leads to the equations

$$\text{Tr} \left\{ \Sigma(\beta) R^T \Psi C_A \Psi^T R \right\} = Z^T \Sigma(\beta)^{-1} R^T \Psi C_\lambda \Psi^T R \Sigma(\beta)^{-1} Z \qquad (107)$$

$$\text{Tr} \left\{ \Sigma(\beta) R^T R \right\} = Z^T \Sigma(\beta)^{-1} R^T R \Sigma(\beta)^{-1} Z \qquad (108)$$

Solving (107) and (108) for α and σ_ν^2 will then give their maximum likelihood estimates. In general this will require a numerical search. However, in [21] Jensen notes that if a symmetric product can be defined on $\{\Sigma(\beta), \beta \in \mathcal{D}\}$ in such a way as to make it a Jordan Algebra then we can just as well substitute $\Sigma(\beta) = I$ into (107) and (108). In this case we obtain an unbiased estimator of the form (70), (71) with $A = R^T \Psi C_\lambda \Psi^T R$, $B = R^T R$. However, the requirement of a Jordan Algebra structure for possible Σ is a harsh one. It can only be satisfied in our case for $R^T \Psi C_\lambda \Psi^T R$ idempotent[7]. This will never by the case. Furthermore, [44] notes that achieving the Jordan Algebra structure is a necessary and sufficient condition for the maximum likelihood

---

[7] centro-cyclosymmetric actually.

estimate to be unbiased and exist in closed form. We must therefore resign
ourselves to finding biased estimates by numerical means.

Analysing further properties of the maximum likelihood estimate proves
to be difficult since standard MLE theory [1] cannot be used because $\{z_k\}$
is not iid. Ljung's prediction error results [26, 28] cannot be used since the
predictor in our case does not depend upon β. Finally, more recent general
results on the asymptotics of MLE's [46, 33, 2] cannot be used either because
we do not satisfy required regularity conditions. To proceed, we introduce
the following assumptions

Assumption Set A

1. $\eta \sim \mathcal{N}(0, \alpha C_\lambda)$, $C_\lambda = \text{diag}_{1 \leq k \leq L} \{\lambda^k\}$
2. The true parameter vector $\beta_0 \in \text{int} \{\mathcal{D}\}$
3. The input $\{u_k\}$ is persistently exciting of all orders [8].
4. L is made a function of N such that $L(N) = \frac{\ln N}{\rho |\ln \lambda_*|}$ for some $\rho < 1$
   where $\lambda_* = \inf_{\lambda \in \mathcal{D}} \lambda$.

Then in the case of orthogonal regressors we have the following asymptotic
result.

Theorem 4. *For the orthogonal regressor case, Assumption set A holding
and the MLE $\hat{\beta}_N$ defined by (105)*

$$\hat{\beta}_N \xrightarrow{a.s.} \beta_0 \quad as \ N \to \infty \tag{109}$$

Proof. See [3]

This is our main result and proves that it is possible to estimate α, λ and σ_ν^2
simultaneously from the available data. Of course, it is an infinite data result
where we have argued we are only interested in the finite data case. However,
the utility of Theorem 4 is in proving that our philosophy of estimating
embedding parameters for noise *and* undermodelling is sound by rigorously
showing that sensible asymptotic properties hold. To try to give some idea
of the finite data properties of $\hat{\theta}_N$ we examine the rate of convergence in
Theorem 4.

Theorem 5. *For the orthogonal regressor case, Assumption set A holding,
the estimates $\hat{\alpha}, \hat{\lambda}, \hat{\sigma}_\nu^2$ found by Maximum Likelihood via (105) and $\hat{\xi}_N$ defined
by*

$$\hat{\xi}_N = \left[\sqrt{\frac{\ln N}{|\ln \lambda|}} \frac{\hat{\alpha}}{\alpha}, \left(\frac{\ln N}{|\ln \lambda|} \right)^{3/2} \frac{\hat{\lambda}}{\lambda}, \sqrt{L(N)} \frac{\hat{\sigma}_\nu^2}{\sigma_\nu^2} \right] \tag{110}$$

we have

$$M_N^{-1/2}(\hat{\xi}_N - \xi_0) \xrightarrow{D} \mathcal{N}(0, I) \quad as \ N \to \infty \tag{111}$$

---

[8] This is required so that an orthogonal sequence $\{\tilde{z}_1, \cdots, \tilde{z}_L, \cdots\}$ extending to ∞
can be generated.

where

$$\lim_{N\to\infty} M_N = 2 \begin{bmatrix} 4 & -6 & 0 \\ -6 & 12 & 0 \\ 0 & 0 & 1 \end{bmatrix} \tag{112}$$

Proof. See [3].

This indicates that estimates of α,λ and σ_ν^2 converge like $1/\ln N, 1/\ln^3 N$ and $1/L(N)$ respectively.

6 Simulation Example

A simulation study was conducted to examine the use of the stochastic embedding paradigm in quantifying the estimation errors when rational, fixed denominator models were fitted to the data. In this case the following continuous time system, sampled with period 1 second, was simulated :

$$G(s) = \frac{e^{-2s}}{(10s + 1)(s + 1)}$$

The test input sequence $\{u_k\}$ was a 0.02 Hz fundamental square wave. The output of this system was corrupted with a noise sequence $\{\nu_k\}$ distributed as $\nu_k \sim \mathcal{N}(0, 0.005)$. One hundred and fifty samples of data were collected, the first one hundred were used to get rid of initial condition effects in the simulated plant and regressor filters, and the last fifty were used for least squares model fitting. A 2nd order model of the form:

$$y_k = \left(\frac{\theta_1 q^{-1}}{(1 + \xi q^{-1})} + \frac{\theta_2 q^{-1}(1 - (2 + \xi)q^{-1})}{(1 + \xi q^{-1})^2} \right) u_k$$

was fitted to the data using least squares. Here $\xi = -0.8$ was chosen (between the true system poles). Note that the unusual regressors are motivated by Laguerre polynomials [39]. The resulting least squares estimates were:

$$\hat{\theta}_1 = 0.0653 \quad \hat{\theta}_2 = -0.1013$$

The response of the estimated model $G(q^{-1}, \hat{\theta}_N)$ to the observed input is shown dashed against the noise corrupted true response in the upper left of Fig. 2. The true (full line) and estimated (dashed line) frequency responses are shown in the upper right corner of Fig. 2.

Next, the parameters of the distributions of the measurement noise and undermodelling were estimated from the data. The stochastic embedding chosen for the undermodelling and noise was

$$\nu_k \sim \mathcal{N}(0, \sigma_\nu^2) \tag{113}$$

$$\eta \sim \mathcal{N}(0, C_\eta) \tag{114}$$

$$C_\eta = \operatorname*{diag}_{1 \leq k \leq L} \{\alpha \lambda^k\} \tag{115}$$

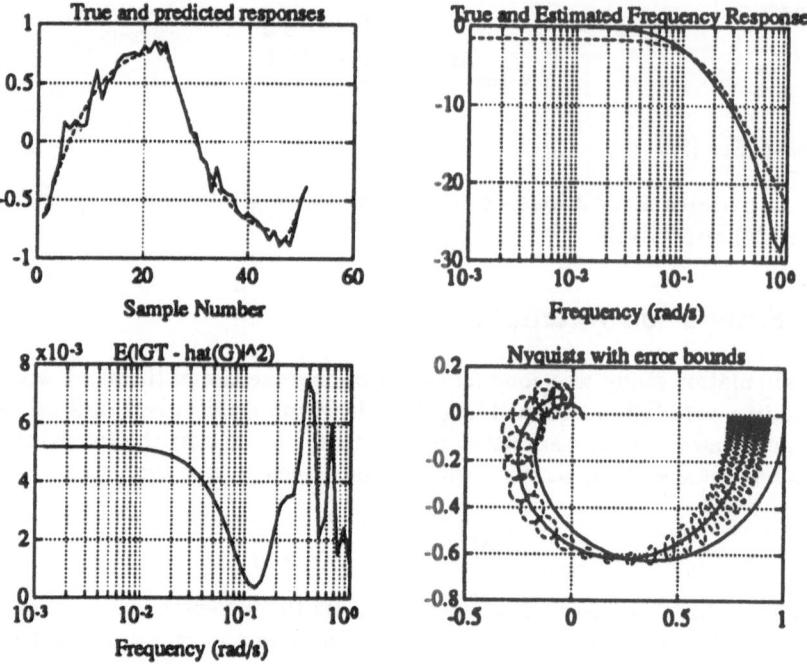

Fig. 2. Results of fitting a 2nd order, rational Laguerre model to the data

$\dim\{\eta\} = L = 40$ was chosen for the FIR model G_Δ. Because of the Gaussian assumptions for the undermodelling, and the Gaussian distribution for the measurement noise, the log likelihood function for the observed data is as given in (103). This was maximized to find the estimates:

$$\hat{\alpha} = 0.217 \quad \hat{\lambda} = 0.488 \quad \hat{\sigma}^2_\nu = 0.0061$$

Substituting these estimates in (113) and (115) then gives estimates of C_η and $C_\nu = \sigma^2_\nu I$. These were then used to derive error bounds for the frequency response estimation error. Error bounds on magnitude estimation were calculated via (55) replacing C_η and C_ν by their estimates and are shown in the lower left of Fig. 2. Uncertainty ellipsoids in the complex plane were calculated similarly and are shown superimposed on the true and estimated Nyquist diagrams in the lower right of Fig. 2. The uncertainty ellipsoids are one standard deviation ellipses; that is, the locus of points satisfying

$$\tilde{g}(f)^T \left[\Gamma(f) \Upsilon \Gamma(f)^T \right]^{-1} \tilde{g}(f) = 2 \tag{116}$$

As can be seen, the error bounds give a very good indication of the true modeling errors in the frequency domain. Note in particular from the lower left diagram in Fig. 2 that the estimated error in the estimation of the system magnitude response is small at the fundamental frequency ($0.02 \times 2\pi$ rad per sec) of the input signal square wave and at the odd harmonics. This concurs with the analysis of [8].

7 Conclusion

In this paper we have examined the question of providing estimated frequency responses from finite noisy data records together with error bounds. We believe that this problem is of fundamental importance for robust control design because of the common need to provide stable closed loop control on the basis of minimal information. We observed that such problems always involve some degree of undermodelling. As a result the estimated frequency responses should also contain bounds accounting for both noise and undermodelling if it is to be useful in Robust Control. The mainstream approach to the problem is to recognise that the effects of undermodelling are deterministic and therefore abandon existing stochastic estimation theory in favour of a simpler theory which forces noise descriptions into the same deterministic setting as undermodelling induced disturbances. The approach we have outlined here still recognises the deterministic nature of undermodelling. However, we choose to proceed by extending stochastic estimation theory to provide a satisfactory description of undermodelling. This entails many advantages both practical and philosophical. The major advantage is that it drastically reduces the amount of prior information that must be injected into the problem. *This is because the parameters describing the assumed class of undermodellings affecting the data may in fact be estimated from the data.* This is only possible because the extension of stochastic theory to handle undermodelling explicitly characterises the ensemble of possible undermodellings; simpler hard bounding work characterises them implicitly. We believe this renders the stochastic approach we describe of both practical and theoretical importance.

Of course, the frequency domain bounds resulting from a stochastic approach are confidence regions rather than hard bounds. However, we argue that this is appropriate since prior assumptions can never be specified with absolute certainty. Indeed, we also suggest that real world control problems are nearly always solved by aiming for high performance in the belief that the set of pathological conditions associated with extreme bounds will rarely, if ever occur. Therefore, control engineers always work with a tradeoff of uncertainty versus performance. Consequently, while the estimation community has a mandate to provide transfer function estimates together with error bounds, the robust control community has a responsibility to accept this information in a realistic format which almost certainly precludes bounds which are absolute.

Appendix A. Proof of Lemma 1

Proof.

$$|y_k - \phi_k^T \theta| < |\nu_k| \tag{117}$$

so feasible $\theta \in \mathbb{R}^p$ satisfy

$$\phi_k^T \theta \in [y_k - |\nu_k|, y_k + |\nu_k|] \tag{118}$$

For the purposes of forming Θ, the best possible realizations are $y_k = \phi_k^T \pm \delta$. Therefore, the best bounds obtainable for one sample are given by the hyperplane delineated region in θ space:

$$\Theta_k \triangleq \{\theta \in \mathbb{R}^p : \phi_k^T \theta \in [\phi_k^T \theta_0 + \delta - |\nu_k|, \phi_k^T \theta_0 - \delta + |\nu_k|]\} \tag{119}$$

But our conservative assumption is $|\nu_k| < \mu$ to give

$$\Theta_k \triangleq \{\theta \in \mathbb{R}^p : \phi_k^T (\theta - \theta_0) \in [\delta - \mu, \mu - \delta]\} \tag{120}$$

So $\tilde{\theta} \triangleq \theta - \theta_0$ lies between a pair of hyperplanes perpendicular to ϕ_k. Also, if β lies on one hyperplane, then $-\beta$ lies on the other, so the distance between the hyperplanes in the direction β is $2\|\beta\|$. But by the Cauchy-Schwartz inequality:

$$|\mu - \delta| = |\phi_k^T \beta| < \|\phi_k\|\|\beta\| \quad \text{for some } \beta \in \mathbb{R}^p \tag{121}$$

so a lower bound on the distance between the two hyperplanes forming Θ_k is

$$2\|\beta\| \geq \frac{2|\mu - \delta|}{\|\phi_k\|} \geq \frac{2(\mu - \delta)}{\sigma_\phi} \tag{122}$$

But this lower bound is valid $\forall k \in \mathbb{N}$ and

$$\Theta_N = \bigcap_{k=1}^N \Theta_k \tag{123}$$

so Θ_N is underbounded by a sphere of diameter d

$$d \geq \frac{2(\mu - \delta)}{\sigma_\phi} \tag{124}$$

regardless of how large we make N. $\square$

References

1. A. Wald, *Asymptotic properties of the maximum likelihood estimate of an unknown parameter of a discrete stochastic process*, The Annals of Mathematical Statistics, 19 (1947), pp. 40–46.

2. B. Hoadley, *Asymptotic properties of maximum likelihood estimators for the independent not identically distributed case*, The Annals of Mathematical Statistics, 42 (1971), pp. 1977–1991.

3. B.M. Ninness, *Estimation of variance components in linear models with applications to stochastic embedding*, Technical Report EE9226, Department of Electrical and Computer Engineering, University of Newcastle, AUSTRALIA, (1992).

4. ——, *Stochastic and deterministic approaches to estimation in H_∞*, Technical Report EE9308, Department of Electrical Engineering, University of Newcastle, Australia, (1993).

5. ——, *Stochastic embedding of undermodelling descriptions:the frequency domain non-stationary case*, Technical Report EE9309, Department of Electrical Engineering, University of Newcastle, Australia, (1993).

6. B.M. Ninness and G.C. Goodwin, *Rapprochment between bounded error and stochastic estimation theory*, International Journal of Adaptive Control and Signal Processing, To Appear (1993).

7. B. Wahlberg, *System identification using laguerre models*, IEEE Transactions on Automatic Control, AC-36 No.5 (1991).

8. B. Wahlberg and L. Ljung, *Design variables for bias distribution in transfer function estimation*, IEEE Trans.Autom.Control,, AC-31 (1986), pp. 134–144.

9. C.R. Rao, *Estimation of variance and covariance components*, Journal of Multivariable Analysis, 1 (1972), pp. 257–275.

10. ——, *Estimation of variance and covariance components in linear models*, Journal of the American Statistical Association, 67 (1972), pp. 112–115.

11. D.A. Harville, *Quadratic unbiased estimation of variance components for the one-way classification*, Biometrika, 56 (1969), pp. 313–326.

12. D. Tse, M. Dahleh, and J. Tsitsiklis, *Optimal and robust identification in the ℓ_1 norm*, Proceedings of the American Control Conference, (1991).

13. D.W. De Vries and P.M.J. Van den Hof, *Quantification of model uncertainty from data:input design, interpolation, and connection with robust control design specifications*, Proceedings of the American Control Conference, Chicago, 4 (1992), pp. 3170–3175.

14. E. Fogel and Y.F. Huang, *On the value of information in system identification-bounded noise case*, Automatica, 18 (1982), pp. 229–238.

15. G.C. Goodwin, M. Gevers, and B.M. Ninness, *Quantifying the error in estimated transfer functions with application to model order selection*, IEEE Transactions on Automatic Control., 37 (1992).

16. G.C. Goodwin and M. Salgado, *A stochastic embedding approach for quantifying uncertainty in the estimation of restricted complexity models*, International Journal of Adaptive Control and Signal Processing, 3(4) (1989), pp. 333–356.

17. G. Gu and P. Khargonekar, *A class of algorithms for identification in H_∞*, Automatica, 28 (1992), pp. 299–312.

18. ——, *Linear and nonlinear algorithms for identification in H_∞ with error bounds*, IEEE Transactions on Automatic Control, 37 (July 1992), pp. 953–963.

19. A. Helmicki, C. Jacobson, and C. Nett, *Control oriented system identification:A worst case/deterministic approach in H_∞*, IEEE Transactions on Automatic Control, 36 (October 1991), pp. 1163–1176.

20. C. Jacobson and C. Nett, *Worst case system identification in ℓ_1:optimal algorithms and error bounds*, Proceedings of the American Control Conference, (1991), pp. 3152–3157.

21. S. Jensen, *Covariance hypotheses with are linear in both the covariance and inverse covariance*, Annals of Statistics, 16 (1988), pp. 302–322.

22. J.P. Norton, *Identification and application of bounded parameter models*, Automatica, 23 (1987), pp. 497–507.

23. ——, *Identification of parameter bounds of armax models from records with bounded noises*, International Journal of Control, 42 (1987), pp. 375–390.

24. W. Larimore, *Accuracy confidence bands including the bias of model underfitting*, Paper presented at teh Workshop on the Modeling of Uncertainty in Control Systems, University of California, Santa Barbara, (June 1992).

25. L. Ljung, *System Identification: Theory for the User*, Prentice-Hall, Inc., New Jersey, 1987.

26. L. Ljung, *Convergence analysis of parametric identification methods*, IEEE Transactions on Automatic Control, AC-23 No.5 (1978), pp. 770–783.

27. ——, *Asymptotic variance expressions for identified black-box transfer function models*, IEEE Transactions on Automatic Control, AC-30 (1985), pp. 834–844.

28. L. Ljung and P.E. Caines, *Asymptotic normality of prediction error estimators for approximate system models*, Stochastics, 3 (1979), pp. 29–46.

29. L. Ljung and B. Wahlberg, *Asymptotic properties of the least squares method for estimating transfer functions and disturbance spectra*, Advances in Applied Probability, 24 (1992), pp. 412–440.

30. L. Ljung and Z.D. Yuan, *Asymptotic properties of black-box identification of transfer functions*, IEEE Transactions on Automatic Control, AC-30 (1985), pp. 514–530.

31. R. L. Maire, L. Valavani, M. Athans, and G. Stein, *A frequency domain estimator for use in adaptive control systems*, Automatica, 27 (1991), pp. 23–38.

32. P. Mäkilä and J. Partington, *Robust approximation and identification in H_∞*, Proceedings of the American Control Conference, (1991), pp. 70–76.

33. M. Crowder, *Maximum likelihood estimation for dependent observations*, Journal of the Royal Statistical Society, Series B, 38 (1987), pp. 45–53.

34. M.E. Salgado, *Issues in Robust Identification*, PhD thesis, University of Newcastle, 1989.

35. M. Milanese and G. Belforte, *Estimations theory and uncertainty intervals evaluation in the presence of unknown but bounded errors:linear families of models and estimators*, IEEE Transactions on Automatic Control, AC-27 (1982), pp. 408–414.

36. P. Parker and R. Bitmead, *Adaptive frequency response estimation*, Proceedings of the Conference on Decision and Control, (1987), pp. 348–353.

37. J. Partington, *Robust identification and interpolation in H_∞*, International Journal of Control, 54 (1991), pp. 1281–1290.

38. P.J. Parker and R.R. Bitmead, *Adaptive frequency response identification*, Proceedings of 26th Conference on Decision and Control, (1987), pp. 348–353.
39. P.M. Mäkila, *Approximation of stable systems by laguerre filters*, Automatica, 26 (1990), pp. 333–345.
40. R.C. Younce and C.E. Rohrs, *Identification with parameteric and non-parametric uncertainty*, IEEE Transactions on Automatic Control, 37 (1992), pp. 715–728.
41. R.L. Kosut, *Adaptive control via parameter set estimation*, International Journal of Adaptive Control and Signal Processing, 2 No.4 (1988), pp. 371–400.
42. ———, *Adaptive robust control via transfer function uncertainty estimation*, Proceedings ACC, Atlanta, (1988).
43. R.L. Kosut, M.K. Lau, and S.P. Boyd, *Set membership identification of systems with parametric and non-parametric uncertainty*, IEEE Transactions on Automatic Control, 37 (1992), pp. 929–941.
44. S. Morgera, *The role of abstract algebra in structured estimation theory*, IEEE Transactions on Information Theory, 38 (1992), pp. 1053–1065.
45. S.R. Searle, *Another look at henderson's methods of estimating variance components*, Biometrics, (1968), pp. 749–778.
46. T.J. Sweeting, *Uniform asymptotic normality of the maximum likelihood estimator*, The Annals of Statistics, 8 (1980), pp. 1375–1381.
47. J. Traub, G. Wasilkowski, and H. Wozniakowski, *Information-Based Complexity*, Academic Press, New York, 1988.
48. T. Söderström and P. Stoica, *System Identification*, Prentice Hall, New York, 1989.
49. B. Wahlberg, *System identification using laguerre models*, IEEE Transactions on Automatic Control, 36 (1991), pp. 551–562.
50. B. Wahlberg and L. Ljung, *Hard frequency-domain model error bounds from least-squares like identification techniques*, IEEE Transactions on Automatic Control, 37 (1992), pp. 900–912.
51. W. Rudin, *Real and Complex Analysis*, McGraw Hill, New York, 1966.

Non-Vanishing Model Errors

*Håkan Hjalmarsson**

Department of Electrical Engineering, Linköping University
S-581 83 Linköping, Sweden

1 Introduction

In system identification one often assumes an idealized "smooth" linear time-invariant world. A consequence of this assumption is that a model of arbitrary accuracy can be obtained provided a sufficiently large data set is available. In practice, however, one often encounters the problem that models based on different batches of data lie outside each others confidence regions. There may be many reasons behind this phenomenon, for example unmodeled dynamics or non-linear effects. Here we consider the case when the cause is time-variations in the system dynamics. It is shown that model validation based on simple correlation tests often fail to detect that a time-invariant model is incorrect. A new correlation test that is especially tailored for this situation is introduced. It is based on the assumption that the true system varies around a nominal system.

The system is assumed to be time-varying but in a stationary way as opposed to the conventional random walk model. The motivation for this is that such a system essentially is time-invariant but with small but arbitrary fast fluctuations of the dynamics and hence likely from a practical point of view to be treated as a time-invariant system. It is, however, important to assess the "size" of the time-varying term if, for example, the model is to be used for control design.

A framework for such time-varying dynamics is defined in Sect. 2. In Sect. 3 we analyze what the result of conventional (time-invariant) system identification techniques will be. The main section, Sect. 4, is devoted to testing whether a time-varying term is present or not in the system dynamics.

---

* This work was made when the author visited the Chemical Engineering Department at the California Institute of Technology during the 1991-92 school year and was partially supported by the Swedish Institute and the Blanceflor Boncompagni-Ludovisi Foundation.

Once a time-varying term has been detected one would of course like to estimate this term. This has been treated in [1]. There a method, consistent with the framework in this paper, is developed that provides the answer to simple but important questions such as in which frequency range the variations are largest and whether it is the phase or the amplitude that varies at a particular frequency.

Since the following sections are rather formal we ask the reader to keep in mind that the purpose of this exercise really is not to detect time-variations. Instead it is an attempt to obtain a methodology to detect whether the "assumption" that with infinitely many data it is possible to get a perfect model is valid or not.

2 A Stochastic Framework

In this section we define a model structure for time-varying systems and examine the spectral properties of such models.

2.1 System Description

Throughout the paper a transfer function with impulse response coefficients $\{g(k)\}$ will be said to be stable if $\sum_{k=0}^{\infty} |g(k)| < \infty$.

Unless otherwise specified, the data will be assumed to be generated according to the following definition throughout the paper.

Definition 1. The system $\tilde{S}$ with scalar input and output is given by

$$y(t) = G_0(q)u(t) + \tilde{\theta}^T(t)F_0(q)u(t) + H_0(q)e_0(t) \tag{1}$$

for some stable transfer functions $G_0(q)$ and $H_0(q)$ where $H_0(q)$ is monic, *i.e.* its first impulse response coefficient is one, and inversely stable. In (1) $\{e_0(t)\}$ is a sequence of independent random variables with zero mean, variance σ_0^2 and bounded moments of order $\gamma > 4$.

The transfer function $F_0(q)$ is vector-valued with dimension p and $\{\tilde{\theta}(t)\}$ is a p-dimensional vector-valued stationary stochastic process generated by $\tilde{\theta}(t) = M(q)m(t)$ where $M(q)$ is a stable $p \times q$ dimensional filter and where $m(t)$ is a q-dimensional sequence of random vectors with the same statistical properties as $\{e_0(t)\}$. The covariance function of $\{\tilde{\theta}(t)\}$ will be denoted by $\tilde{P}(\cdot)$.

The scalar input signal $\{u(t)\}$ is generated by $u(t) = W(q)w(t)$ where $\{w(t)\}$ is a sequence of random variables with the same statistical properties as $\{e_0(t)\}$. The spectrum of $\{u(t)\}$ will be denoted by $\Phi_{uu}(e^{i\omega})$.

Remarks

- The term $W(q)w(t)$ should be interpreted as a model of an exogenous input such as a reference signal to the system.

- The assumption that $\{u(t)\}$ and $\{\tilde{\theta}(t)\}$ are independent rules out feedback in the system.
- This description arises when a finite-dimensional time-varying linear system with transfer function $G(q, \theta(t))$ is approximated by a first order Taylor expansion.

Introducing $\tilde{G}(q, t) = \tilde{\theta}^T(t)\mathbf{F}_0(q)$ we may rewrite (1) more suggestively as

$$y(t) = G_0(q)u(t) + \tilde{G}(q, t)u(t) + H_0(q)e(t) \ . \tag{2}$$

For slow time-variations this can be interpreted as that the total system dynamics $G_0(q) + \tilde{G}(q, t)$ varies around an "average" represented by $G_0(q)$.

2.2 Spectral Properties

We know that a linear time-invariant system retains stationary properties of the input signal. This turns out to be true also for (1). Define $\tilde{y}(t) = \tilde{G}(q, t)u(t) = \varphi^T(t)\tilde{\theta}(t)$. It can be shown that $\{\tilde{y}(t)\}$ is stationary and that its spectrum is

$$\Phi_{\tilde{y}\tilde{y}}(e^{i\omega}) = \text{Tr}\left\{\mathbf{F}_0(e^{-i\omega})\mathbf{F}_0^T(e^{i\omega})\Phi_{uu}(e^{i\omega}) * \Phi_{\tilde{\theta}\tilde{\theta}}(e^{i\omega})\right\} \tag{3}$$

where $\Phi_{\tilde{\theta}\tilde{\theta}}(e^{i\omega})$ is the spectrum of $\{\tilde{\theta}(t)\}$ and $*$ is the convolution operator.

3 Identification of Time-Varying Systems

In this section we analyze a prediction error method (PEM) when the data is generated by $\tilde{\mathcal{S}}$.

3.1 Prediction Error Methods

Let the model structure be given by

$$y(t) = G(q, \theta)u(t) + H(q, \theta)e(t)$$

where $\theta \in D_{\mathcal{M}}$, a connected open subset of $\mathbb{R}^p$. The corresponding prediction error is given by

$$\varepsilon(t, \theta) = -H^{-1}(q, \theta)G(q, \theta)u(t) + H^{-1}(q, \theta)y(t) \ .$$

The model structure is assumed to be uniformly stable which means that the filter that generates the prediction error and its second derivatives are uniformly stable (see [2]).

Let $\mathbf{Z}^N$ be a data set consisting of N measurements of the output and the input of a system. A prediction error method (PEM) using the model structure above and a quadratic criterion identifies a model by finding the parameter value $\hat{\theta}(N)$ that minimizes

$$V_N(\theta, \mathbf{Z}^N) = \frac{1}{N} \sum_{t=1}^{N} \frac{1}{2} \varepsilon^2(t, \theta) \ . \tag{4}$$

Under mild conditions (see [2])

$$\hat{\theta}(N) \rightarrow D_c = \{\theta^* : \bar{V}(\theta^*) = \inf_\theta \bar{V}(\theta)\} \quad \text{w.p. 1 as } N \rightarrow \infty \tag{5}$$

where

$$\bar{V}(\theta) = \frac{1}{2} \mathrm{E}[\varepsilon^2(t, \theta)] \ . \tag{6}$$

Under some identifiability conditions (see [2]) $D_c = \{\theta^*\}$ is a single point.

3.2 Time-Varying Data

The key to being able to analyze the behavior of a PEM when $u(t)$ and $y(t)$ are measurements from (1) is the following theorem.

Theorem 2. *The spectrum of $\{[y(t), u(t)]\}$ is the same as if the data was generated by*

$$S_e: \qquad y(t) = G_0(q)u(t) + \tilde{H}_0(q)\tilde{e}_0(t) \tag{7}$$

where $G_0(q)$ and $\{u(t)\}$ are as in Definition 1, where $\{\tilde{e}_0(t)\}$ and $\tilde{H}_0(q)$ are subject to the same conditions as $\{e_0(t)\}$ and $H_0(q)$, respectively, in Definition 1 and where $\tilde{\sigma}_0^2 = \mathrm{E}[\tilde{e}_0^2(t)]$ and $\tilde{H}_0(q)$ are defined by

$$|\tilde{H}_0(e^{i\omega})|^2 \tilde{\sigma}_0^2 = |H_0(e^{i\omega})|^2 \sigma_0^2 + \Phi_{\tilde{y}\tilde{y}}(e^{i\omega}) \tag{8}$$

where $\Phi_{\tilde{y}\tilde{y}}(e^{i\omega})$ is defined in (3).

The theorem is proved using that $\tilde{G}(t)u(t)$ is uncorrelated with both $u(s)$ and $e_0(s)$ for any s.

Since it is the spectrum of the data that determines the parameter estimate we are now in position to formulate the following theorem.

Theorem 3. *The limiting parameter set D_c is the same as if the data set was generated by the second order equivalent system S_e in Theorem 2.*

Proof. By exploiting the independence between $\{\tilde{\theta}(t)\}$ and $\{u(t)\}$ it can be shown that (5) is true when the data is generated by S using the same method as the one that is used in [2] to show (5) when the data is generated by (7).

The limiting set D_c is defined by (5) and (6). The quadratic criterion $\bar{V}(\theta)$ can be expressed in the frequency domain using Parseval's formula. Let $\Phi_{\varepsilon\varepsilon}(e^{i\omega}, \theta)$ be the spectrum of $\{\varepsilon(t, \theta)\}$, then

$$\bar{V}(\theta) = \frac{1}{2\pi} \int_{-\pi}^{\pi} \frac{1}{2} \Phi_{\varepsilon\varepsilon}(e^{i\omega}, \theta) d\omega \ .$$

Now $\Phi_{\epsilon\epsilon}(e^{i\omega}, \theta)$ is a filtered version of the spectrum of $[y(t), u(t)]^T$ which is the same for $\tilde{S}$ and S_e. Hence, the criterion $\bar{V}(\theta)$ is identical for the two systems under consideration which proves the theorem. $\Box$

The essence of the theorem is that

$$w_0(t) = \tilde{G}(q,t)u(t) + H_0(q)e_0(t) \tag{9}$$

is treated as noise in the identification.

Theorem 3 implies that (7) is the best possible time-invariant model that can be obtained when using conventional system identification. With this observation, it is easy to state when the estimated transfer function, $G(q, \hat{\theta}(t))$, will converge to the "average" transfer function $G_0(q)$. Replace $\tilde{S}$ by S_e as the "true" system. Then all results concerning parameter convergence will hold if a quadratic criterion is used.

3.3 Model Validation

Since, the systems $\tilde{S}$ and S_e gives the same limiting estimate one may wonder if they can be distinguished in the model validation phase. It is standard to use the test statistics

$$\hat{r}_{\epsilon\epsilon}^N(\tau) = \frac{1}{\sqrt{N}} \sum_{t=1}^{N-\tau} \epsilon(t)\epsilon(t+\tau) \quad \text{and} \quad \hat{r}_{\epsilon u}^N(\tau) = \frac{1}{\sqrt{N}} \sum_{t=\tau}^{N} \epsilon(t)u(t-\tau) \tag{10}$$

to test whiteness of the residuals and cross-correlation between the input and residuals, respectively. These tests are based on the idea that if there is correlation between the factors in the sums, these statistics grows as $\sqrt{N}$ and if not they are asymptotically normal distributed.

The next lemma shows how these statistics will behave when the output of the system contains a time-varying part.

Lemma 4. *Suppose that the data set Z^∞ has been generated by the system $\tilde{S}$ given by Definition 1. Assume that the model structure is such that*

$$G(e^{i\omega}, \theta^*) \equiv G_0(e^{i\omega}); \qquad H(e^{i\omega}, \theta^*) \equiv \tilde{H}_0(e^{i\omega}) \tag{11}$$

where $\tilde{H}_0(e^{i\omega})$ is defined in (8).

Then $\epsilon(t, \theta^*)$ and $\epsilon(t - \tau, \theta^*)$ are uncorrelated for $\tau \neq 0$. Furthermore $\epsilon(t, \theta^*)$ and $u(s)$ are uncorrelated for all t and s.

Proof. We have

$$\epsilon(t, \theta^*) = \tilde{H}_0^{-1}(q) \left[\varphi^T(t)\tilde{\theta}(t) + H_0(q)e_0(t) \right]. \tag{12}$$

From (3), (11) and (8) it follows that the spectrum of $\{\epsilon(t, \theta^*)\}$ is constant. Hence

$$E\left[\epsilon(t, \theta^*)\epsilon(t+\tau, \theta^*)\right] = \frac{1}{2\pi} \int_{-\pi}^{\pi} Ce^{i\tau\omega} d\omega = 0 \tag{13}$$

which proves the first proposition. The second proposition follows from (12) using that $u(t)$ is uncorrelated with $e_0(s)$ and $\tilde{\theta}(s)$ for any s. $\Box$

Lemma 4 shows that the means of the test statistics in (10) are zero. It also can be shown that they are asymptotically normal distributed. However, the variances of these normal distributions depend on the fourth order properties of $\{\varepsilon(t)\}$ and $\{u(t)\}$ (see, for example [3]). Even though $\tilde{S}$ and S_e are equivalent with respect to the second order properties, their higher order moments will not be equal. Hence, in general the test statistics above will have different properties depending on whether the residuals are from the time-varying system or from the second-order equivalent time-invariant system. But let us emphasize that since the mean is zero there is an upper bound, which is independent of the number of data, on the probability that the hypothesis of a correct model will be rejected.

We conclude this section by showing an example when tests based on (10) fail.

Example 8. The true system is given by

$$y(t) = G(q, \theta_0)u(t) + \tilde{\theta}^T(t)G_\theta(q, \theta_0)u(t) + H_0(q)e_0(t) \qquad (14)$$

where $\theta = [\theta_1\ \theta_2]^T$, $\theta_0 = [0.3\ -0.7]^T$ and

$$G(q, \theta) = \frac{\theta_1 q^{-1}}{1 + \theta_2 q^{-1}}; \quad H_0(q) = \frac{1}{1 - 0.7q^{-1}}; \quad \tilde{P} = 10^{-4}\begin{bmatrix} 3.2\ 1.1 \\ 1.1\ 5.3 \end{bmatrix} .$$

The noise sequence $\{e_0(t)\}$ is Gaussian white noise with zero mean and variance 0.002. The input is Gaussian white noise with unit variance. The variance of the time-varying term is approximately one fourth of the variance of the noise.

The time-varying parameters, $\tilde{\theta}(t) + \theta_0$, are shown in Fig. 1. Figure 2 shows the Nyquist diagram of the frozen transfer function together with the "average" system $G(e^{i\omega}, \theta_0)$. Included also are the true 95% confidence bounds at some frequencies. Since the dynamics of $\{\tilde{\theta}(t)\}$ is much slower than the time-invariant dynamics, the spectrum of $\tilde{\theta}^T(t)G_\theta(q, \theta_0)u(t)$ will have almost the same characteristics as $G(q, \theta_0)u(t)$. This is confirmed by using the ARX-model

$$y(t) + a_1 y(t - 1) = b_1 u(t - 1) + e(t) . \qquad (15)$$

Estimating a_1 and b_1 using 1000 measurements gives $\hat{a}_1 = -0.70$ and $\hat{b}_1 = 0.30$. Let us analyze the model using residual tests on new data. Figure 3, shows the correlation function of the residuals when using 1000 new measurements. The dotted lines are confidence intervals (corresponding to 3 standard deviations) when the residuals are independent, hence the hypothesis that the residuals are white noise can not be rejected. The cross correlation function between input and residuals (see Fig. 3) also indicates that the dynamic model is reasonable. Here the dotted lines denotes confidence intervals (corresponding to 3 standard deviations) when input and residuals are independent.

Compare also Fig. 3 with Fig. 4 where the data has been generated by (15) with $a = -0.7$ and $b = 0.3$ and the variance of $e(t)$ adjusted so that it corresponds to the variance of sum of the time-varying term and the noise term in (14).

The variance of the parameter estimates is low, supporting the idea of having obtained a good model. There is nothing in the above analysis that indicates that the system is time-varying. Thus, the conventional analysis will lead us totally astray in this case.

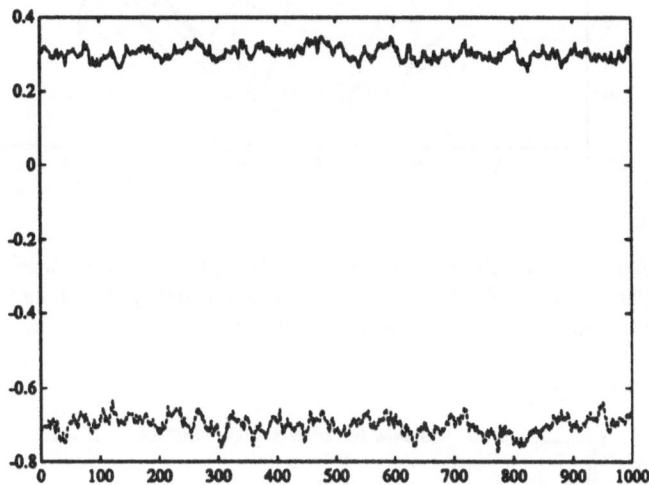

Fig. 1. Example 1: $\tilde{\theta}(t)$ as a function of time

4 Detecting Time-Variability

It is of course desirable to be able to determine if the term $\tilde{G}(q,t)u(t)$ is present at all in the system.

We saw in the previous section that the hypothesis tests based on (10) fail because their means are zero even when the system is time-varying. The reason was that a time-varying system is equivalent to a time-invariant system in a second order sense. Hence, a useful test statistic must be of higher order.

A candidate is suggested by observing the following. Let $v(t)$ be a function of u^t. Then

$$\mathrm{E}\left[\left(w_0^2(t) - \mathrm{E}\left[w_0^2(t)\right]\right)\left(|v(t)|^2 - \mathrm{E}\left[|v(t)|^2\right]\right)\right]$$

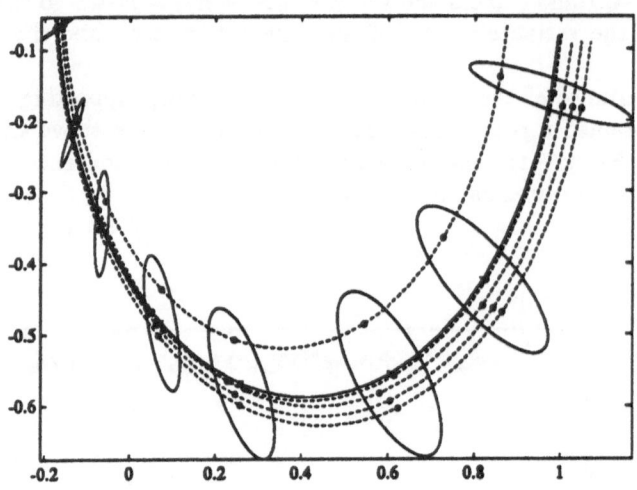

Fig. 2. Example 1: Dashed lines: Nyquist plot of the true system at $t = 100, 200, \ldots, 500$; solid line: "average" system $G_0(e^{i\omega})$. The true 95% ($\alpha = 6$) confidence bounds are shown for some frequencies

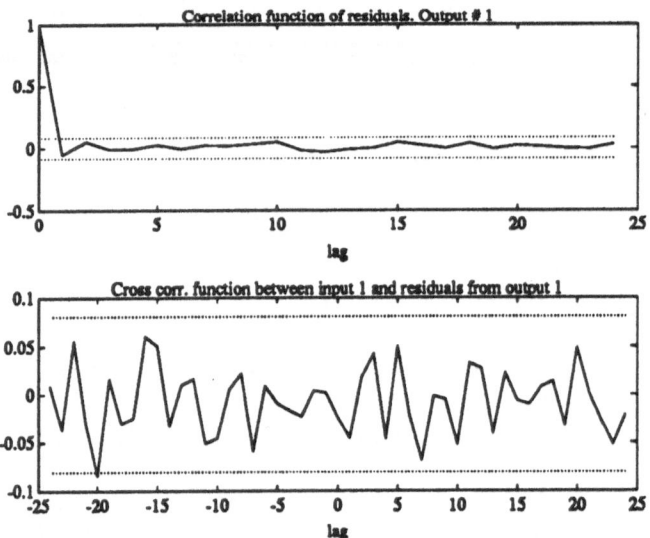

Fig. 3. Example 1: Residual analysis of time-invariant model using 1000 new measurements. The data is generated by (14)

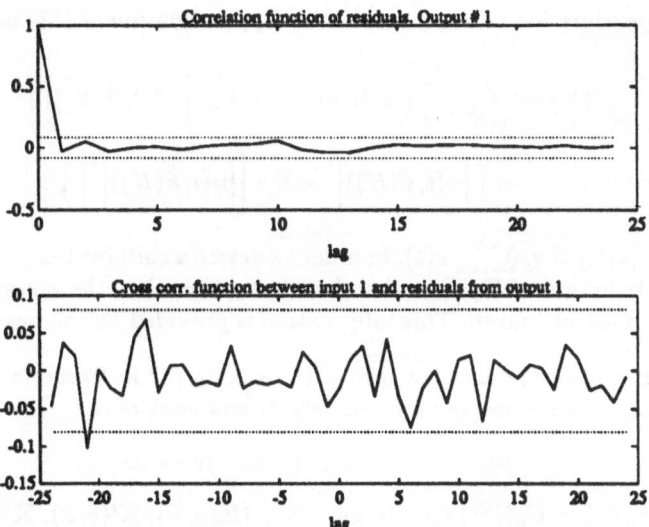

Fig. 4. Example 1: Residual analysis of time-invariant model using 1000 new measurements. The data is generated by (15)

$$= \text{Tr}\left\{ \text{E}\left[\left(|v(t)|^2 - \text{E}\left[|v(t)|^2 \right] \right) \left(\varphi(t)\varphi^T(t) - \text{E}\left[\varphi(t)\varphi^T(t) \right] \right) \right] \cdot \tilde{\text{P}} \right\} \quad (16)$$

where $\tilde{\text{P}}$ is the covariance matrix of $\tilde{\theta}(t)$. This mean will thus always be zero if $\tilde{\text{P}} = 0$ and in general non-zero otherwise.

Hence

$$\frac{1}{\sqrt{N}} \sum_{t=1}^{N} \left(w_0^2(t) - \text{E}\left[w_0^2(t) \right] \right) \left(|v(t)|^2 - \text{E}\left[|v(t)|^2 \right] \right) \quad (17)$$

should be a suitable statistic. However, neither $w_0(t)$ nor $\text{E}[w_0^2(t)]$ are available and hence we shall have to use approximations for these quantities.

In the sequel we will restrict the choice of $v(t)$ to be linear in u^t but we will allow $v(t)$ to depend on θ so that $v(t, \theta) = \text{K}(q, \theta)u(t)$ where $\text{K}(q, \theta)$ is a stable filter. Later on in this section we will discuss the choice of $v(t, \theta)$. But first we must discuss how to obtain an approximation of $w_0(t)$. The answer is given by Theorem 3 which shows that we can estimate $G_0(q)$ consistently if we have a flexible enough model structure. Hence, we estimate $G_0(q)$ using a uniformly stable model structure which we will assume allows us to estimate $G_0(q)$ consistently. Using the transfer function estimate $G(q, \hat{\theta}(N))$ we then take

$$w(t, \hat{\theta}(N)) = y(t) - G(q, \hat{\theta}(N))u(t) \quad (18)$$

as an approximation of $w_0(t)$. A suitable approximation of (17) is then

$$T(N) = \frac{1}{\sqrt{N}} \sum_{t=1}^{N} \left(w^2(t, \hat{\theta}(N)) - \hat{E}_N \left[w^2(t, \hat{\theta}(N)) \right] \right)$$
$$\times \left(\left| v(t, \hat{\theta}(N)) \right|^2 - \hat{E}_N \left[\left| v(t, \hat{\theta}(N)) \right|^2 \right] \right) \tag{19}$$

where $\hat{E}_N[x(t)] = \frac{1}{N} \sum_{t=1}^{N} x(t)$. In order to device a suitable test, the (asymptotic) distribution of $T(N)$ under the assumption that the system is time-invariant must be known. This information is provided by the next theorem.

Theorem 5. *Suppose that the system is given by (7) in Theorem 2. Assume that the model structure is uniformly stable and such that*

$$\hat{\theta}(N) \to \theta^* \quad \text{w.p.1} \quad \text{as} \quad N \to \infty \tag{20}$$

and $G(e^{i\omega}, \theta^*) = G_0(e^{i\omega}) \forall \omega$. *Assume that* $\{K(q, \theta), K'(q, \theta), K''(q, \theta); \theta \in D_\mathcal{M}\}$ *is uniformly stable. Let* $v^*(t) = K(q, \theta^*)u(t)$.
  Then $T(N) \sim \text{AsN}(0, P_T)$ *where*

$$P_T = \sum_{\tau=-\infty}^{\infty} R_{v_0^2}(\tau) R_{|v^*|^2}(\tau) \tag{21}$$

where $R_{v_0^2}(\cdot)$ *and* $R_{|v^*|^2}(\cdot)$ *are the covariance functions of* $v_0^2(t) - E[v_0^2(t)]$ *and* $|v^*(t)|^2 - E[|v^*(t)|^2]$, *respectively.*

The proof of this theorem is omitted.
  Based on Theorem 5 a family of hypothesis tests is given as follows. Let $\hat{P}_T(N)$ be a consistent estimate of P_T. Reject the null hypothesis $\mathcal{H}_0 : \tilde{P} = 0$ if $T(N)/\sqrt{\hat{P}_T(N)} \in R_\alpha^c$ where R_α^c is the complement of R_α, a subset of the real line such that if X is a $N(0,1)$ distributed random variable then

$$P(X \in R_\alpha) = \alpha . \tag{22}$$

The level of significance α is the probability that the null hypothesis will not be rejected if it is true.
  The power of the test, *i.e.* the probability to reject $\mathcal{H}_0$ if $\tilde{P}$ is non-zero depends on the data length and how large $\tilde{P}$ is but also on R_α and on $v(t, \theta^*)$. R_α should be chosen such that the probability is high that the test statistic is in R_α^c when the null hypothesis is false. Suppose that $v(t, \theta)$ is such that

$$E\left[\left(|v(t, \theta^*)| - E\left[|v(t, \theta^*)|^2 \right] \right) \left(\varphi(t)\varphi^T(t) - E\left[\varphi(t)\varphi^T(t) \right] \right) \right] > 0 . \tag{23}$$

When this is true it follows from (16), using the positive definiteness of $\tilde{P}$, that the mean of $T(N)$ is positive and proportional to $\sqrt{N}$ asymptotically. Furthermore the difference between $T(N)$ and its mean will be asymptotically normally distributed and thus concentrated around the mean. Hence, in this case it is advantageous to choose a one-sided region $R_\alpha = (-\infty, N_\alpha]$. Thus, it should be desirable to choose $\{v(t, \theta)\}$ such that (23) holds. Notice that, although this mean in general is unknown we can estimate it by replacing all means by sample means.

A natural choice is $v(t, \theta) = |\varphi(t, \theta)|$. Notice, however, that this choice does not automatically ensure (23) but that this, nevertheless, is the generic situation.

To complete the test it remains to provide a consistent estimate of the asymptotic variance of $T(N)$. We notice that this amounts to estimating the spectrum of $\left(v_0^2(t) - \mathrm{E}[v_0^2(t)]\right)\left(|v(t, \theta^*)|^2 - \mathrm{E}\left[|v(t, \theta^*)|^2\right]\right)$ evaluated at the zero frequency. Taking advantage of the independence between $v_0(t)$ and $u(t)$ a natural candidate is

$$\sum_{\tau = -\gamma(N)}^{\gamma(N)} w_{\gamma(N)}(\tau) \hat{R}_{v_0^2}^N(\tau) \hat{R}_{|v^*|^2}^N(\tau) \tag{24}$$

where

$$\hat{R}_{v_0^2}^N(\tau) = \frac{1}{N} \sum_{t=1}^N \left(w^2(t, \hat{\theta}(N)) - \hat{\mathrm{E}}_N \left[w^2(t, \hat{\theta}(N)) \right] \right)$$
$$\times \left(w^2(t - \tau, \hat{\theta}(N)) - \hat{\mathrm{E}}_N \left[w^2(t, \hat{\theta}(N)) \right] \right) \tag{25}$$

where $\hat{R}_{|v^*|^2}^N(\tau)$ is defined analogously and where $w_\gamma(\tau)$ is a so called *lag-window* such as for example a Hamming window. For a thorough treatment of spectral estimation see [4] or [2]. The length $\gamma(N)$ of the window determines the accuracy of the estimate. A too small $\gamma(N)$ does not take advantage of all the information about the spectrum in the data set. But on the other hand if $\gamma(N)$ grows too fast the accumulated error in the sum will not tend to zero and estimate will not be consistent. The next theorem provides an upper bound on the growth rate of $\gamma(N)$ in order for the estimate to be consistent.

Theorem 6. *Suppose that the system is given by (1) in Definition 1 but with $\tilde{\theta}(t) \equiv 0$. Assume that the model structure is uniformly stable and such that*

$$\hat{\theta}(N) \to \theta^* \quad \text{w.p.1} \quad \text{as} \quad N \to \infty \tag{26}$$

and $G(e^{i\omega}, \theta^*) = G_0(e^{i\omega}) \quad \forall\omega$. Assume that $\{\mathrm{K}(q, \theta), \mathrm{K}'(q, \theta), \mathrm{K}''(q, \theta); \theta \in D_\mathcal{M}\}$ is uniformly stable. Let $\{\gamma(N)\}$ be a positive, monotonously increasing sequence such that for some $\delta > 0$

$$\frac{\gamma(N)}{\sqrt{N}/\log^{2+\delta}(N)} \to C \quad \text{as} \quad N \to \infty . \tag{27}$$

Let $|w_\gamma(\tau)| < C_W$ for $\gamma = 1, 2, \ldots; \tau = 0, \pm1, \ldots$ and let $\lim_{\gamma \to \infty} w_\gamma(\tau) = 1 \, \forall \tau$.

Then (24) converges to P_T with probability one as $N \to \infty$.

Proof. This follows from a minor modification of Lemma 3.6 in [5].   □

We conclude the section with an example where we illustrate how the ideas above can be used and compare them with the cross-correlation test in (10).

Example 9. (Example 8 continued.) In this example Monte Carlo simulations were used to see if the hypothesis test outlined above could detect the time-variations of the system (14). Recall that the variance of the time-varying part is one fourth of the variance of the variance of the noise term. Figure 5 shows a comparison between the output of the time-varying system (14) and the time-invariant second order equivalent (15). Hopefully this plot convinces the reader that it is not a trivial problem to discriminate between the two systems.

To estimate $w_0(t)$ and $\varphi(t)$ an output error model structure

$$y(t) = \frac{b}{1 + aq^{-1}} u(t - 1) + e(t) \tag{28}$$

was used since it is known to give a consistent estimate of $G_0(q)$ if the noise is uncorrelated with $\{u(t)\}$. We chose $v(t, \theta) = |\varphi(t, \theta)|$ and the window length $\gamma(N)$ was approximately $\log(N)$.

Table 1a shows the percentage of the trials that rejected the hypothesis of time-invariance for different data lengths. The total number of trials was 1000 for each data length. The level of significance used was approximately 95% ($N_{0.95} = 1.65$). To check the sensitivity of the test with respect to the distribution simulations where all random quantities were uniformly distributed was also done. The results from these simulations also are presented in Table 1a. From this it seems as if the test is very insensitive to the underlying distributions. The corresponding results for the cross-correlation test in (10) with $\tau = 1$ is also shown.

Table 1b shows the corresponding results when the data is generated by (15). In this case we also tested the normality of $T(N)/\hat{P}_T^{1/2}$ using a χ^2-test (see [6]). This test indicated that the statistic is far from normal except when $N = 5000$. From this and the results in Table 1b we conclude that the level of significance seems to be robust with respect to the distribution of $T(N)$.

5 Final Discussion

From a formal standpoint we have in this contribution discussed a way to characterize and detect the variability of linear time-varying systems. We showed that in that case the time-varying system has a time-invariant counterpart with respect to the second order properties. Example 8 showed that it is by no means obvious that standard model validation techniques will distinguish these systems from each other. Hence, one may be mislead to believe

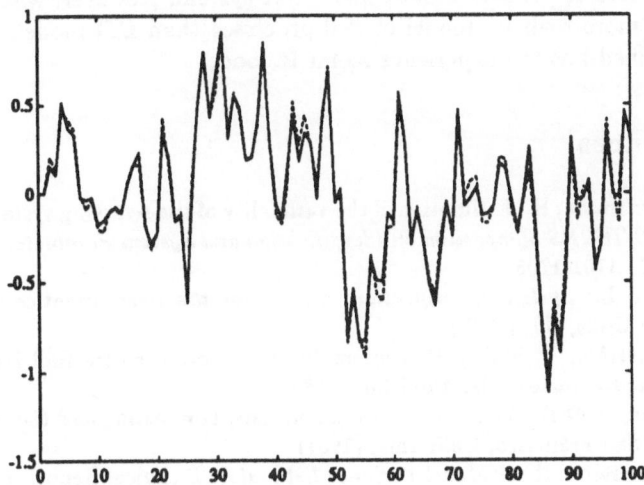

Fig. 5. Output of time-varying and second order equivalent time-invariant systems.
Solid line: time-varying system; dashed line: time-invariant system

Distribution	Number of data points			
	200	500	1000	5000
Normal	29/28	55/35	82/32	100/38
Uniform	32/30	54/32	82/34	100/34

a

Distribution	Number of data points			
	200	500	1000	5000
Normal	4.2	5.8	6.2	5
Uniform	5.6	5.2	5.6	6.1

b

Table 1. Percentage of rejected trials in hypothesis test in Example 9. Numbers below a slash correspond to the cross-correlation test based on (10). a, Data generated
by (14). b, Data generated by (15).

that a time-varying system is time-invariant. A remedy for this was to use
higher order statistics in the test statistic.

However, the method assumes that no unmodelled time-invariant dynamics are present (recall that $w_0(t)$ in (9) only depends on the input through
the time-varying dynamics). Presence of unmodelled dynamics may introduce
bias in the estimate. Another limitation is that no feedback is allowed.

Let us close this section by returning to what was our primary concern.
The fact that with the conventional linear time-invariant (LTI) assumption
one may estimate an arbitrarily good model if only enough money and time
is spent. A property that can be questioned for many real world processes. By
adding a "nice" time-varying term to the system we prevented this from being
possible but kept the system behaviour very close to that of a LTI system. In
fact, it turned out that with respect to conventional (second order) system

identification it behaves almost like a LTI system. However, whether or not this is a more realistic model of real processes than LTI models remains to be examined but the experience so far is good.

References

1. Hjalmarsson, H.: Estimation of the variability of time-varying systems. In *Proc. IFAC/IFORS Symposium on Identification and System Parameter Estimation,* (1991) 1701-1706
2. Ljung, L.: *System Identification: Theory for the User.* Prentice-Hall, Englewood Cliffs, NJ, (1987)
3. Söderström, T. Stoica, P.: *System Identification.* Prentice-Hall International, Hemel Hempstead, Hertfordshire, (1989)
4. Brillinger, D.R.: *Time Series: Data Analysis, Forecasting and Control.* Holden-Day, San Francisco, California, (1981)
5. Hjalmarsson, H.: *A Model Variance Estimator.* Technical Report, Dep. of Electrical Engineering, Linköping University, S-581 83 Linköping, Sweden. (1992)
6. Rao, C.R.: *Linear Statistical Inference and Its Applications.* John Wiley & Sons, New York, (1973)

Accuracy Confidence Bands Including the Bias of Model Under-Fitting*

Wallace E. Larimore

Adaptics, Inc, 40 Fairchild Drive, Reading, MA 01867

Abstract:
New results are developed that give accuracy confidence bands based upon the parameter estimation error of system identification including the bias of model under-fitting. In this context, bias means the underestimation of model accuracy due to selection of the model state order less than the true system order. It is shown that the optimal accuracy model is given by the model minimizing the Akaike information criterion (AIC), and that the bias is indicated by the behavior of the AIC as a function of model order beyond the minimum order. The resulting confidence bands are simultaneous so that they guarantee the system frequency response function lies entirely within the band for all frequencies with the stated probability. Such bands are then appropriate for guaranteeing robust stability and performance of robust controllers.

1 Introduction

The primary problem addressed in this paper concerns the determination of accuracy confidence bands on models identified from observational data via system identification procedures. It is assumed that the system identification procedure is asymptotically efficient in that it approaches the minimum possible error in parameter estimation as the sample size becomes large. It is widely recognized that in system identification, there is a optimum selection of model order, such as state order or AR and MA orders, beyond which the model accuracy increases. At such a point, it is preferable to truncate the additional process model structure because the estimation of these additional parameters introduces more error in the model than truncating the higher order structure, i.e. setting the corresponding parameters to zero. Currently available procedures for assessing the accuracy of the identified model include the variability of the estimated parameters, but do not assess the error resulting from the truncating of higher order structure. Several recent papers have

---

* This work was supported in part by the National Science Foundation under contract with Computational Engineering, Inc., presently a part of Coleman Research Corporation.

attempted to assess the truncation or bias error using Bayesian procedures that put a Bayesian prior probability on the parameters that are truncated, i.e. are not estimated. While this does allow computation of model accuracy resulting from the truncated parameters, the statistical justification for such a procedure is lacking and the choice of such a prior is arbitrary.

In the previously developed theory, the confidence bands express the error due to the parameters estimated in the model fitting but ignore the bias error due to truncating additional structure. The bands are expressed in terms of a central Chi-squared distribution where the degrees of freedom are related to the number of estimated parameters. In the theory developed in this paper, this is replaced by a noncentral Chi-squared distribution where the noncentrality parameter is related to the truncated model structure. The noncentrality parameter is given by the Kullback discrimination information [7] between the fitted model and the true model of the process. This is a measure of the error in approximation of the true process by the fitted model including the truncation of model structure beyond the chosen model order. Since the Akaike information criterion (AIC) is an asymptotically unbiased estimate of the Kullback information, it can be used to estimate the bias of model under-fitting. An estimate of the noncentrality parameter is given by the difference between the observed AIC curve and the theoretical average AIC curve assuming no additional model structure present beyond the chosen model order.

2 Approximation of Stochastic Systems

In this paper, we are concerned with the approximation of dynamical systems including stochastic noise disturbance models. Only linear, time invariant, gaussian processes will be considered. The system outputs y_t and inputs u_t are assumed to be defined at equal spaced times on the integers $t : 1 \leq t \leq N$. The model is considered in the state space form

$$x_{t+1} = \Phi x_t + G u_t + w_t \tag{1}$$

$$y_t = H x_t + A u_t + B w_t + v_t \tag{2}$$

where x_t is a k-order Markov state and w_t and v_t are white noise processes that are independent with covariance matrices Q and R respectively. These state equations are more general than typically used since the noise $B w_t + v_t$ in the output equation is correlated with the noise w_t in the state equation. This more general structure with the term $B w_t$ present is necessary if a minimal order Markov state for the system is required. The coefficient matrices Φ, G, H, A, and B along with the covariance matrices Q and R specify the state space model. These coefficients are not unique, but can be chosen so that the covariance structure of the process is in a canonical form [10] The coefficients are assumed to be constant, however nonstationary noise results in the output if the system is unstable. This allows the consideration of general nonstationary gaussian Markov processes.

Any procedure for system identification can be viewed as a model approximation procedure. A major issue arises concerning the measure of approximation and the sense in which a given system identification procedure is a good approximation. These issues have been largely resolved in the past decade and are reviewed here briefly. Suppose that $p_*(Y)$ is the true stochastic process model for the observed system outputs Y of the form (1) and (2) of a particular state order and with particular values of the parameters matrices. Suppose further that some system identification procedure is used that produces an identified model $p_1(Y)$ of the form (1) and (2), and that an alternative identification procedure produces the model $p_2(Y)$ of the same form. Then the most general and statistically justifiable measure will involve a measure of the adequacy of the identified models $p_1(Y)$ or $p_2(Y)$ in describing future observations Z of the process, where Z just denotes another set of observations Y from the future.

A formal discussion of this problem is given in Larimore [8] and Larimore and Mehra [9] using the concepts of predictive inference and statistical sufficiency. We digress to give a short derivation of the Kullback information as the natural measure of model approximation error since it turns out to be the fundamental quantity in both model order selection and quantifying the error in model under-fitting. The Kullback information appears repeatedly in the various quantities as a fundamental quantity because fundamental principles of statistical inference require the comparison of models in terms of likelihood ratios that lead directly to the Kullback information.

To develop these concepts, a more precise notation is needed. Consider, then, the hypothetical future observations Z as a second sample from the process. A system identification procedure fitted on the first sample Y is judged by the adequacy of the model for describing a future sample Z. In particular let $p_1(Z|Y)$ denote a probability model for the future sample Z obtained from fitting a model using the sample Y. The model $p_1(Z|Y)$ has the interpretation of a conditional density or predictive density (Larimore [8], Larimore and Mehra [9]). The problem is as follows: what measure of approximation of the model $p_1(Z|Y)$ to the true process $p_*(Z|Y)$ can be inferred by the hypothetical second sample Z? From the fundamental statistical principle of sufficiency, if such a measure of approximation is to loose no information, then it must be a function of a sufficient statistic. Such a sufficient statistic for comparing the models $p_1(Z|Y)$ and $p_2(Z|Y)$ on the hypothetical sample Z is the likelihood ratio

$$\frac{p_1(Z|Y)}{p_2(Z|Y)} \tag{3}$$

If the above hypothetical experiment of drawing the two samples (Y, Z) were repeated R times in independent experiments with the samples from the i-th repetition denoted (Y_i, Z_i), then the likelihood ratio Λ_R comparing the two procedures p_1 and p_2 on the joint outcome $(Z_1, \ldots, Z_R)$ is

$$\Lambda_R = \frac{p_1(Z|Y) \cdots p_1(Z|Y)}{p_2(Z|Y) \cdots p_2(Z|Y)} . \tag{4}$$

The likelihood ratio Λ_R grows exponentially with R so that its behavior is best studied by looking at $\log(\Lambda_R)$.

$$\lim_{R \to \infty} \log(\Lambda_R) = \frac{1}{R} \sum_{i=1}^{R} \log \frac{p_1(Z_i|Y_i)}{p_2(Z_i|Y_i)}$$

$$= \int p_*(Z|Y) \log \frac{p_1(Z|Y)}{p_2(Z|Y)} dZ p_*(Y) dY$$

$$= E\{I_Y(p_*, p_2) - I_Y(p_*, p_1)\} \tag{5}$$

where $E\{\ \}$ denoted the expected value or average operation and $I_Y(p_*, p_1)$ is defined below. Thus if the true probability density is p_* and an approximating density is p_1, then the measure of approximation of the model p_1 to the truth p_* is given by the expected value of the Kullback discrimination information [7],

$$I_Y(p_*, p_1) = \int p_*(Z|Y) \log \frac{p_*(Z|Y)}{p_1(Z|Y)} dZ . \tag{6}$$

The Kullback information is an extremely general measure of model approximation error based upon the very general statistical principles of sufficiency and repeated sampling in a predictive inference setting. This measure is appropriate for the measure of any departures in the model including the input-output dynamics, biases or trends in the process, or changes in the correlation structure of the noise disturbances.

3 Optimal Order Selection

The Kullback information provides a theoretical measure of the error in an approximating model relative to the true process. Its actual evaluation would require knowledge of the true process model which of course is not known. To proceed in system identification, it is necessary to estimate the Kullback information from the observed data for various tentative model orders or structures, and devise decision procedures based upon such an estimate.

To make the discussion precise, let $p(Y, \theta)$ be a parameterized family of probability densities on the observations Y with parameter vector θ

$$Y^T = (y_1^T, \ldots, y_N^T) \tag{7}$$

$$\theta^T = (\psi_1^T, \psi_2^T, \ldots) \tag{8}$$

where N is the sample size and ψ_i is a subvector associated with the additional parameters estimated for order i. The constrained model of order i is parameterized by

$$\theta_i^T = (\psi_1^T, \ldots, \psi_i^T, 0, \ldots) \tag{9}$$

and $M_i = \dim(\theta_i)$ is the number of estimated parameters in θ_i.

The term $p_*(Z|Y) \log p_*(Z|Y)$ in the Kullback information is a constant that does not depend upon the choice of the identified model $p_1(Z|Y)$. Thus for the purpose of model selection, it can be ignored. The remaining term can be estimated by using the log likelihood function $\log p_1(Y, \hat{\theta}_k)$. This turns out to be a biased estimate which, when corrected for bias, gives the Akaike information criterion (AIC) [1]. The AIC for each order k is defined by

$$AIC(k) = -2 \log p(Y^N, U^N; \hat{\theta}_k) + 2M_k \qquad (10)$$

where p is the likelihood function based on the observations (Y^N, U^N) at N time points, and where $\hat{\theta}_k$ is the maximum likelihood parameter estimate using a k-order model with M_k parameters. The model order k is chosen corresponding to the minimum value of AIC(k).

The number of parameters in the state space model (1) and (2) is

$$M_k = k(2n + m) + mn + n(n + 1)/2 \qquad (11)$$

where k is the number of states, n is the number of outputs, and m is the number of inputs to the system. This result is developed by considering the size of the equivalence class of state space models having the same input/output and noise characteristics (Candy et al. [3]). Thus the number of functionally independent parameters in a state space model is far less than the number of elements in the various state space matrices.

A small sample correction to the AIC has been recently developed for model order selection (Hurvich and Tsai [5], Hurvich, Shumway and Tsai [4], Hurvich and Tsai [6]). The corrected quantity denoted $AIC_C(k)$ is given by

$$AIC_C(k) = -2 \log p(Y^N, U^N; \hat{\theta}_k) + 2fM_k \qquad (12)$$

where the small sample correction factor f is

$$f = \frac{\overline{N}}{\overline{N} - \left(\frac{M(k)}{n} + \frac{n+1}{2} \right)} . \qquad (13)$$

The effective sample size $\overline{N}$ is the number of time points at which one-step predictions are made using the identified model. The small sample correction has been shown to produce model order selection that is close to the optimal as prescribed by the Kullback information measure of model approximation error.

The AIC provides an optimal tradeoff between the increase in variability introduced by the selection of too high an order and the bias error introduced by the selection of too low an order. The AIC chooses the order where the bias error of truncation is smaller than the variability that would be introduced if a higher order model were estimated.

It can be shown that for large sample the AIC is an optimal estimator of the Kullback information and achieves optimal decisions on model order (Shibata [12]). Also it has been shown that order consistent procedures for order selection are necessarily suboptimal in minimizing the Kullback information (Shibata [13]).

4 Distribution of the AIC

The behavior of the AIC and its statistical distribution provide the key in assessing the bias of model under-fitting. Consider the difference of AICs for the two orders $j < k$ given by

$$AIC(j) - AIC(k) = -2\log\frac{p(Y,\hat{\theta}_j)}{p(Y,\hat{\theta}_k)} - 2(M_k - M_j) . \qquad (14)$$

From the large sample theory of maximum likelihood estimators (Cox & Hinkley, 1974), the $-2\log\lambda_{j,k}$ of the likelihood ratio $\lambda_{j,k}$ is asymptotically a noncentral Chi-squared distribution $\chi^2(\nu,\delta)$ on $\nu = (M_k - M_j)$ degrees of freedom. The expected value of a noncentral Chi-squared random variable with noncentrality parameter δ and degrees of freedom ν is $\delta^2 + \nu$ so that

$$E\{-2\log\lambda_{j,k}\} = M_k - M_j + \delta_{j,k}^2 = E\{AIC(j) - AIC(k) + 2(M_k - M_j)\} . \qquad (15)$$

Since the AIC is asymptotically an unbiased estimator of the Kullback information, we have that

$$\delta_{j,k}^2 = E\{AIC(j) - AIC(k)\} + (M_k - M_j) \qquad (16)$$
$$= E\{I_Y(p_*,p_j) - I_Y(p_*,p_k)\} + (M_k - M_j) . \qquad (17)$$

Thus we have as an unbiased estimate $\hat{\delta}_{j,k}^2$ of the noncentrality parameter $\delta_{j,k}^2$ for the Chi-squared random variable $-2\log\lambda_{j,k}$

$$\hat{\delta}_{j,k}^2 = AIC(j) - AIC(k) + M_k - M_j . \qquad (18)$$

The standard deviation of the estimate $\hat{\delta}_{j,k}^2$ is

$$\sigma(\hat{\delta}_{j,k}^2) = \sqrt{2(M_k - M_j) + 4\delta_{j,k}^2} . \qquad (19)$$

When the true process order is j so that the additional structure in model order k is unnecessary, the noncentrality parameter $\delta_{j,k} = 0$ so that

$$E\{AIC(k)\} = E\{AIC(j)\} + (M_k - M_j) . \qquad (20)$$

Thus $AIC(k)$ as a function of k would be expected to grow on the average as the number of additional parameters estimated in the higher order models.

Observing such behavior in the AIC is then the basis for inferring that there is no additional structure in the process beyond the fitted order j.

It will prove useful below to have a confidence interval on the noncentrality parameter. From (14), we have that the quantity

$$-2\log \lambda_{j,k} = AIC(j) - AIC(k) + 2(M_k - M_j) \qquad (21)$$

is distributed as $\chi^2(M_k - M_j, \delta_{j,k})$, so that confidence intervals can be constructed for the noncentrality $\delta_{j,k}$ parameter. Suppose that we want the interval to be such that $\delta_{j,k}$ is less than the interval with probability α_1 and greater with probability $1 - \alpha_2$. This is done by finding $\delta_{\alpha_1,j,k}$ and $\delta_{\alpha_2,j,k}$ such that

$$P(-2\log \lambda_{j,k} < \chi^2(M_k - M_j, \delta_{\alpha_i,j,k})) = \alpha_i \ . \qquad (22)$$

A rough indication of the variability of the estimate $\delta^2_{j,k}$ is given by the variance (19). If the true $\delta^2_{j,k}$ is zero, then the standard deviation is $\sqrt{2(M_k - M_j)}$. Since the Chi-squared distribution is asymptotically normally distributed, we can expect that the confidence on the noncentrality parameter estimate $\hat{\delta}^2_{j,k}$ will grow proportional to $\sqrt{M_k - M_j}$.

5 Finite Sample Approximation

The above distribution theory assumes that the sample size is large compared to the number of parameters estimated in the model fitting. In many applied problems, to apply the above methods requires that high order models be fitted. For moderate sample sizes, this results in a poor asymptotic approximation. In this section, finite sample size modifications to the above asymptotic theory is developed.

From multivariate analysis (Anderson, 1984 p. 317), a more accurate approximation to the distribution of the likelihood ratio is that $-2\log \lambda_{j,k}$ is distributed as $C_{j,k}\chi^2(M_k - M_j, \delta^2_{j,k})$ were the constant is very nearly the small sample correction factor (13). In the following, we will use the approximation $C_{j,k} = f_j$ and will need that $k - j << j$ which we will insure by considering $k = j + 1$. Then the results of the previous section become

$$AIC(j) - AIC(j+1) = -2\log \frac{p(Y, \hat{\theta}_j)}{p(Y, \hat{\theta}_{j+1})} - 2(f_{j+1}M_{j+1} - f_j M_j) \qquad (23)$$

$$E\{-2\log \lambda_{j,j+1}\} = f_{j+1}M_{j+1} - f_j M_j + f_j \delta^2_{j,j+1}$$
$$= E\{AIC(j) - AIC(j+1) + 2(f_{j+1}M_{j+1} - f_j M_j)\} \qquad (24)$$
$$\delta^2_{j,j+1} = E\{AIC(j) - AIC(j+1)\}/f_j + (M_{j+1} - M_j) \ . \qquad (25)$$

Thus we obtain an unbiased estimate of the incremental change in the noncentrality parameter. From the distribution theory, the first term is a noncentral Chi-squared random variable with degrees of freedom $M_{j+1} - M_j$. This can then be used in testing if the noncentrality parameter is incrementally zero beyond some order K.

6 Confidence Bands on Model Accuracy

The above provides the basis for simultaneous confidence bands including an estimate for the bias of model under-fitting. Suppose that $f(\omega, \theta)$ is a complex vector function of frequency ω for a system parameterized by θ. $f(\omega, \theta)$ could be the elements of the transfer function matrix collected in a vector. Let $f_\theta(\omega, \theta)$ denote the matrix of partial derivatives of f with respect to θ^T and $F(\theta)$ denote the Fisher information matrix.

Parameter errors relate to transfer function errors as

$$f(\omega, \hat{\theta}) - f(\omega, \theta) = f_\theta(\omega, \theta)[\hat{\theta} - \theta] \tag{26}$$

where $f_\theta(\omega, \theta)$ denotes the matrix of partial derivatives of f with respect to θ^T and $F(\theta)$ denotes the Fisher information matrix. At a fixed frequency ω, the covariance matrix of the frequency response vector, if the true order is θ_k, is asymptotically

$$Cov[f(\omega, \hat{\theta}_k), f(\omega, \hat{\theta}_k)] = f_\theta(\omega, \theta_k) F^{-1}(\theta_k) f_\theta(\omega, \theta_k)^* . \tag{27}$$

This gives a point-wise confidence ellipse at each frequency that is simultaneous for all elements of the transfer function matrix. It holds only for selection of the true model order k.

For robust control design, the above confidence bands have two deficiencies. To guarantee robustness for a class of models, the confidence bands need to be simultaneous for all frequencies, i.e. hold for the *function* of ω. Also confidence bands need to include the error due to selecting the model order less than the true order. Such a result has been recently developed in Larimore [11].

Theorem 1 *For suitable regularity conditions and for an asymptotically normal and efficient estimator $\hat{\theta}$, as $N \to \infty$, the probability is at least $1 - \alpha$ that simultaneously for all $\omega \in \Omega$ the true vector transfer function $f(\omega, \theta)$ is bounded by*

$$[f(\omega, \hat{\theta}_j) - f(\omega, \theta_k)]^* [f_\theta(\omega, \theta_k) F^{-1}(\theta_k) f_\theta(\omega, \theta_k)^*]^\dagger [f(\omega, \hat{\theta}_j) - f(\omega, \theta_k)] \tag{28}$$

$$\leq M_j \chi^2(\alpha, M_j, \delta_{j,k}) \tag{29}$$

where $\chi^2(\alpha, M_j, \delta_{j,k})$ is the α point of the noncentral χ^2 distribution.

To use this Theorem 1, in principle an estimate of $\delta_{j,\infty}$ would be required. This is not possible with a finite sample. However, if the growth in $\delta_{j,k}$ is zero or small for k beyond some order K, then an estimate or bound on $\delta_{j,K}$ will suffice in practice. To determine that the growth is small, then setting (21) to zero we need to determine that

$$AIC(k) = AIC(K) + M_k - M_K + e_{K,k} \tag{30}$$

where $e_{K,k}$ behaves as a central chi-squared random variable. In such a case, a confidence bound on $\delta_{j,K}$ is provided by (22) with α_2 set to the value of α in the Theorem 1.

7 Examples

With a simulation example, the behavior of the AIC and its use in estimating the noncentrality parameter can be illustrated. An example that has
been previously considered in the literature (Larimore [10]) is a 2-input, 2-
output, 6-state system including state disturbances and feedback. For model
approximation, we consider fitting autoregressive models with exogenous inputs (ARX models). A 6-state system with zeros in the transfer function
cannot be expressed exactly by a finite order ARX model. Thus we can only
approximate the true system by such an ARX model.

The AIC as a function of ARX order is shown in Fig. 1. Note that the
minimum appears to be somewhat flat from order 6 through 12 however there
is a fair amount of variability.

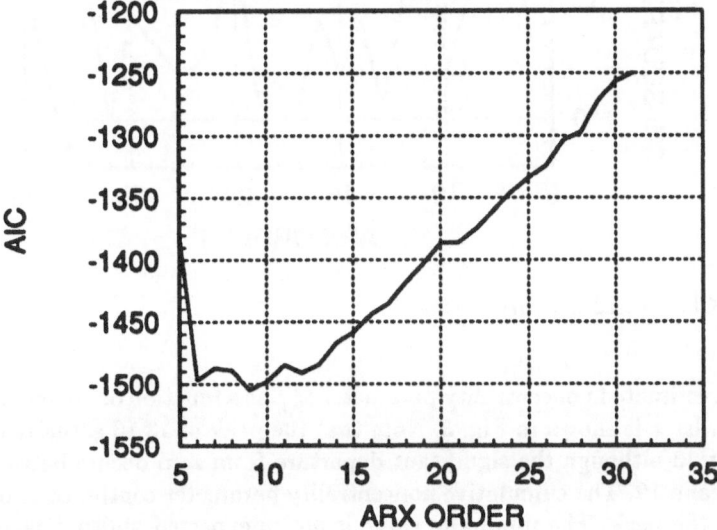

Fig. 1. AIC versus ARX order

Figure 2 shows the Chi-squared variable (21) calculated from the likelihood ratio $\lambda_{j,j+1}$ for a change of ARX order of one. This should be a Chi-
squared random variable on 8 degrees of freedom since for each increase in
ARX order there are two 2×2 matrices to be estimated. If the true ARX
order is 9 where the minimum AIC occurs, then the noncentrality parameter
would be zero for $j \geq 9$. The peak at order 11 of 21.8 is significant at the
0.01 level so that the probability is less than 0.01 that such a change in the
AIC occurred by chance. On the other hand, the peak at order 20 of 15.7 is
just barely significant at the 0.05 level. Since there are 21 orders greater than
order 9 being computed, it is not surprising that one of these is significant at

the 0.05 level. In the same light, the peak at order 11 could be considered as marginally significant except that it occurs very close to the order 9 minimum and the AIC is rather flat from order 6 through 12. In this respect, the order 11 case is an unlikely peak in only a few possible orders close to order 9.

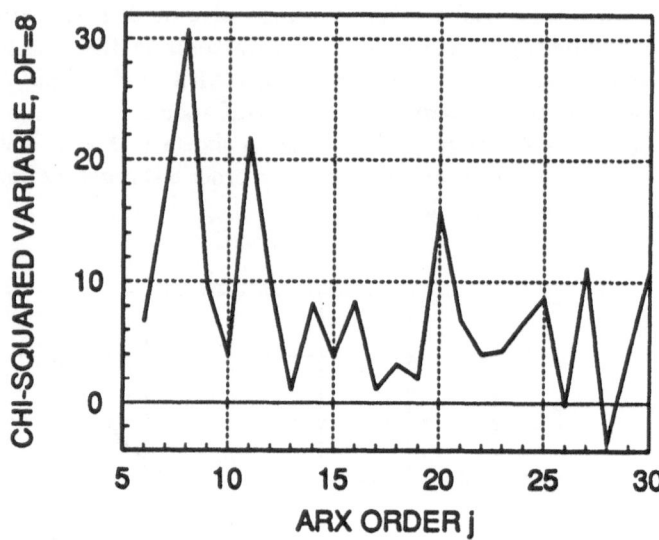

Fig. 2. Chi-squared measure of bias

The estimated noncentrality parameter $\delta_{9,k}^2$ as a function of assumed 'true' ARX order k is shown in Fig. 3. Note that the peak of 13.16 actually occurs at order 13 although the significant departure from zero occurs between orders 11 and 12. The cumulative noncentrality parameter continues to decline beyond the peak. The minimum value is not unexpected and indicates that there is no systematic nonzero value for high orders of the ARX model. It is only the jump from order 11 to 12 that indicates the likely under modeling of the ARX model.

Including the estimated noncentrality in the confidence band estimates will have little effect since the Chi-squared variable involved in (22) has 79 degrees of freedom and the noncentrality parameter is only 13.16. Plots of confidence bands on one of the input output pairs of the magnitude and phase frequency response is shown in Figs. 4 and 5. These confidence bands include the bias of model under fitting assuming that the true system order is order 12 while the fitted model order is 9.

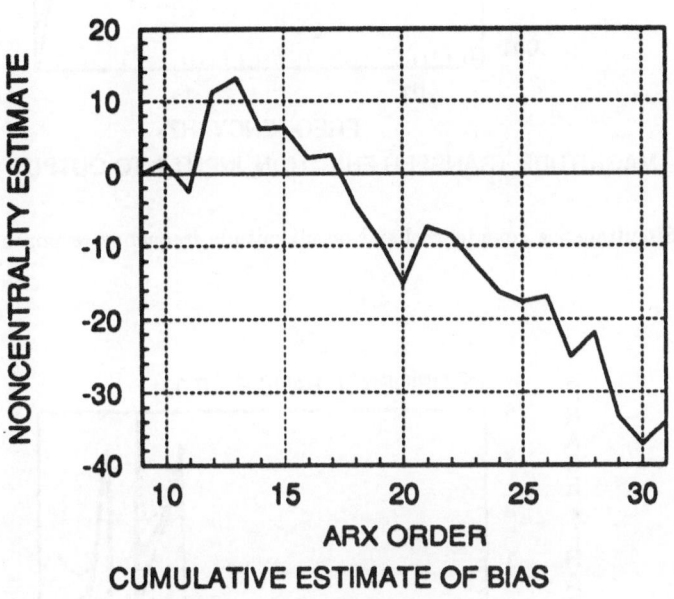

Fig. 3. Estimate of noncentrality parameter $\delta^2_{9,k}$

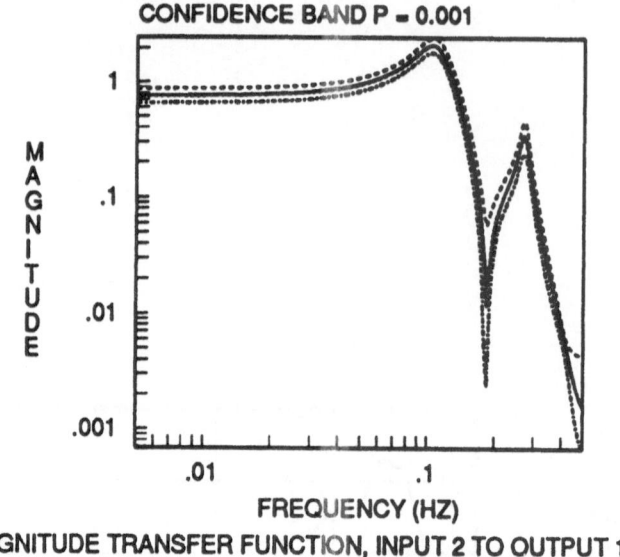

MAGNITUDE TRANSFER FUNCTION, INPUT 2 TO OUTPUT 1

Fig. 4. Simultaneous confidence band on magnitude frequency response function

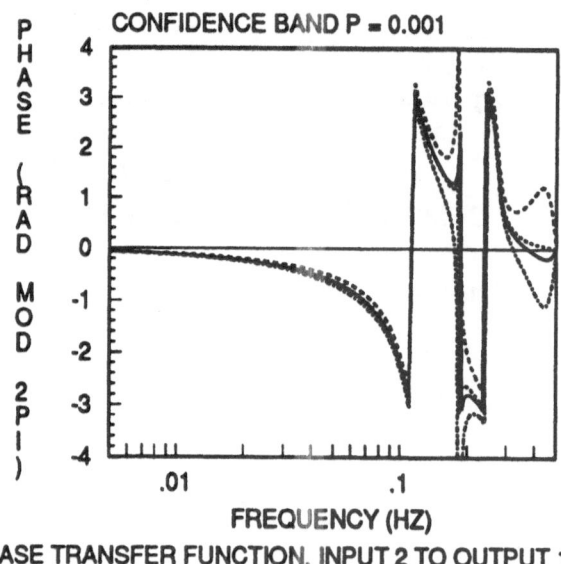

PHASE TRANSFER FUNCTION, INPUT 2 TO OUTPUT 1

Fig. 5. Simultaneous confidence band on phase frequency response function

References

1. Akaike, H.: Information Theory and an Extension of the Maximum Likelihood Principle, 2nd International Symposium on Information Theory, Eds. B.N. Petrov and F. Csaki, Budapest: Akademiai Kiado 267-281.

2. Anderson, T.W.: An Introduction to Multivariate Statistical Analysis, New York: Wiley.

3. Candy, J.V., Bullock, T.E., and Warren, M.E.: Invariant Description of the Stochastic Realization, Automatica, 15 (1979) 493-5

4. Hurvich, C.M., R. Shumway and C.L. Tsai: Improved Estimators of Kullback-Leibler Information for Autoregressive Model Selection in Small Samples, Biometrika 77 (1990) 709-720

5. Hurvich, C.M. and C.L. Tsai: Regression and Time Series Model Selection in Small Samples, Biometrika 76 (1989) 297-307

6. Hurvich, C.M. and C.L. Tsai: Bias of the Corrected AIC Criterion for Under-fitted Regression and Time Series Models, Biometrika 78 (1991) 499-510

7. Kullback, S.: Information Theory and Statistics, Dover

8. Larimore, W.E.: Predictive Inference, Sufficiency, Entropy, and an Asymptotic Likelihood Principle, Biometrika 70 (1983) 175-81

9. Larimore, W.E. and R.K. Mehra: The Problem of Overfitting Data, Byte 10 (1985) 167-80

10. Larimore, W.E.: Canonical Variate Analysis for System Identification, Filtering, and Adaptive Control, Proc. 29th IEEE Conference on Decision and Control, 1, Honolulu, Hawaii, December (1990) 635-9

11. Larimore, W.E.: Simultaneous Confidence Bands for Efficient Parametric Multivariate Spectral Estimation, Submitted for publication.

12. Shibata, R.: An Optimal Autoregressive Spectral Estimate, Ann. Statistics 9 (1981) 300-6

13. Shibata, R.: A Theoretical View of the Use of AIC, Proc of the International Time Series Meeting, O. D. Anderson, Ed.

Iterative Identification and Control Design: A Worked Out Example

Ruud Schrama and Paul Van den Hof

Mechanical Engineering Systems and Control Group,
Delft University of Technology, Mekelweg 2, 2628 CD Delft, The Netherlands.

1 Introduction

Recently it has been motivated that the problem of designing a high performance control system for a plant with unknown dynamicics through separate stages of (approximate) identification and model based control design requires iterative schemes to solve the problem [7, 11, 13, 16]. The underlying idea is that there actually is a joint problem of finding an appropriate model $\hat{P}$ of the plant P, and a controller $C_{\hat{P}}$ based on $\hat{P}$, such that $C_{\hat{P}}$ achieves a high performance for the modelled plant P and a similar performance for the nominal model $\hat{P}$. The former is the true control objective; the latter is needed in order that we are confident about the compensator $C_{\hat{P}}$. Simultaneous high performances are accomplished, if the feedback system composed of the nominal model $\hat{P}$ and its own high performance compensator $C_{\hat{P}}$ approximately describes the feedback system containing the plant P and the same compensator $C_{\hat{P}}$. The quality of each candidate nominal model depends on its own compensator and vice versa. Hence the problem of designing a high performance compensator for an imprecisely known plant boils down to a *joint problem* of approximate identification and model-based control design. Solving this joint problem through separate stages of identification and control design can be done, only if these procedures are embedded in an iterative scheme. We elaborate an iterative scheme, in which each identification is based on new data collected while the plant is controlled by the latest compensator. Each new nominal model is used to design an improved compensator, which replaces the old compensator.

A few iterative schemes proposed in literature have been based on the prediction error identification method, together with LQG/LTR control design [2] and with LQG control design [5, 6, 16]. Alternatively, in [8] the identification and control design are based on covariance data. In [7] the IMC-design method is employed, and the identification step is replaced by a model reduction based on full plant knowledge. Alternatively, in [10] an iteration is used to build prefilters for a control-relevant prediction error identification from one open-loop dataset.

Our iterative scheme is composed of a robust control design method and a frequency domain identification technique that are conceived in terms of coprime factorizations. We will discuss the iterative scheme, and show an extensive simulation example. For more details on the approach presented, the reader is referred to [11].

2 A Link Between Identification and Control Design

We adopt the following control design paradigm from [1, 9]. The feedback configuration of interest is the interconnection $H(\hat{P}, C)$, which is depicted in Fig. 1. The transfer matrix $T(\hat{P}, C)$ defined as

$$T(\hat{P}, C) \doteq \begin{bmatrix} \hat{P}(I+C\hat{P})^{-1}C & \hat{P}(I+C\hat{P})^{-1} \\ (I+C\hat{P})^{-1}C & (I+C\hat{P})^{-1} \end{bmatrix} \tag{1}$$

maps $\mathrm{col}(r_2, r_1)$ into $\mathrm{col}(\hat{y}, \hat{u})$. This transfer matrix is called the nominal *feedback matrix*, because it embodies all feedback properties like disturbance and noise attenuation, sensitivity, stability and robustness margins. The model-based controller $C_{\hat{P}}$ is derived from the nominal model $\hat{P}$ according to

$$C_{\hat{P}} = \arg\min_C \|T(\alpha\hat{P}, C/\alpha)\|_\infty \tag{2}$$

with $\alpha \in \Re$ a scalar weight. The resulting controller is optimally robust against stable perturbations of the normalized right coprime factors of $\alpha\hat{P}$ (see [1, 15] for details). At the same time this controller $C_{\hat{P}}$ pursues some traditional control objectives like a small sensitivity at the lower frequencies and a small complementary sensitivity at the higher frequencies. C_i optimizes robustness for a nominal performance level associated with α. The resulting designed feedback system has its bandwidth close to the cross-over frequency of $\alpha\hat{P}$ [9], and thus a large α corresponds to a high nominal performance.

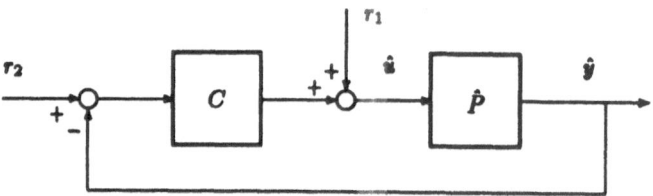

Fig. 1. Feedback configuration $H(\hat{P}, C)$ for control design

Conformably to (2) the nominal performance is high, if $\|T(\alpha\hat{P}, C_{\hat{P}}/\alpha)\|_\infty$ is small. We examine the performance norm of the actual plant P by the triangular inequalities:

$$\|T(\alpha\hat{P}, C_{\hat{p}}/\alpha)\| - \|T(\alpha P, C_{\hat{p}}/\alpha) - T(\alpha\hat{P}, C_{\hat{p}}/\alpha)\| \leq \|T(\alpha P, C_{\hat{p}}/\alpha)\|$$
$$\|T(\alpha P, C_{\hat{p}}/\alpha)\| \leq \|T(\alpha\hat{P}, C_{\hat{p}}/\alpha)\| + \|T(\alpha P, C_{\hat{p}}/\alpha) - T(\alpha\hat{P}, C_{\hat{p}}/\alpha)\| \qquad (3)$$

The middle term reflects the performance of the controlled plant. The nominal performance norm $\|T(\alpha\hat{P}, C_{\hat{p}}/\alpha)\|_\infty$ is minimized by the design of (2); and $\|T(\alpha P, C_{\hat{p}}/\alpha) - T(\alpha\hat{P}, C_{\hat{p}}/\alpha)\|$ is the 'worst-case' performance degradation due to the fact that $C_{\hat{p}}$ has been designed for the nominal model $\hat{P}$ rather than for the plant P. With the above inequalities we can make more precise the implications of the high performance control design problem. The point is to find a nominal model $\hat{P}$ with an induced controller $C_{\hat{p}}$ such that

$$1) \quad \|T(\alpha\hat{P}, C_{\hat{p}}/\alpha)\| \text{ is small} \qquad (4)$$

$$2) \quad \|T(\alpha P, C_{\hat{p}}/\alpha) - T(\alpha\hat{P}, C_{\hat{p}}/\alpha)\| \ll \|T(\alpha\hat{P}, C_{\hat{p}}/\alpha)\| . \qquad (5)$$

The requirement of (4) pertains to a high *nominal* performance. The strong inequality of (5) embodies the demand of a *robust* performance: if (5) is satisfied, then there is only a relatively small difference between the feedback properties of the designed and actual feedback systems $T(\hat{P}, C_{\hat{p}})$ and $T(P, C_{\hat{p}})$.

As the control design of (2) pursues a small nominal performance norm $\|T(\alpha\hat{P}, C_{\hat{p}}/\alpha)\|_\infty$, the remaining task for the approximate identification would be to find such a nominal model $\hat{P}$ that the performance degradation

$$\|T(\alpha P, C_{\hat{p}}/\alpha) - T(\alpha\hat{P}, C_{\hat{p}}/\alpha)\|_\infty$$

is relatively small. This approximate identification problem cannot be solved straightforwardly, because the compensator $C_{\hat{p}}$ is not available prior to the identification. This explains once more that the problems of approximate identification and model-based control design have to be treated as a joint problem.

Note that the bounds of (3) are used to express the identification objective in terms of the control objective of (2). The same approach applies to any other control design method that optimizes a norm or a distance function of the nominal feedback matrix $T(\hat{P}, C_{\hat{p}})$. As explained in [11] these methods include LQ control design and the H_∞-optimization of a weighted sensitivity.

As the choice of α refers to a required nominal performance level, we will not fix its value a priori, e.g. aiming at a very high but unachievable performance, but we will gradually increase α during the iteration process. A motivation for this will be given later on.

We propose the following iterative scheme to tackle the joint problem of approximate identification and model-based control design.

Step i. Given $\hat{P}_{i-1}, C_{i-1}, \alpha_{i-1}$

(a) Obtain data from the plant, while it operates under feedback by C_{i-1}. The nominal model $\hat{P}_i$ is identified with an identification scheme that asymptotically obtains

$$\hat{P}_i = \arg \min_{P \in \mathcal{P}(\theta)} \|T(\alpha_{i-1}P, C_{i-1}/\alpha_{i-1}) - T(\alpha_{i-1}\bar{P}, C_{i-1}/\alpha_{i-1})\|_2 \quad (6)$$

where $\mathcal{P}(\theta)$ is the set of parameterized candidate models.

(b) Determine $\alpha_i > \alpha_{i-1}$ and design a new controller C_i according to

$$C_i = \arg \min_C \|T(\alpha_i \hat{P}_i, C/\alpha_i)\|_\infty . \quad (7)$$

such that the performance degradation $\|T(\alpha_i P, C_i/\alpha_i) - T(\alpha_i \hat{P}_i, C_i/\alpha_i)\|_\infty$ is kept small.

(c) Perform a robust stability test to verify whether the plant P will be stabilized by the new controller C_i, prior to implementing it.

Note that we have replaced the infinity-norm with a 2-norm in (6). This is done since there exists no identification technique yet that can handle an H_∞ (or L_∞) approximation. The rationale for this replacement is that the L_2 approximation yields a reasonably good nominal model in an L_∞ sense, provided that the error-term is sufficiently smooth. This observation is backed up by the result in [3] on the L_∞ consistency of L_2 estimators.

Since the control design scheme does not take account of model uncertainties directly, the design weight is used to tune the design. We intend to gradually increase the design weight during the iteration in order to keep the performance degradation small at each iteration step. In this way we guarantee that in the control design step, there remains a good resemblance between the feedback properties of $H(\hat{P}_i, C_i)$ and $H(P, C_i)$.

The different steps in the iterative scheme will be described in more detail in the next sections.

3 Control-Relevant Identification

We consider the feedback configuration of Fig. 2, in which the plant P is stabilized by the controller C_{i-1}. The feedback system is driven by the exogenous inputs r_1 and r_2 and the additive output noise v. The noise v is uncorrelated with r_1 and r_2 and it is modelled as $v = P_{yw}w$, where w is a white noise.

The problem of concern is to identify a nominal model $\hat{P}_i$ from measurements of u and y such that the asymptotic identification criterion reflects

$$\hat{P}_i = \arg \min_{P \in \mathcal{P}(\theta)} \|T(P, C_{i-1}) - T(\bar{P}, C_{i-1})\|_2 . \quad (8)$$

Only for notational simplicity we use $\alpha_{i-1} = 1$. We recall from the previous sections that we actually use system identification to find an approximate description of the feedback properties of $H(P, C_{i-1})$. Therefore we concentrate on the so-called "asymptotic bias distribution" due to undermodelling.

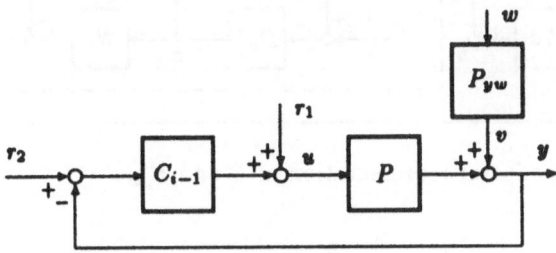

Fig. 2. Feedback configuration for identification

Since $P \notin \mathcal{P}(\theta)$ the minimization in (8) from u and y combines all problems that are encountered in approximate identification and in closed-loop identification. The desired $\hat{P}_i$ cannot be derived by a direct application of some standard identification method to u and y. In order to obviate this problem we first represent the plant P by a right coprime factorization (definitions are provided in [15]).

The plant P is known to be stabilized by the latest controller C_{i-1}. As P belongs to the set of all systems that are stabilized by C_{i-1}, it can be represented by a coprime factorization that is dual to the (Youla-) parameterization of all stabilizing compensators [15]. This dual parameterization can be extended to incorporate the "noise filter" P_{yw}, [14, 11]. A similar parameterization has been used by Hansen [4] for closed-loop experiment design.

This parametrization of P is sketched in Fig. 3, with P_0 an auxiliary model that is stabilized by C_{i-1}; $P_o = N_0(D_o)^{-1}$, and $C_{i-1} = N_c(D_c)^{-1}$ are right coprime factorizations, and R, S any stable transfer functions. For notational convenience we define

$$N \doteq N_o + D_c R; \quad D \doteq D_o - N_c R, \tag{9}$$

so that $P = ND^{-1}$, which is the dual of the Youla parameterization. The following result can now be employed.

Proposition 1. *[11, 12] Consider the notation and parametrization as presented above. Then*

(a)

$$\begin{pmatrix} y \\ u \end{pmatrix} = \begin{bmatrix} N \\ D \end{bmatrix} x + \begin{bmatrix} D_c \\ -N_c \end{bmatrix} Sw \tag{10}$$

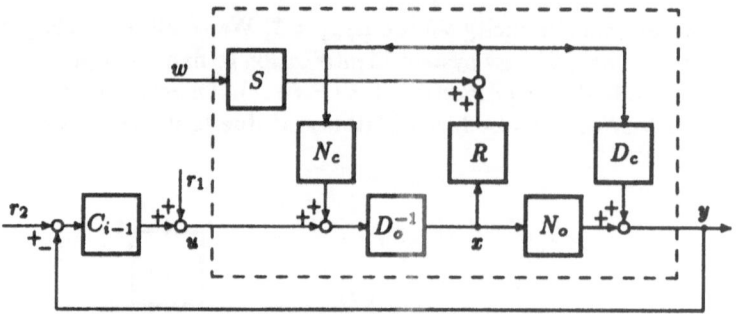

Fig. 3. Coprime factor representation of P and P_{yw}.

(b)

$$x = (D_o + C_{i-1}N_o)^{-1}(u + C_{i-1}y) \tag{11}$$
$$= (D_o + C_{i-1}N_o)^{-1}(r_1 + C_{i-1}r_2) \tag{12}$$

and x is uncorrelated with w provided that w is uncorrelated with r_1, r_2;
(c)

$$T(P, C_{i-1}) = \begin{bmatrix} N \\ D \end{bmatrix} (D_o + C_{i-1}N_o)^{-1} \begin{bmatrix} C_{i-1} & I \end{bmatrix} . \tag{13}$$

The proposition shows that the signal x can be reconstructed from the measure signals u and y, and that x is "designed" through appropriate choice of r_1 and/or r_2. Since x is uncorrelated with w, (10) shows that identification of the coprime factors (N, D) actually is an open loop identification problem. This creates the possibility of approximate identification of $P = ND^{-1}$ with an asymptotic identification criterion that is not influenced by w.

Application of an output error identification algorithm to (10) with a prefilter L, shows an asymptotic identification criterion (for SISO systems):

$$\int_\omega [|N(jws) - \hat{N}(jws)|^2 + |D(jws) - \hat{D}(jws)|^2]\Phi_x(\omega)|L(jws)|^2 d\omega . \tag{14}$$

Through the choice of a prefilter L, satisfying

$$|L(jws)|^2 = \frac{1 + |C_{i-1}(jws)|^2}{\Phi_{r_1}(\omega) + |C_{i-1}(jws)|^2\Phi_{r_2}(\omega)} \tag{15}$$

the expression (14) can be shown to be equal to $\|T(P, C_{i-1}) - T(\hat{P}, C_{i-1})\|_2^2$.

In the example shown later on, we will employ (10) to construct an estimate of the frequency response of $N(jws)$, $D(jws)$; this estimate is substituted for P in (8), in order to construct $\hat{P}_i$ through nonlinear optimization. This latter minimization problem is all but trivial, because the nominal model $\bar{P}_i$ appears in $T(\bar{P}_i, C_{i-1})$ in a multiple and non-linear fashion. The problem is attacked by the Newton-Raphson method in [11]. Due to its highly non-linear character the utility of this particular optimization hinges on a good initial estmate. In [11] such an estimate is obtained by parametrizing $\bar{P}_i$ in (8) in terms of its coprime factors.

Finally we have to verify whether the estimated model $\hat{P}_i$ induces a performance degradation

$$\|T(\alpha_{i-1}P, C_{i-1}/\alpha_{i-1}) - T(\alpha_{i-1}\hat{P}_i, C_{i-1}/\alpha_{i-1})\|_\infty$$

that is sufficiently small with respect to the nominal performance. If this is not true, we might have to increase the order of the model. The evaluation of this degradation is done by replacing P by its estimated frequency response, and evaluating the norm for the available frequency response data points.

4 Enhancement of the Controller

We have a system $H(\hat{P}_i, C_{i-1})$ that provides a good description of $H(P, C_{i-1})$ in view of the weighted performance norm. We may expect that this holds also if C_{i-1} is slightly changed. Hence we design an improved controller C_i for $\hat{P}_i$ in such a way that C_i does not differ too much from the old controller C_{i-1}. The change of the compensator will be moderate if the performance requirements are increased moderately. Hence we may choose α_i a bit larger than α_{i-1}. We outline how, in essence, this selection of α_i is guided by a frequency response estimate of P (details can be found in [11]). We build this estimate from the frequency response estimates of N and D used in the previous section. Then we evaluate the ratio of maximum singular values

$$\frac{\bar{\sigma}\{T(\alpha_i P, C_{i-1}/\alpha_i)(jws)\}}{\bar{\sigma}\{T(\alpha_i P, C_i/\alpha_i)(jws)\}} \left/ \frac{\bar{\sigma}\{T(\alpha_i \hat{P}_i, C_{i-1}/\alpha_i)(jws)\}}{\bar{\sigma}\{T(\alpha_i \hat{P}_i, C_i/\alpha_i)(jws)\}} \right.$$

and a similar ratio for the upper bound of (3). We choose α_i such that these ratios are bounded for every frequency response sample of P. Thereby C_i changes $H(P, C_{i-1})$ similarly to $H(\hat{P}_i, C_{i-1})$. In the example shown later on, these bounds are chosen to be 0.7–1.3.

As the choice of α_i is based on a "prediction" of the frequency response of $T(P, C_i)$, the feedback systems $H(\hat{P}_i, C_i)$ and $H(P, C_i)$ are expected to be similar in an L_∞-sense. However, stability still has to be ascertained.

5 Robustness Analysis

Before the enhanced compensator C_i is actually applied to the plant P, we have to ascertain the stability of the new control system $H(P, C_i)$. This stability test is necessary anyway, because the optimization of robustness against stable coprime factor perturbations is an unconstrained optimization, without the guarantee of obtaining sufficient robustness. We employ the following result from [14].

Proposition 2. *Let $\hat{P}_i = \hat{N}\hat{D}^{-1}$ be stabilized by $C_i = \tilde{Y}^{-1}\tilde{X}$, the latter being a normalized coprime factorization, such that $\tilde{Y}\hat{D} + \tilde{X}\hat{N} = I$. Denote $P_\Delta = N_\Delta D_\Delta^{-1}$ with $N_\Delta = \hat{N} + \Delta N$, $D_\Delta = \hat{D} + \Delta D$. Then $H(P_\Delta, C_i)$ is stable for all P_Δ such that $\left\| \begin{bmatrix} \Delta_N \\ \Delta_D \end{bmatrix} \right\|_\infty < 1$.*

The result is used as follows. Knowing $\hat{P}_i$ and C_i, we can construct $\hat{N}, \hat{D}$. For N_Δ, D_Δ we substitue the estimated frequency response of the coprime factors (N, D) of P. If there exists a stable filter Q such that

$$\left\| \begin{bmatrix} \hat{N} - N_\Delta Q \\ \hat{D} - D_\Delta Q \end{bmatrix} \right\|_\infty < 1,$$

the stability of $H(P, C_i)$ can be concluded. In practice the 2-norm of the expression is minimized over stable Q, and the inequality is verified for the ∞-norm [11, 14].

6 Simulation Study

We apply the iterative approach to a simulation example. The data consist of 100 frequency response samples that are uniformly distributed over a logarithmic interval ranging from 0.1 to 100 rad/s. We use exact frequency response data in order to stress the effects due to undermodelling. We merely list the results of this iterative high performance control design procedure, which is investigated in much more detail in [11].

The continuous-time plant P under investigation has a transfer function $n(s)/d(s)$ with

$$n(s) = 30s^6 + 3020s^5 + 30538s^4 + 40373s^3 + 74041s^2 + 41972s + 12467$$
$$d(s) = s^8 + 26.023s^7 + 321.70s^6 + 2635.9s^5 + 10412s^4$$
$$+ 3091.4s^3 + 11032s^2 + 306.81s + 986.86 .$$

In order to simulate a real application we pretend that the plant P is imprecisely known. Accordingly we do not use any knowledge of the plant's number of poles or (unstable) zeros; we just know that P is open-loop stable. Hence we cannot tell a priori how complex a compensator must be in order to obtain some performance. Conversely we do not know what performance is achievable with a compensator of constrained complexity.

The iteration commences with an open-loop identification of $\hat{P}_1$. Figure 4 shows the Bode log-magnitude plots of P (—) and $\hat{P}_1$ (--). The nominal model $\hat{P}_1$ provides an accurate description of the low frequency behavior of P. The mismatch at the higher frequencies hardly contributes to the identification criterion of (8) with $C_0 = 0$, because this criterion measures an additive error on a linear scale.

From $\hat{P}_1$ we design the compensator C_1 as in (7) with $\alpha_i = 0.113$. We apply C_1 to P, we obtain new data, and we subsequently derive several nominal models and compensators. The iteration ends with the nominal model $\hat{P}_5(s) = \hat{n}(s)/\hat{d}(s)$, where

$$\hat{n}(s) = 8.8 \cdot 10^{-4} s^5 - 4.77 \cdot 10^{-2} s^4 + 34.7 s^3 + 2494 s^2 + 1663 s + 6028$$
$$\hat{d}(s) = s^5 + 13.3 s^4 + 156.3 s^3 + 712.4 s^2 + 131.3 s + 369.4,$$

with the compensator $C_5(s) = n_c(s)/d_c(s)$, where

$$n_c(s) = 71.407 s^4 + 2182.1 s^3 + 28718 s^2 + 23854 s + 68457$$
$$d_c(s) = s^4 + 129.16 s^3 + 4829.0 s^2 + 3344.1 s + 11571,$$

and with $\alpha_5 = 20$. The evolution of the nominal models and of the controllers are illustrated respectively in Fig. 4 and Fig. 5. The latter figure displays the gradual increase of control action. The former figure reveals that during the iteration the accuracy of the nominal model is improved in the high frequency range at the expense of a large mismatch for the lower frequencies. Despite the large open-loop mismatch between P and $\hat{P}_5$ (see again Fig. 4), the nominal model $\hat{P}_5$ is suited for high performance control design. This is illustrated in Fig. 6, which shows the log-magnitudes of $T(P, C_5)$ and of

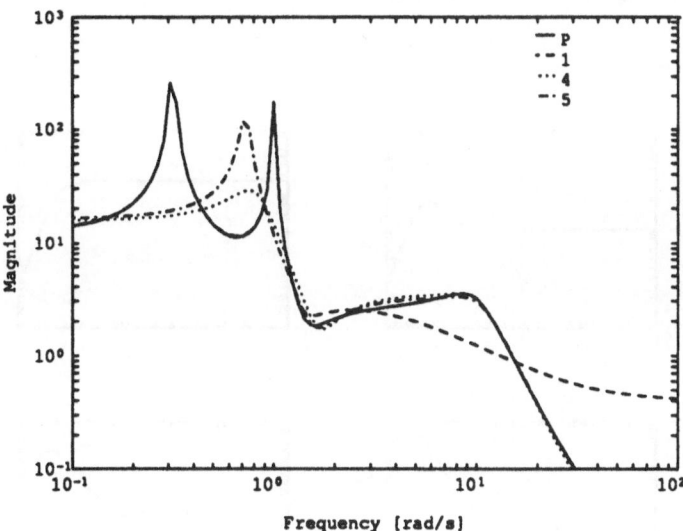

Fig. 4. Log-magnitudes of P (—), $\hat{P}_1$ (--), $\hat{P}_4$ (⋯) and $\hat{P}_5$ (-·-)

$T(\hat{P}_5, C_5)$. Considering the logarithmic scale we may conclude that P and $\hat{P}_5$ have very similar high performances under feedback by C_5. Hence the couple $\hat{P}_5, C_5$ is a solution to the joint problem of approximate identification and model-based control design.

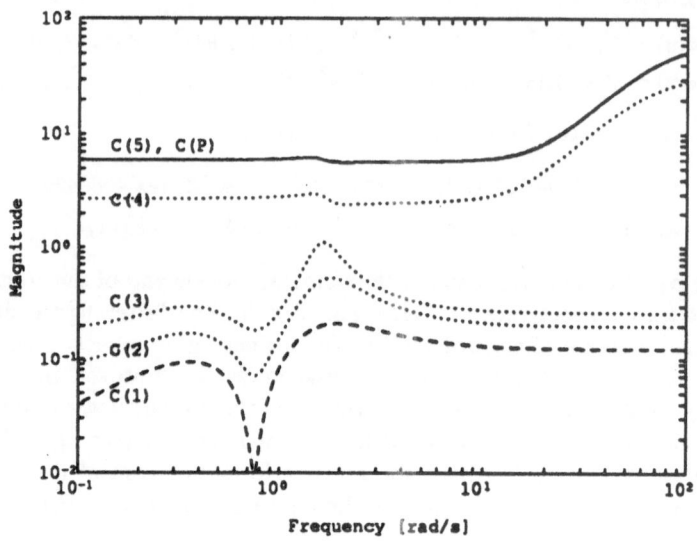

Fig. 5. Log-magnitudes of the designed controllers

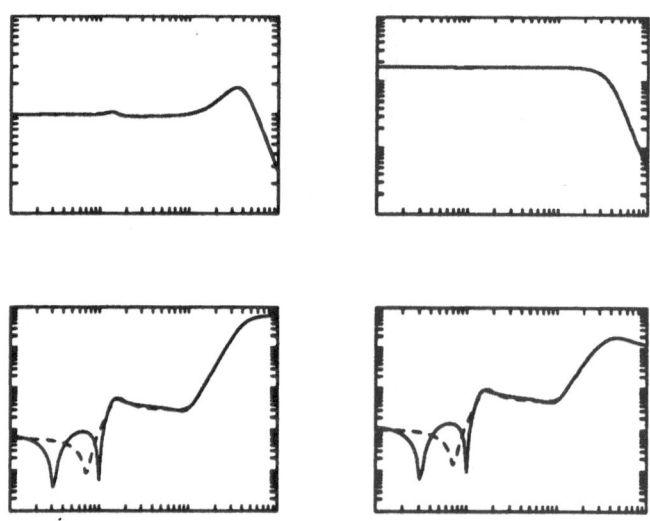

Fig. 6. Log-magnitudes of $T(P, C_5)$ (—) and $T(\hat{P}_5, C_5)$ (--)

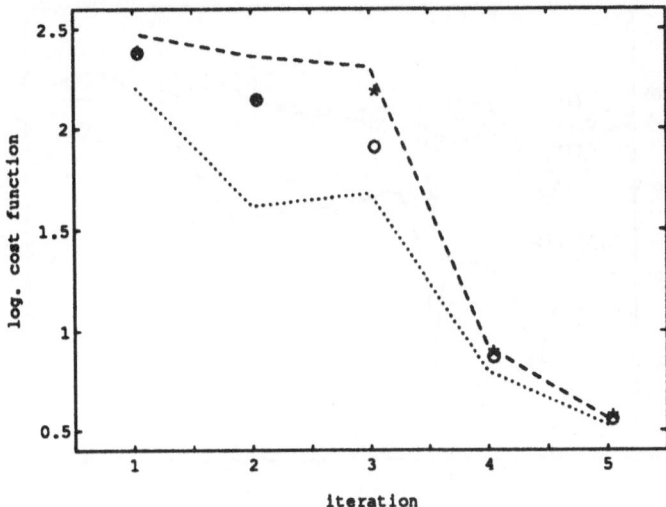

Fig. 7. Logarithm of the performance norms

Note that the model error that appears in the low frequency range is due to the fact that we have chosen a model order that is too small to capture all system dynamics. If one additionally to the high performance control requirements, would require the estimated model to have a similar open loop response as the plant, one has to "pay" for that in terms of a higher model order.

We evaluate the performance norms for all pairs of nominal models and compensators for α_5. That is, we determine for instance $\|T(\alpha_5 P, C_i/\alpha_5)\|$ as the maximum singular value over all frequency reponse samples. These performance norms have been plotted in Fig. 7. The indices at the horizontal axis indicate the iteration step. The performance norms corresponding to $T(\hat{P}_i, C_i)$ and $T(P, C_i)$ are marked respectively by 'o' and '$\star$'. The upper bound of (3), indicated by (--), and the analogous lower bound (····) disclose that the approximation of $T(P, C_i)$ by $T(P_i, C_i)$ is relatively accurate. This is a direct consequence of the frequency reponse based controller enhancement of Section 4. The figure also displays that the "worst-case" performance (--) is improved in each step of the iteration. Finally Fig. 8 shows the evolution of the sensitivity that is achieved for the plant P.

We complete the evaluation by using the method of (7) and α_5 to design also the compensator C_P of order 4 directly from the plant P. In regard of α_5 this C_P is the optimal compensator of order 4 that can be designed for and from the plant P. In Fig. 5 we see that the frequency responses of C_P and C_5 are indiscernible, which produces indiscernible sensitivities for P (see Fig. 8). Thus the iteratively designed high performance compensator C_5 is

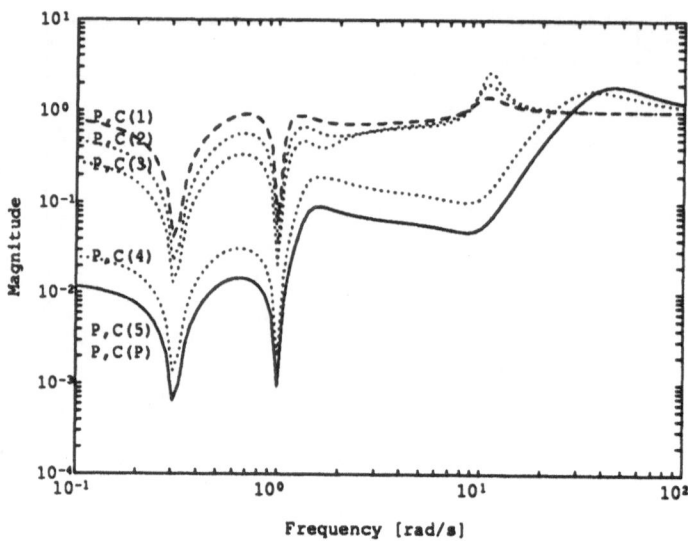

Fig. 8. Sensitivities achieved for the plant P

almost identical with the optimal plant-based compensator C_P, even though no exact knowledge of P nor any information of C_P has been used to achieve C_5.

Lastly we elucidate the need of an iteration to solve the joint problem of approximate identification and model-based control design. The left upper term of $T(P, C_5) - T(\hat{P}_5, C_5)$, which equals $PC_5(I+PC_5)^{-1} - \hat{P}_5 C_5(I+\hat{P}_5 C5)^{-1}$, can be rewritten to $(I+PC_5)^{-1}(P-\hat{P}_5)C_5(I+\hat{P}_5 C5)^{-1}$. Similar expressions can be derived for the other elements of $T(P, C_5)$ and $T(\hat{P}_5, C_5)$. Hence $\hat{P}_5, C_5$ make a *couple* that produces a small mismatch

$$W_L(P, C_5) \ (P - \hat{P}_5) \ W_R(\hat{P}_5, C_5),$$

where W_L and W_R are weighting functions depending on P, $\hat{P}_5$ and C_5. It is tempting to suggest that $\hat{P}_5$ could have been obtained directly from a weighted open-loop identification. However, $W_L(P, C_5)$ and $W_R(\hat{P}_5, C_5)$ depend on the outcome of the iteration, and thus the required weighting functions are not available at the outset.

7 Concluding Remarks

We addressed the problem of designing a high performance compensator for an imprecisely known plant. We tackled this problem by an iterative scheme of repeated identification and control design. At each stage of the iteration data is obtained from the plant while it is controlled by the latest compensator.

As the iterative design procedure evolves, it learns about the control-relevant dynamics of the plant in question. The resulting nominal model is accurate near the cross-over frequency and, at least as important, the large mismatch at other frequencies does not impair the control design. In addition the iteration reveals the performance that is attainable for the imprecisely known plant.

References

1. P.M.M. Bongers and O.H. Bosgra (1990). Low order robust H_∞ controller synthesis. *Proc. 29th IEEE Conf. Decision and Control*, Honolulu, HI, pp. 194-199.

2. R.R. Bitmead, M. Gevers and V. Wertz (1990). *Adaptive Optimal Control, The Thinking Man's GPC*. Prentice Hall, Englewood Cliffs, NJ.

3. P.E. Caines and M. Baykal-Gürsoy (1989). On the L_∞ consistency of L_2 estimators. *Syst. Control Lett.*, vol.12, pp. 71-76.

4. F.R. Hansen (1989). *A Fractional Representation Approach to Closed Loop System Identification and Experiment Design*. Ph.D.-Thesis, Stanford University, Stanford, CA, March 1989.

5. R.G. Hakvoort (1990). Optimal experiment design for prediction error identification in view of feedback design. *Selected Topics in Identification, Modelling and Control*, vol. 2. Delft University Press, pp. 71-78.

6. R.G. Hakvoort, R.J.P. Schrama and P.M.J. Van den Hof (1992). Approximate identification in view of LQG feedback design. *Proc. 1992 American Control Conf.*, June 26-28, 1992, Chicago, IL, pp. 2824-2828.

7. W.S. Lee, B.D.O. Anderson, R.L. Kosut and I.M.Y. Mareels (1992). On adaptive robust control and control-relevant system identification. *Proc. 1992 American Control Conf.*, June 26-28, 1992, Chicago, IL, pp. 2834-2841.

8. K. Liu and R.E. Skelton (1990). Closed loop identification and iterative controller design. *Proc. 29th IEEE Conf. Decision and Control*, Honolulu, HI, pp. 482-487.

9. D. McFarlane and K. Glover (1988). An H_∞ design procedure using robust stabilization of normalized coprime factors. *Proc. 27th IEEE Conf. Decision and Control*, Austin, TX, pp. 1343-1348.

10. D.E. Rivera (1991). Control-relevant parameter estimation: a systematic procedure for prefilter design. *Proc. 1991 American Control Conf.*, Boston, MA, pp. 237-241.

11. R.J.P. Schrama (1992). *Approximate Identification and Control Design with Application to a Mechanical System*. Dr. Dissertation, Delft University of Technology, May 1992.

12. R.J.P. Schrama and P.M.J. Van den Hof (1992). An iterative scheme for identification and control design based on coprime factorizations. *Proc. 1992 American Control Conf.*, June 24-26, 1992, Chicago, IL, pp. 2842-2846.

13. R.J.P. Schrama (1992). Accurate models for control design: the necessity of an iterative scheme. *IEEE Trans. Automat. Contr.*, AC-37, pp. 991-994.

14. R.J.P. Schrama, P.M.M. Bongers and D.K. de Vries (1992). Assessment of robust stability from experimental data. *Proc. 1992 American Control Conf.*, Chicago, IL, pp. 266-270.

15. M. Vidyasagar (1985). *Control System Synthesis: A Factorization Approach.* M.I.T.-Press, Cambridge, MA.

16. Z. Zang, R.R. Bitmead and M. Gevers (1991). H_2 iterative model refinement and control robustness enhancement. *Proc. 30th IEEE Conf. Decision and Control*, Brighton, UK, pp. 279-284.

Frequency Domain Identification for Robust Control Design

David S. Bayard and Yeung Yam

Jet Propulsion Laboratory, California Institute of Technology,
4800 Oak Grove Drive, Pasadena, California 91109

1 Introduction

This paper demonstrates an approach to frequency domain identification for the explicit purpose of designing robust H_∞ controllers. The approach transforms raw experimental data into a plant set estimate directly usable by modern robust control design software (e.g., Matlab Robust Control Toolboxes [2, 11]). A key issue in control design from raw data is the question of whether the controller will work when applied to the true system. The main feature of this approach is that the resulting controller is guaranteed to work as designed (when applied to the true system) to a prescribed statistical confidence. While the overall methodology addresses key theoretical issues, it has at the same time been specifically designed to support practical implementations.

The basic approach put forth in this paper has already been reported in an earlier 1991 conference paper [29]. However, due to the brevity of the original paper (3 pages), and recent interest in control relevant identification (as evidenced by the present UCSB workshop), it was felt that an expanded version of the original paper would be warranted. This task is simplified by the fact that each of the constituent algorithms have since appeared in the literature, cf., [3, 4, 6, 8, 23, 24]. This fact has simplified the present treatment by allowing all of the essential elements behind the algorithms to be discussed, while referring to the literature for further details.

There is presently a great deal of excellent research going on in the areas of plant set estimation, control-relevant identification, robust identification etc.. Due to the present focus as discussed above, there will be no attempt to survey other methods or compare various approaches. In this respect, this paper is very limited to one particular point of view. It is emphasized that this is not the only point of view nor necessarily the best approach. In the end, the main justification for the present approach may be that it is specifically designed to support practical implementations, and has already been demonstrated successfully in a number of large flexible structures studies.

For the purposes of this paper, a linear multivariable plant $\mathcal{P}(z^{-1})$ will be identified in the representation shown in Fig. 1. Many modern robust control methods are applicable to uncertainty expressed in this form. Here, $\hat{P}(z^{-1})$ is a nominal estimate of the true plant $\mathcal{P}(z^{-1})$; Δ_A is the additive uncertainty defined as $\Delta_A = \mathcal{P} - \hat{P}$; $C(z^{-1})$ is the digital controller under consideration, d is a disturbance, and $W_d(z^{-1})$ is a frequency weighting filter which characterizes the effect of $d(k)$ on the open-loop plant output $y(k)$.

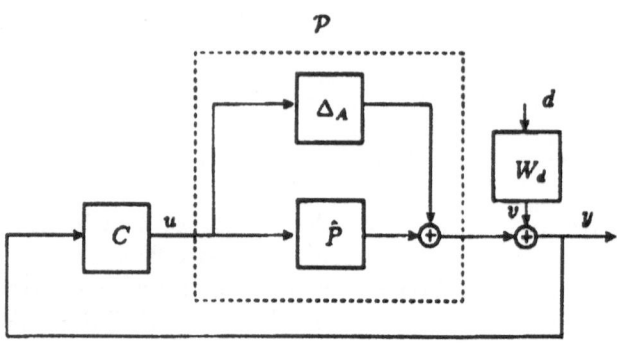

Fig. 1. Canonical representation for identification and robust control

For control design purposes, it is desirable to represent the additive uncertainty in the form $\Delta_A = \Delta W_A$ such that Δ is norm bounded, i.e., such that $||\Delta||_\infty \leq 1$. The filter W_A is then typically incorporated into the control design, to ensure robustness properties over the additive uncertainty set.

The goal of this paper will be to identify a nominal plant estimate $\hat{P}$, and a weighting filter W_A such that the relation $\mathcal{P} = \hat{P} + \Delta W_A$ holds (for some $||\Delta||_\infty \leq 1$) to a specified statistical confidence $1 - \kappa$ specified by the designer. Robust control synthesis methods can then be used to find a compensator C that has desirable stability/performance properties for all plants in the uncertainty set defined by $\hat{P}$ and W_A.

An overview of the basic approach is given in Sect. 2. Frequency domain identification methods to determine a nominal plant estimate $\hat{P}$ are given in Sect. 3. Plant set estimation is discussed in Sect. 4, leading to an additive uncertainty weighting W_A. Robust control design methods incorporating quantities $\hat{P}$ and W_A are discussed in Sect. 5. A flexible structures control example is given in Sect. 6 and conclusions are postponed until Sect. 7.

2 Background and Overview

2.1 Overview

The overall identification and robust control design approach is outlined schematically in Fig. 2. For simplicity, the approach will be described here for a single-input single-output (SISO) plant. Extensions to the multivariable case are given in the body of the paper.

The approach begins with the gathering of input-output time-domain data. Here, a Schroeder-phased multisinusoidal input u_s is applied to the plant until the output y_s reaches steady-state. Input/output data are gathered in steady-state, and averaged using the spectral estimation processing scheme of Fig. 3 to give the plant spectral estimate P_s. The spectral estimate P_s is curve fitted by minimizing a weighted 2-norm error criteria using the algorithms developed in [3], to obtain a transfer function B/a. The transfer function matrix is then realized/balanced/reduced to give the state-space model $\hat{P} = (A, B, C, D)$.

An additive uncertainty ball $\ell_A^{1-\kappa}(\omega) \geq \overline{\sigma}(\mathcal{P} - \hat{P})$ is then characterized in nonparametric form to within statistical confidence $(1 - \kappa) \times 100\%$ (specified by the designer) using the statistical plant set estimation methods found in [4]. The profile $\ell_A^{1-\kappa}(\omega)$ is overbounded (tightly) using the linear programming spectral overbounding and factorization (LPSOF) algorithm from [23, 24] to give a parametric minimum-phase transfer function W_A of specified order such that the additive uncertainty has the form $\Delta_A = \Delta W_A$ where $||\Delta||_\infty \leq 1$ is norm bounded.

Using the nominal state-space estimate $\hat{P}$ and the additive uncertainty set parametrized by W_A, available robust control methods can be used to design a compensator which will ensure some specified stability/performance for all plants within the uncertainty set. For example, it will be shown here that the robust control design problem can be posed in terms of a weighted mixed sensitivity H_∞ problem and solved using existing software from [11].

2.2 A Priori Knowledge

Plant set estimation is not possible without some initial knowledge about the plant and environment. The a priori knowledge required in the present scheme is stated precisely in the set of assumptions below. Additional assumptions will be stated as needed in the body of the paper.

First, the following definition will be needed.

Definition 1. A MIMO transfer function $G(z^{-1})$ is said to be in $\mathcal{D}(M, \rho)$ if the impulse response matrix sequence $\{g(kT)\}_{k=0}^{\infty}$ defined by the Z-transform relation $\sum_{i=0}^{\infty} g(kT)z^{-1} = G(z^{-1})$ satisfies,

$$\overline{\sigma}(g(kT)) \leq M\rho^k \qquad (2.1)$$

for some $\infty > M > 0$ and $1 > \rho \geq 0$.

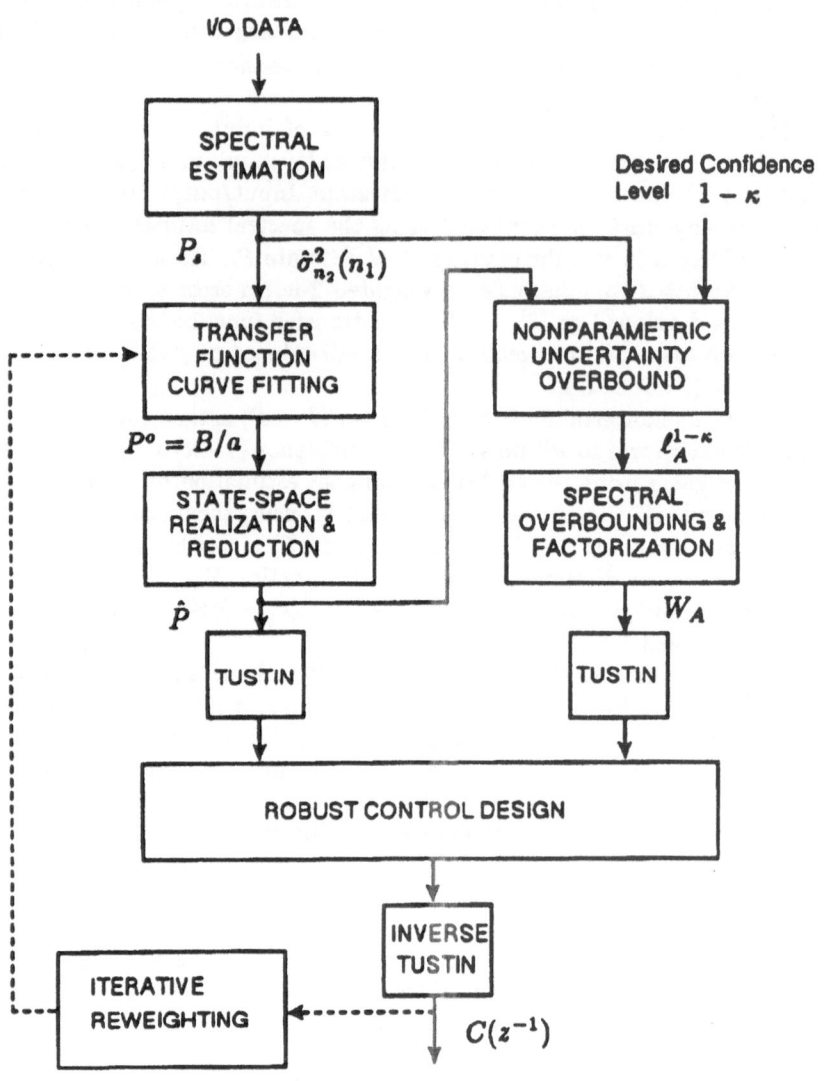

Fig. 2. Frequency domain identification and robust control approach

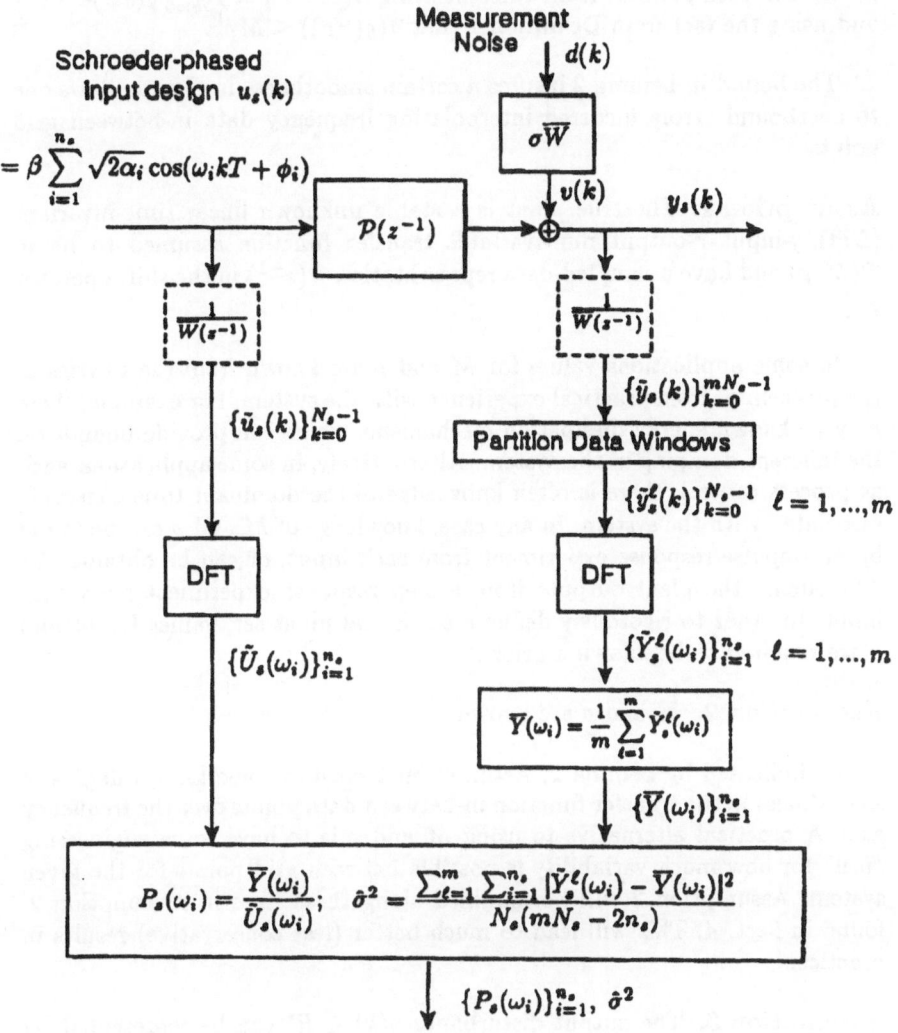

Fig. 3. Spectral estimation and noise variance estimation using Schroeder-phased multisinusoidal input design

The main usefulness of Definition 1 is due to the next lemma.

Lemma 2. *Let* $G(z^{-1}) \in \mathcal{D}(M, \rho)$. *Then the modulus of the derivative of* G *on the unit circle can be uniformly bounded from above as follows,*

$$\overline{\sigma}\left(\frac{dG(e^{-j\omega T})}{d\omega}\right) \leq \frac{TM\rho}{(1-\rho)^2}$$

Proof. The result follows from differentiating $G(e^{-j\omega T}) = \sum_{k=0}^{\infty} g(kT)e^{-j\omega kT}$ and using the fact from Definition 1 that $\bar{\sigma}(g(kT)) \leq M\rho^k$.   □

The bound in Lemma 2 insures a certain smoothness in G and allows one to overbound errors incurred interpolating frequency data in-between grid points.

Assumption 1. The true plant is a stable unknown linear time-invariant (LTI) q-input r-output multivariable transfer function assumed to be in $\mathcal{D}(M, \rho)$ and have a sampled-data representation $\mathcal{P}(z^{-1})$ in the shift operator z^{-1}.

In some applications values for M and ρ are known from the physics of the problem or from practical experience with the system. For example, there may be known energy dissipation mechanisms which can provide bounds on the inherent damping in the system. Alternatively, in some applications such as process control, there is often knowledge of the dominant time constants associated with the system. In any case, knowledge of M and ρ can be found by an impulse response experiment from each input, or can be obtained by differencing the plant outputs from a step response experiment from each input. In order to rigorously define a statistical plant set, values for M and ρ are required to be known a priori.

Assumption 2. M and ρ are known.

As indicated by Lemma 2, Assumption 2 ensures some known degree of smoothness in the transfer function in-between data points over the frequency axis. A practical alternative to using M and ρ is to have some engineering "feel" for how much variability is possible between grid points for the given system. Assumption 2 will be modified along these lines in Assumption 2' found in Sect. 4. This will lead to much better (less conservative) results in practice.

Assumption 3. The output disturbance $v(k) \in R^r$ can be represented by $v(k) = W_d d(k)$ where $d(k) \in R^r$ is a white Gaussian zero-mean vector noise sequence normalized such that $E[d(j)d^T(k)] = \delta_{jk} \cdot I$; W_d is a *diagonal matrix* of stable filters

$$W_d(z^{-1}) = \text{diag}\left\{\sigma_1 \overline{W}_1(z^{-1}), ..., \sigma_r \overline{W}_r(z^{-1})\right\} \qquad (2.2)$$

where $\sigma_i < \infty$ is a scalar (possibly unknown) and $\overline{W}_i(z^{-1})$ is a *known* minimum phase transfer function $i = 1, ..., r$.

Most generally, if W_d is not known, it can be estimated using a separate *disturbance identification* experiment. Typically this would proceed by,

1. setting $u = 0$, measuring $y = v$;

2. using standard spectral estimation methods for computing the output PSD $P_{yy} = P_{vv}$;

3. fitting W_d to the data such that $W_d^* W_d$ approximates P_{vv}.

In modern parametric methods for spectral estimation, steps 2 and 3 are combined. A general reference on this problem is [12].

Assumption 3 is made for convenience to restrict the scope of the present paper. In particular, the diagonal structure of W_d simplifies plant set estimation in the multivariable case. Furthermore, while estimates of filters $\overline{W}_i$ $i = 1, ..., r$ can be bootstrapped from the spectral analysis, expressions for the plant set become very complicated. The assumption that these filters are known avoids such complications. In contrast, it is emphasized that the noise variances σ_i^2 in (2.2) are *not required to be known a priori*, and will be estimated as part of the overall approach.

3 Frequency Domain Identification

In this section a systematic procedure is given for computing $\hat{P}$. Before starting, it is assumed that a sampling period T has already been chosen commensurate with desired control bandwidth, and that all anti-aliasing filters are in place and are not modified in the course of identification experiments and control implementation.

3.1 Schroeder-Phased Sinusoidal Input Design

A method for nonparametric frequency domain estimation is presented based on a Schroeder-phased multisinusoidal input design [4]. To simplify presentation, the approach is first described on a single-input single-output system of the form,

$$y = P(z^{-1})u + \sigma \overline{W} d . \tag{3.1}$$

A signal processing diagram is given in Fig. 3 for the nonparametric frequency domain identification scheme applied to (3.1).

Consider the periodic input design composed of a harmonically related sum of sinusoids,

$$u_s(k) = \beta \sum_{i=1}^{n_s} \sqrt{2\alpha_i} \cos(\omega_i kT + \phi_i) \tag{3.2}$$

where $\omega_i = 2\pi i/T_p$, $T_p/T = N_s$, $n_s \leq N_s/2$. The power is normalized as,

$$\sum_{i=1}^{n_s} \alpha_i = 1 \tag{3.3}$$

where the relative power in each component $\{\alpha_i > 0, i = 1, ...n_s\}$ is assumed specified. In order to minimize peaking in time domain the sinusoids are phased according to Schroeder [25] as,

$$\phi_i = 2\pi \sum_{j=1}^{i} j\alpha_j \ . \tag{3.4}$$

(Here, a slightly modified form of the Schroeder phase is used in (3.4), as derived in Young and Patton [31]). For example, assuming equal powers $\alpha_i = 1/n_s$, gives, $\phi_i = \frac{\pi}{n_s}(i^2 + i)$. Linear terms which appear in i can always be dropped since they correspond to pure time shifts. A complete statistical analysis of Schroeder-phased harmonic signals for system identification is given in Bayard [4].

For technical reasons, the following assumption will be made,

Assumption 4. The system is driven by Schroeder-phased sinusoidal input (3.2), and allowed to reach steady-state before experimental data is taken.

At steady-state, the plant response to u_s is denoted as y_s and is given by,

$$y_s(k) = \sum_{i=1}^{n_s} \left(b_i \beta \sqrt{2\alpha_i} \cos(\omega_i kT + \phi_i) - a_i \beta \sqrt{2\alpha_i} \sin(\omega_i kT + \phi_i) \right) + v(k) \tag{3.5}$$

where,

$$a_i = \Im\{P(e^{-j\omega_i T})\}, \quad b_i = \Re\{P(e^{-j\omega_i T})\} \ . \tag{3.6}$$

For notational convenience, the index k starts from 0 in (3.5) even though we are in steady-state. Since the goal is to estimate the quantities a_i and b_i it is convenient to collect these quantities in a single vector θ defined as follows,

$$\theta = [a^T, b^T]^T \tag{3.7a}$$

$$a = [a_1, ..., a_{n_s}]^T; \quad b = [b_1, ..., b_{n_s}]^T \ . \tag{3.7b}$$

In order to "whiten" the effect of the noise in (3.5), the time domain input u_s and output y_s will be inverse filtered by $\overline{W}$ to give filtered signals $\tilde{u}_s$ and $\tilde{y}_s$ as follows,

$$\overline{W}(z^{-1})\tilde{y}_s(k) = y_s(k); \quad \overline{W}(z^{-1})\tilde{u}_s(k) = u_s(k) \ . \tag{3.8}$$

Since the frequencies in u_s are harmonically related, both the input $\tilde{u}_s$ and deterministic part of the output $\tilde{y}_s$ at steady-state will be periodic with period T_p. Assume that m periods of filtered input/output data $\tilde{u}_s, \tilde{y}_s$ are collected at steady-state. Denote the output data from the ℓth period as,

$$\tilde{y}_s^\ell(k) = \tilde{y}_s(k + (\ell - 1)N_s) \tag{3.9}$$

for $k = 0, ..., N_s - 1$ and $\ell = 1, ..., m$.

Remark 3.1. It is noted that when inverse filtering by $\overline{W}$ is used, the steady-state assumption (i.e., Assumption 4), requires that the filter transient settles out in addition to the plant transient.

Frequency domain estimates P_s, $\hat{a}_i$, $\hat{b}_i$ are now constructed by taking DFT's on the filtered time-domain data.

DFT Frequency Domain Estimator

$$P_s(\omega_i) = \frac{\frac{1}{m} \sum_{\ell=1}^{m} \tilde{Y}_s^\ell(\omega_i)}{\tilde{U}_s(\omega_i)} \qquad (3.10)$$

$$\hat{a}_i = \Im\{P_s(\omega_i)\}, \quad \hat{b}_i = \Re\{P_s(\omega_i)\} \qquad (3.11)$$

where,

$$\tilde{Y}_s^\ell(\omega_i) = \frac{1}{N_s} \sum_{k=0}^{N_s-1} \tilde{y}_s^\ell(k) e^{-j\omega_i kT}; \quad \tilde{U}_s(\omega_i) = \frac{1}{N_s} \sum_{k=0}^{N_s-1} \tilde{u}_s(k) e^{-j\omega_i kT} \qquad (3.12)$$

Remark 3.2. It is emphasized that the DFT is evaluated precisely on the points of support of the Schroeder-phased input (3.2). The use of an FFT to implement (3.12) requires using the full number of sinusoids ($n_s = N_s/2$) in the sum (3.2), choosing the frequency separation in the input design $2\pi/(N_sT)$ such that the number of samples N_s in one period of u_s is some power of 2.

Remark 3.3. The DFT estimator is conveniently computed recursively in the number of data windows m since one can keep track of the running sum, $\sum_{\ell=1}^{m} \tilde{Y}_s^\ell(\omega_i)$ where each $\tilde{Y}_s^\ell$ is computed using an FFT of fixed size.

It has been assumed that $W = \sigma\overline{W}$, where $\overline{W}$ is assumed known, and σ may be either known or unknown. If σ is unknown, it can be estimated as follows,

Noise Variance Estimator

$$\hat{\sigma}^2 = \frac{\sum_{\ell=1}^{m} \sum_{i=1}^{n_s} |\tilde{Y}_s^\ell(\omega_i) - \overline{Y}(\omega_i)|^2}{N_s(mN_s - 2n_s)} \qquad (3.13)$$

where,

$$\overline{Y}(\omega_i) = \frac{1}{m} \sum_{\ell=1}^{m} \tilde{Y}_s^\ell(\omega_i) . \qquad (3.14)$$

3.2 Statistical Analysis

A detailed statistical analysis of frequency domain estimator (3.10) and noise estimator (3.13) is given in [4]. The results are summarized below. To aid subsequent discussion, a vector $\hat{\theta}$ of estimated quantities $\hat{a}_i$ and $\hat{b}_i$ in (3.11) is defined as follows,

$$\hat{\theta} = [\hat{a}_1, ..., \hat{a}_{n_s}, \hat{b}_1, ..., \hat{b}_{n_s}]^T \ . \tag{3.15}$$

The following theorem summarizes results for the input design (3.2) applied to the single-input single-output plant (3.1).

Theorem 3. *Assume that the Schroeder-phased sinusoidal input u_s defined in (3.2) is applied to exponentially stable plant $\mathcal{P}(z^{-1})$ (Assumption 1), giving rise to the steady-state output y_s defined in (3.5, 3.6). Let the measurement noise coloring filter be given by $W = \sigma\overline{W}$ (Assumption 3) and implement inverse filtering of u_s and y_s by $\overline{W}$ (cf., (3.8)) giving rise to filtered input $\tilde{u}_s$ and filtered output $\tilde{y}_s$. Let spectral estimates P_s defined in (3.10) and $\hat{\theta}$ defined in (3.11, 3.15), be computed based on $m > 1$ periods of the filtered steady-state data $\{\tilde{y}_s^\ell\}_{\ell=1}^m$ in response to the Schroeder-phased input (3.2) (Assumption 4). Then,*

(i) *If σ is known, the exact error probability distributions are given as,*

$$\frac{|\mathcal{P}(e^{-j\omega_i T}) - P_s(\omega_i)|^2}{\sigma^2 c_{ii}} \sim \chi^2(2); \quad \theta - \hat{\theta} \sim N(0, \Sigma) \tag{3.16}$$

$$\Sigma = \sigma^2 \begin{pmatrix} C & 0 \\ 0 & C \end{pmatrix} \tag{3.17}$$

$$C = \text{diag}[c_{11}, ..., c_{n_s n_s}]; \quad c_{ii} = |\overline{W}(e^{-j\omega_i T})|^2/(\beta^2 \alpha_i m N_s) \tag{3.18}$$

where $\chi^2(\nu)$ denotes a Chi-Squared distribution with ν degrees of freedom.
(ii) *If σ is unknown, and estimated using (3.13), the exact error probability distributions are given as,*

$$(mN_s - 2n_s)\frac{\hat{\sigma}^2}{\sigma^2} \sim \chi^2(mN_s - 2n_s) \tag{3.19}$$

$$\frac{|\mathcal{P}(e^{-j\omega_i T}) - P_s(\omega_i)|^2}{2\hat{\sigma}^2 c_{ii}} \sim F(2, mN_s - 2n_s) \tag{3.20}$$

$$\frac{\hat{a}_i - a_i}{\hat{\sigma}\sqrt{c_{ii}}} \sim t(mN_s - 2n_s); \quad \frac{\hat{b}_i - b_i}{\hat{\sigma}\sqrt{c_{ii}}} \sim t(mN_s - 2n_s) \tag{3.21}$$

where $F(\nu_1, \nu_2)$ denotes a Fisher distribution with ν_1 and ν_2 degrees of freedom, and $t(\nu)$ denotes a Student t distribution with ν degrees of freedom.

Proof. See [4].   □

The following corollary to Theorem 3 is useful when statistical confidence regions are desired.

Corollary 4. *Under the conditions of Theorem 3, the* $(1-\alpha)\cdot100\%$ *confidence bounds associated with the DFT estimates (3.10, 3.11) are summarized below,*

$$|\mathcal{P}(e^{-j\omega_i T})-P_s(\omega_i)|^2 \leq \begin{cases} c_{ii}\sigma^2\chi^2_{1-\alpha}(2) & \text{for } \sigma^2 \text{ known} \\ c_{ii}\hat\sigma^2 2F_{1-\alpha}(2, mN_s - 2n_s) & \text{for } \sigma^2 \text{ estimated by } \hat\sigma^2 \end{cases}$$
$$(3.22)$$

$$|\hat a_i - a_i|^2 \leq \begin{cases} c_{ii}\sigma^2\eta^2_{1-\alpha} & \text{for } \sigma^2 \text{ known} \\ c_{ii}\hat\sigma^2 t_{1-\alpha}(mN_s - 2n_s) & \text{for } \sigma^2 \text{ estimated by } \hat\sigma^2 \end{cases} \quad (3.23)$$

$$|\hat b_i - b_i|^2 \leq \begin{cases} c_{ii}\sigma^2\eta^2_{1-\alpha} & \text{for } \sigma^2 \text{ known} \\ c_{ii}\hat\sigma^2 t^2_{1-\alpha}(mN_s - 2n_s) & \text{for } \sigma^2 \text{ estimated by } \hat\sigma^2 \end{cases} \quad (3.24)$$

$$c_{ii} = |\overline{W}(e^{-j\omega_i T})|^2/(\beta^2\alpha_i mN_s) \qquad (3.25)$$

where $\eta_{1-\alpha}$, $t^2_{1-\alpha}(\nu)$, $\chi^2_{1-\alpha}(\nu)$ *and* $F_{1-\alpha}(\nu_1, \nu_2)$ *denote the* $(1-\alpha)\cdot100$ *percentiles for the Gaussian distribution, the Student t distribution with ν degrees of freedom, the Chi-Squared distribution with ν degrees of freedom, and the Fisher distribution with ν_1 over ν_2 degrees of freedom, respectively.*

Remark 3.4. To avoid confusion, it is pointed out that percentiles for symmetric densities (Gaussian and Student) used in this paper are assumed to be two-sided e.g., for x Gaussian, the percentile $\eta_{1-\alpha}$ is defined as,

$$Prob\{-\eta_{1-\alpha} \leq x \leq \eta_{1-\alpha}\} = 1 - \alpha .$$

Under the conditions of Theorem 3, properties of the frequency domain estimates are summarized below [4],

P.1 $\hat P(\omega_i)$, $\hat a_i$, and $\hat b_i$, are unbiased and consistent estimators of $\mathcal{P}(e^{-j\omega_i T})$, $\Im\{\mathcal{P}(e^{-j\omega_i T})\}$, and $\Re\{\mathcal{P}(e^{-j\omega_i T})\}$, respectively for $i = 1, ..., n_s$.

P.2 $\hat a_i$ is statistically independent of $\hat a_j$ for $i \neq j$

P.3 $\hat b_i$ is statistically independent of $\hat b_j$ for $i \neq j$

P.4 $\hat a_i$ is statistically independent of $\hat b_j$ for all i and j

P.5 $P_s(\omega_i)$ is statistically independent of $P_s(\omega_j)$ for $i \neq j$

P.6 $\hat\sigma^2$ is an unbiased and consistent estimator of σ^2

For the purpose of visualization, the estimate $P_s(\omega_i)$ and its confidence region are depicted in a Nyquist plot in Fig. 4. Here the confidence region for the case of σ^2 estimated by $\hat\sigma^2$ is seen as a perfect circle centered at $P_s(\omega_i) = \hat b_i + j\hat a_i$ of radius δ_i where from (3.22),

$$\delta_i^2 = \frac{\hat\sigma^2|\overline{W}(e^{-j\omega_i T})|^2 2F_{1-\alpha}(2, mN_s - 2n_s)}{\beta^2\alpha_i mN_s} \qquad (3.26) .$$

Noting that $F_{1-\alpha}(2, \nu)$ is bounded as ν becomes large (e.g., $F_{1-\alpha}(2, \nu) \leq 9$ for $1 - \alpha = .999$ and $\nu > 30$) the uncertainty region increases with noise-to-signal ratio $\hat\sigma^2\overline{W}/\alpha_i$ and decreases with the amount of measurement data mN_s.

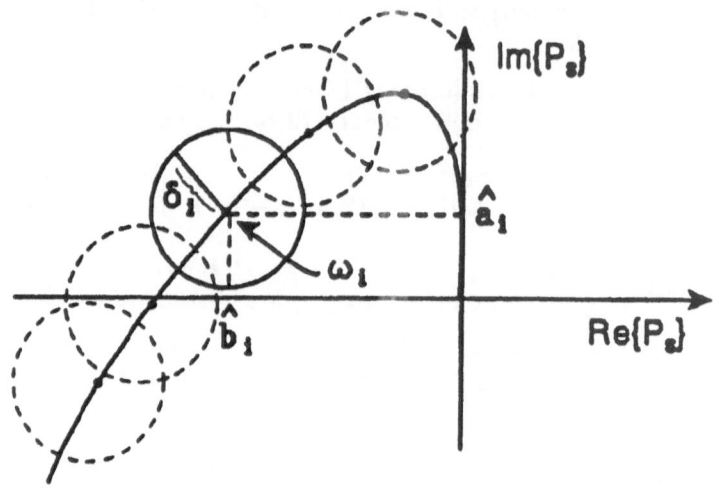

Fig. 4. Nyquist plot showing spectral estimate $P_s(\omega_i)$ with $(1-\alpha)\cdot 100\%$ confidence radius δ

Remark 3.5. Under the conditions stated in Theorem 3, the error distributions (3.16)-(3.21) and confidence regions (3.22)-(3.25) are exact. However, they become approximate if any of the conditions of Theorem 3 are violated,

C.1 The system has not reached steady-state before the data is taken
C.2 The noise $v(k)$ is not Gaussian
C.3 Inverse filtering by $\overline{W}$ is omitted

The effect of C.1 is to create a bias in the estimate, and the previous expressions must be correspondingly modified. In contrast, the effect of C.2 is mild due to the Central Limit Theorem for m reasonable large. Furthermore, the effect of C.3 can be neglected if the data window of length $T_p = TN$, is large compared to the time constants of filter $\overline{W}^{-1}$. The fact that inverse filtering can be removed, significantly simplifies the implementation of the spectral analysis, and is depicted by the dashed lines in Fig. 3.

3.3 Extension to Multivariable Case

The results of Sect. 3.2 can be extended to the multivariable case by performing q separate single-input multiple-output (SIMO) experiments.

Consider a single-channel input design $u_s(k, n_1)$, with design weights $\beta(n_1)$, $\alpha_i(n_1)$, $n_1 = 1, ..., q$ of the form (3.2) into the n_1th actuator. The system is allowed to reach steady-state, at which time the plant response at the n_2th sensor is denoted as $y_s(k, n_2, n_1)$ for each sensor $n_2 = 1, ..., r$. By Assumption 3 the noise on the n_2th output is colored by a filter $\sigma_{n_2}\overline{W}_{n_2}$.

Hence, the noise on the n_2th output channel is whitened by inverse filtering as follows,

$$\overline{W}_{n_2}(z^{-1})\tilde{y}_s(k,n_2,n_1) = y_s(k,n_2,n_1); \quad \overline{W}_{n_2}(z^{-1})\tilde{u}_s(k,n_1) = u_s(k,n_1) \ .$$

For each output $n_2 = 1,...,r$, assume that $m(n_1)$ periods of input/output data $u_s(k,n_1), y_s(k,n_2,n_1)$ are collected at steady-state, and spectral estimates $P_s^{n_2,n_1}$ are computed using the SISO processing approach of Sect. 3.2 for each filtered input/output data set $\tilde{u}_s(k,n_1), \tilde{y}_s(k,n_2,n_1), n_2 = 1,...,r$. This provides spectral estimates $P_s^{n_2 n_1} \ n_2 = 1,...,r$. The entire process is then repeated for each actuator $n_1 = 1,...,q$, (i.e., q SIMO experiments altogether) giving a full multivariable spectral estimate P_s.

Consider estimation of noise variances $\sigma_{n_2}^2 \ n_2 = 1,...,r$. Using the SISO formula (3.13), there will be a separate estimate, denoted as $\hat{\sigma}_{n_2}^2(n_1)$, of the noise variance in the n_2th output for each input experiment $u_s(k,n_1), n_1 = 1,...,q$. Formulas (3.16)-(3.21) are only valid if used with these separate estimates of the noise variances. This is particularly important to remember when using the plant set estimation methods to be discussed in Sect. 4. However, for purposes other than plant set estimation it may be desirable to have only a single estimate of each output noise variance. In this case, one can combine estimates as follows,

$$\hat{\sigma}_{n_2}^2 = \frac{\sum_{n_1=1}^{q}\left(m(n_1)N_s - 2n_s\right)\hat{\sigma}_{n_2}^2(n_1)}{\overline{m}N_s - 2n_s}$$

where $\overline{m} = \sum_{n_1=1}^{q} m(n_1)$. Using properties of the Chi-Squared distribution, and the fact that the noise sequences in separate experiments are statistically independent, it can be shown that this estimate is distributed as

$$(\overline{m}N_s - 2n_s)\frac{\hat{\sigma}_{n_2}^2}{\sigma^2} \sim \chi^2(\overline{m}N_s - 2n_s)$$

and is unbiased, and consistent as $\overline{m} \to \infty$.

3.4 Transfer Function Curve Fitting

In this section, a frequency domain identification approach will be discussed based on the algorithms contained in [3] and [6]. Here, one considers the problem of finding a q-input/r-output rational transfer function matrix $P^o(z^{-1})$ which minimizes the 2-norm of the error between itself and specified frequency domain data P_s, i.e.,

$$\min_{P^o} \sum_{i=1}^{n_s} \mathrm{w}^2(\omega_i)\left\| P_s(\omega_i) - P^o(e^{-j\omega_i T}) \right\|_f^2 \ . \tag{3.28}$$

Here, $\mathrm{w}(\omega_i)$ is a specified weighting function of frequency; and the Frobenious norm is defined as,

$$\|X\|_f^2 = Tr\{X^*X\} \tag{3.29}$$

with " $*$ " denoting the complex conjugate transpose. For optimization purposes, the transfer function matrix $P^o(z^{-1})$ is considered to be in the form of the ratio of a matrix numerator polynomial $B(z^{-1})$ and an nth-order monic scalar denominator polynomial $a(z^{-1})$, i.e.,

$$P^o(z^{-1}) = \frac{B(z^{-1})}{a(z^{-1})} \tag{3.30}$$

$$B(z^{-1}) = B_0 + B_1 z^{-1} + ... + B_n z^{-n} \tag{3.31a}$$

$$a(z^{-1}) = 1 + a_1 z^{-1} + ... + a_n z^{-n} \tag{3.31.b}$$

where $B_k \in \Re^{r \times q}$, $k = 0, ..., n$.

Several algorithms are presently available for solving (3.28). For example, a simple but approximate algorithm found in the work of Sanathanan and Koerner [22] is given by the following fixed-point iteration, (denoted here as the SK iteration),

$$a^{k+1}, B^{k+1} =$$

$$arg \min_{a,B} \sum_{i=1}^{m} w^2(\omega_i) \left\| \frac{1}{a^k(e^{-j\omega_i T})} \left(P_s(\omega_i) a(e^{-j\omega_i T}) - B(e^{-j\omega_i T}) \right) \right\|_f^2$$

$$\tag{3.32}$$

with initial condition $a^0 = 1, B^0 = 0$. With a^k fixed at each iteration, the cost function in (3.32) is quadratic in the coefficients of a and B. Hence, the SK iteration is implemented as a sequence of linear least squares problems.

The original work of Sanathanan and Koerner [22] is formulated in the Laplace s domain. Details of the formulation in the z-domain with some practical improvements can be found in [9, 28] for single-input/single-output systems and in [3] for multivariable systems.

By appropriate definition of parameter vector θ, vector y, regression matrix Φ and weighting matrix $W(a)$, the SK iteration can be written in standard weighted least squares regression form as (see [3]),

SK Iteration:

$$\theta^{k+1} = arg \min_{\theta} \left\| W(a^k)(y - \Phi\theta) \right\|_2^2 \tag{3.33}$$

with initial condition

$$\theta^0 = 0 . \tag{3.34}$$

With no noise and correct choice model order, the SK iteration converges to the optimal solution of (3.28) in a *single* iteration since the error in (3.32) vanishes identically. In the presence of noise and/or incorrect model order, the SK iteration converges to a value which is not generally optimal in any sense. However, it has been shown by Whitfield [26] that the algebraic condition satisfied by the fixed point of the SK iteration differs from a vanishing gradient only by a second-order term in the residual. Hence, while the SK iteration is not generally optimal, it is often close. This motivates the use of the SK iteration as an initializer for a Gauss-Newton (GN) algorithm which forces the gradient to vanish. The GN algorithm can be written in least squares vector form as (see [3]),

Gauss-Newton Iteration:

$$\theta^{k+1} = arg \min_{\theta} \left\| W(a^k)(y_G(\theta^k) - \Phi_G(\theta^k)\theta) \right\|_2^2 \qquad (3.35)$$

where,

$$\Phi_G(\theta^k) = (\Phi + \tilde{\Phi}(\theta^k)) \qquad (3.36a)$$

$$y_G(\theta^k) = (\tilde{\Phi}(\theta^k)\theta^k + y) \ . \qquad (3.36b)$$

In contrast to the SK iteration (3.33) which only updates the weighting function, it is seen that the GN iteration also updates the vector y_G and regression matrix Φ_G at each iteration. Full details and sparse matrix methods for multivariable implementations the SK and GN iterations can be found in Bayard [3]. Extension of these methods for identification of wide-band models using data composition methods can be found in [5].

In summary, the overall approach is to curve fit the data using the SK iteration, followed by a GN iteration. One common difficulty in minimizing cost (3.28) by complex curve fitting, is that an estimate of plant order n is required to be known a priori. Since this is generally not the case in practice, a tedious cut and try process is usually required to find the best model order. This difficulty is overcome in the next section by allowing the polynomials B and a to be *overspecified*, and then determining model order based on Hankel singular values.

3.5 State-Space Realization

Since a state-space model is required for using most modern control design software, a method is given in this section for determining a balanced state-space model from $P^o = B/a$. The algorithm is taken directly from [6]. The key idea is to overparametrize the polynomials B and a when minimizing (3.28), and then to determine the system Markov parameters through a linear algebraic relation. A state-space realization is then determined from the Markov parameters.

Given P^o, one can divide $a(z^{-1})$ into $B(z^{-1})$ to give the Markov parameter sequence $\{H_i\}$,

$$\frac{B(z^{-1})}{a(z^{-1})} = \sum_{i=0}^{\infty} H_i z^{-i} \qquad (3.37)$$

which gives upon cross-multiplying,

$$B_0 + B_1 z^{-1} + ... + B_n z^{-n} = (1 + a_1 z^{-1} + ... + a_n z^{-n}) \sum_{i=0}^{\infty} H_i z^{-i} \ . \qquad (3.38)$$

Equating coefficients of the first N powers of z^{-1} in (3.38) gives the following system of linear equations,

$$
\begin{bmatrix}
I & 0 & \cdots & \cdots & \cdots & \cdots & 0 \\
a_1 I & I & \ddots & & & & \vdots \\
\vdots & a_1 I & I & \ddots & & & \\
a_n I & \ddots & a_1 I & I & \ddots & & \vdots \\
0 & \ddots & \ddots & \ddots & \ddots & \ddots & \\
\vdots & \ddots & \ddots & \ddots & \ddots & \ddots & 0 \\
0 & \cdots & 0 & a_n I & \cdots & a_1 & I
\end{bmatrix}
\begin{bmatrix}
H_0 \\
H_1 \\
\vdots \\
H_n \\
H_{n+1} \\
\vdots \\
\vdots \\
H_N
\end{bmatrix}
=
\begin{bmatrix}
B_0 \\
B_1 \\
\vdots \\
B_n \\
0 \\
\vdots \\
\vdots \\
0
\end{bmatrix} .
\qquad (3.39)
$$

Given the estimated polynomial $a(z^{-1})$ and polynomial matrix $B(z^{-1})$, the multivariable Markov parameters $H_i \in \Re^{r \times q}$, $i = 0, ..., N$ can be calculated by solving the above system of equations. Since the matrix to be inverted is lower triangular with ones on the diagonal, it is always invertible and a solution always exists. Furthermore, since this system of equations is block triangular it can be solved recursively by backsubstitution.

Finally, a balanced state-space realization is determined from the Markov parameters $\{H_i\}$ using any one of a number of realization methods based on the singular-value decomposition [17, 20]. With this approach, the model reduction is performed systematically in terms of the Hankel singular values, and leads to a desired reduced-order balanced state-space realization $\hat{P} = (A, B, C, D)$.

Full details on the state-space identification algorithm outlined above as well as some numerical examples are given in [6].

4 Statistical Plant Set Estimation

In this section, the additive error $\Delta_A(z^{-1}) = \mathcal{P}(z^{-1}) - \hat{P}(z^{-1})$ is computed to characterize the mismatch between the true and estimated plant. As mentioned earlier, it is desired to represent the additive uncertainty as $\Delta_A = \Delta W_A$ where Δ is norm bounded and W_A is a minimum-phase transfer function. The filter W_A will be obtained in two steps. First in Sect. 4.1 a nonparametric overbound is given on the additive error which is valid to a prescribed statistical confidence. Then in Sect. 4.2 a tight parametric overbound will be obtained on the nonparametric bound using an algorithm taken from [23, 24].

4.1 Nonparametric Overbounds

The error $\mathcal{P} - \hat{P}$ at each grid point can be overbounded by using the following inequality,

$$
\bar{\sigma}(\mathcal{P}(\omega_i) - \hat{P}(\omega_i)) \leq \bar{\sigma}(\mathcal{P}(\omega_i) - P_s(\omega_i)) + \bar{\sigma}(P_s(\omega_i) - \hat{P}(\omega_i)) .
\qquad (4.1)
$$

The second term on the right side of (4.1) can be calculated exactly since P_s and $\hat{P}$ are known. The first term is probabilistic and can be overbounded using the statistics of the spectral estimation error (3.16, 3.22), and the facts that,

F.1 Data is taken in a series of SIMO experiments as outlined in Sect. 3.3.

F.2 For a given experiment, the noise sequences in each output channel are statistically independent by Assumption 3. Hence, the errors in the elements of any *column* of P_s are statistically independent.

F.3 The noise sequences from different experiments are statistically independent. Hence, the errors in the elements of any *row* of P_s are statistically independent.

Taken together, facts F.1–F.3 imply that the errors in all elements of P_s are statistically independent. Consider the following inequality involving the Frobenious norm,

$$\overline{\sigma}(\mathcal{P}(\omega_i) - P_s(\omega_i))^2 \leq \|\mathcal{P} - P_s\|_f^2 = \sum_{n_1=1}^{q} \sum_{n_2=1}^{r} |\mathcal{P}^{n_1 n_2}(\omega_i) - P_s^{n_1 n_2}(\omega_i)|^2 \ . \quad (4.2)$$

Note that each term of the sum in (4.2) can be overbounded using the statistical confidence bounds in (3.22). Since each of the terms are statistically independent by facts F.1–F.3, the probability of all terms being overbounded simultaneously is the product of the individual probabilities. We have just proved the following proposition.

Proposition 5. *Let Assumptions 1,3,4 hold, and let multivariable data be gathered in a series of multisinusoidal SIMO experiments (as outlined in Sect. 3.3) so that facts F.1–F.3 hold. Then the event*

$$E_i : \quad \overline{\sigma}(\Delta_A(\omega_i)) \leq \epsilon_i \quad (4.3)$$

is true with at least probability $1 - \alpha$, *where,*

$$\epsilon_i = \left(\sum_{n_1=1}^{q} \sum_{n_2=1}^{r} \beta_i^{n_1 n_2} \right)^{\frac{1}{2}} + \overline{\sigma}(P_s(\omega_i) - \hat{P}(\omega_i)) \quad (4.4)$$

$$\beta_i^{n_1 n_2} =$$

$$\begin{cases} c_{ii}(n_1, n_2)\sigma_{n_2}^2 \chi_{1-\alpha'}^2(2) & \text{for } \sigma_{n_2}^2 \text{ known} \\ c_{ii}(n_1, n_2)\hat{\sigma}_{n_2}^2(n_1)2F_{1-\alpha'}(2, mN_s - 2n_s) & \text{for } \sigma_{n_2}^2 \text{ estimated by } \hat{\sigma}_{n_2}^2(n_1) \end{cases}$$

$$c_{ii}(n_1, n_2) = |\overline{W}_{n_2}(\omega_i)|^2 / \left(\beta^2(n_1)\alpha_i(n_1)m(n_1)N_s \right) \quad (4.5)$$

$$1 - \alpha' = (1 - \alpha)^{\frac{1}{q \cdot r}} \ . \quad (4.6)$$

Alternatively, consider the case where the output noise variances σ_{n_2}, $n_2 = 1, ..., r$ are known. Then by rearranging (4.2),

$$\bar{\sigma}(\mathcal{P}(\omega_i) - P_s(\omega_i))^2 \leq \max_{n_1, n_2}\{\sigma_{n_2}^2 c_{ii}(n_1, n_2)\} \sum_{n_1=1}^{q} \sum_{n_2=r}^{n} \frac{|\mathcal{P}^{n_1 n_2}(\omega_i) - P_s^{n_1 n_2}(\omega_i)|^2}{\sigma_{n_2}^2 c_{ii}(n_1, n_2)}$$

$$(4.7)$$

Using (3.16) each term of the sum in (4.7) is distributed as $\chi^2(2)$. Since by facts F.1–F.3 all terms in the sum are statistically independent it follows that the full summation of terms in (4.7) is distributed as $\chi^2(2qr)$. We have just proved the following proposition.

Proposition 6. *Let Assumptions 1,3,4 hold, and let multivariable data be gathered in a series of multisinusoidal SIMO experiments (as outlined in Sect. 3.3) so that facts F.1–F.3 hold. Then if the output variances $\sigma_{n_2}^2$, $n_2 = 1, ..., r$ are known the event*

$$E_i : \quad \bar{\sigma}(\Delta_A(\omega_i)) \leq \epsilon_i \tag{4.8}$$

is true with at least probability $1 - \alpha$, where,

$$\epsilon_i = \beta_i^{\frac{1}{2}} + \bar{\sigma}(P_s(\omega_i) - \hat{P}(\omega_i)) \tag{4.9}$$

$$\beta_i = \max_{n_1, n_2}\{\sigma_{n_2}^2 c_{ii}(n_1, n_2)\} \cdot \chi_{1-\alpha}^2(2qr) \tag{4.10}$$

Depending on the application, the additive uncertainty bound from Proposition 6 may be less conservative than the bound from Proposition 5. In practice, one can compute both bounds and use the smaller of the two.

The above propositions provide the statistical confidence in overbounding the additive uncertainty at each grid point separately. The next step is to overbound the uncertainty at all n_s grid points *simultaneously*. In the case where the output variances $\sigma_{n_2}^2$ are known, this is made simple by the fact that the errors in P_s are statistically independent at each frequency point. Since the events $E_i, i = 1, ..., N$ in Propositions 5 and 6 are statistically independent, and each has probability $1 - \alpha$ of being true, the probability of all events being satisfied simultaneously is given by

$$1 - \kappa = (1 - \alpha)^{n_s} . \tag{4.11}$$

Alternatively, if the output variances $\sigma_{n_2}^2$ are estimated, the events $E_i, i = 1, ..., N$ in Propositions 5 and 6 are not statistically independent (actually, it can be shown that they are still uncorrelated [4]). In this case, let $\overline{E}_i$ denote the complementary event of E_i. Then one can derive a useful Bonferroni inequality (cf., Feller [16], pp. 110) as follows,

$$Prob\left[\bigcap_{i=1}^{n_s} E_i\right] = 1 - Prob\left[\bigcup_{i=1}^{n_s} \overline{E}_i\right] \geq 1 - \sum_{i=1}^{n_s} Prob[\overline{E}_i] = 1 - \alpha n_s . \tag{4.12}$$

In this case, the probability of all events being satisfied simultaneously is seen from (4.12) to be at least

$$1 - \kappa = 1 - \alpha n_s . \tag{4.13}$$

We have just proved the following result.

Theorem 7. *Let Assumptions 1,3,4 hold, and let multivariable data be gathered in a series of multisinusoidal SIMO experiments (as outlined in Sect. 3.3) so that facts F.1–F.3 hold. Let*

$$\ell_A^{1-\kappa}(\omega_i) = \epsilon_i \quad i = 1, ..., n_s \tag{4.14}$$

where ϵ_i can be calculated from either Proposition 5 or Proposition 6; and $1 - \kappa$ is given by,

$$1 - \kappa = \begin{cases} (1 - \alpha)^{n_s} & \text{for } \sigma_{n_2}^2 \text{ known} \\ 1 - \alpha n_s & \text{for } \sigma_{n_2}^2 \text{ estimated by } \hat{\sigma}_{n_2}^2(n_1) \end{cases} \tag{4.15}$$

Then, $\ell_A^{1-\kappa}$ is an overbound on the additive uncertainty set at all grid points ω_i $i = 1, ..., n_s$ simultaneously with at least probability $1 - \kappa$, i.e.,

$$Prob \left[\bigcap_{i=1}^{n_s} \left\{ \bar{\sigma}(\Delta_A(\omega_i)) \leq \ell_A^{1-\kappa}(\omega_i) \right\} \right] > 1 - \kappa . \tag{4.16}$$

Actually, it is shown in [4] that as $mN_s - 2n_s$ becomes large, one can choose $1 - \kappa = (1 - \alpha)^{n_s}$ with little error whether the σ^2 are estimated or not. This is a consequence of the fact that a multivariate Student t distribution starts to approximate a multivariate Normal distribution as the number of degrees of freedom becomes large.

In Theorem 7, the additive uncertainty set has been statistically overbounded at all grid points simultaneously. A rigorous treatment of additive uncertainty estimation must further ensure overbounding *in-between* grid points. The basic argument goes as follows. Since $\mathcal{P} \in \mathcal{D}(M, \rho)$ and $\hat{P} \in \mathcal{D}(M_1, \rho_1)$ are stable by assumption it can be shown that $\mathcal{P} - \hat{P} \in \mathcal{D}(M_A, \rho_A)$ where $\rho_A = \min(\rho, \rho_1)$ and $M_A = M + M_1$. Hence, Δ_A is stable and its slope is bounded on the unit circle. Using Lemma 2 it is an easy matter to overbound the interpolation error in-between grid points by an expression involving M_A and ρ_A. A complete treatment along these lines can be found in Bayard [4]. However, it has been found from practical experience that bounds on interpolation error (i.e., based on the slope from Lemma 2 in terms of M and ρ) can be very conservative (e.g., several orders of magnitude) in many problems of practical interest. The main problem is the term $1/(1 - \rho)^2$ which can become extremely large for lightly damped poles. In order to simplify the analysis and avoid overly conservative uncertainty bounds, the following engineering assumption will be made to replace Assumption 2 from this point on.

Assumption 2′ The frequency grid has been chosen sufficiently fine so that Δ_A is overbounded in-between grid points if it is overbounded at the grid points with some small margin.

This assumption is mathematically equivalent to having a bound on the slope of the transfer function $\mathcal{P}$ based on engineering intuition, rather than M, ρ information. Results will generally involve fewer grid points and lead to less conservative control designs compared to using M, ρ bounds.

4.2 Spectral Overbounding and Factorization

At this point, an overbound $\ell_A^{1-\kappa}$ is available from Theorem 7 which overbounds the additive uncertainty at all grid points simultaneously with prescribed statistical confidence $1 - \kappa$. While such a bound is useful, it is not in a form directly usable by modern robust control software. Instead, a parametric description is required.

In this section, the LPSOF algorithm introduced in [23, 24] is discussed for determining a minimum-phase transfer function W_A such that $\overline{\sigma}(W_A)$ is a tight overbound on $\ell_A^{1-\kappa}(\omega_i)$ $i = 1, ..., n_g$. The basic idea is to find a spectrally factorizable rational function $W(z)W(z^{-1})$ whose magnitude tightly overbounds $(\ell_A^{1-\kappa})^2$. The uncertainty weighting is then taken as the scalar matrix $W_A = W(z^{-1}) \cdot I$. In this case, the additive uncertainty is overbounded with a scalar matrix. Weightings which preserve more structure to the uncertainty can be developed but are beyond the scope of this paper.

Forming the quantity $W(z)W(z^{-1})$ and evaluating on the unit circle gives an expression of the form,

$$W^*W = \frac{\beta(\omega)}{\alpha(\omega)} \tag{4.17}$$

where,

$$\beta(\omega) = \beta_0 + \beta_1 cos(\omega T) + ... + \beta_m cos(m\omega T) \tag{4.18a}$$

$$\alpha(\omega) = 1 + \alpha_1 cos(\omega T) + ... + \alpha_m cos(m\omega T) \ . \tag{4.18b}$$

The requirement that $\overline{\sigma}(W)$ be an overbound on some specified function of frequency $\ell(\omega)$ is equivalent to the requirement that $\overline{\sigma}(W)^2$ is an overbound on the square of the function ℓ^2 and can be expressed as,

$$\frac{\beta(\omega)}{\alpha(\omega)} \geq \ell^2(\omega) \text{ for all } \omega \in [0, \pi/T] \ . \tag{4.19}$$

The requirement that $\overline{\sigma}(W)^2$ be a "tight" overbound can be expressed as,

$$\min_{\alpha, \beta} \delta \tag{4.20}$$

where,

$$\delta = \max_{\omega} \left\{ \left(\frac{\beta(\omega)}{\alpha(\omega)} - \ell^2(\omega) \right) q^{-1}(\omega) \right\} \ . \tag{4.21}$$

Here, the criterion minimizes a worst-case error δ, which is frequency weighted by the quantity $q^{-1}(\omega)$. The requirement that the overbound β/α admits spectral factor W can be satisfied by ensuring that (cf., Astrom [1]),

$$\beta(\omega)/\alpha(\omega) > 0 \text{ for all } \omega \in [0, \pi/T] \qquad (4.22a)$$

$$\alpha(\omega) > 0 \text{ for all } \omega \in [0, \pi/T] \qquad (4.22b)$$

Note that condition (4.22a) is implied by (4.19), and condition (4.22b) can be enforced explicitly by the constraint, $\alpha(\omega) \geq \underline{\alpha} > 0$ for some small $\underline{\alpha}$. For technical reasons, a similar constraint is enforced on β as $\beta(\omega) \geq \underline{\beta} > 0$ for some small $\underline{\beta}$. The constrained optimization problem above can be written on the frequency grid as,

$$\min_{\delta, \alpha_j, \beta_j} \delta \qquad (4.23)$$

subject to

$$\beta(\omega_i) - \ell^2(\omega_i)\alpha(\omega_i) \geq 0 \qquad (4.24a)$$

$$\beta(\omega_i) - \ell^2(\omega_i)\alpha(\omega_i) \leq \delta q(\omega_i)\alpha(\omega_i) \qquad (4.24b)$$

$$\beta(\omega_i) > \underline{\beta}; \quad \alpha(\omega_i) > \underline{\alpha} \qquad (4.24c)$$

$$\text{for all } \omega_i, \ i = 1, ..., n_s$$

where $\alpha(\omega)$ and $\beta(\omega)$ are defined by (4.18a). A key observation from (4.23, 4.24) is that for fixed δ the optimization over α, β is simply a linear programming problem to find a *feasible solution* for the coefficients $\alpha_i \ \beta_i$. Hence, the joint optimization problem can be solved by a nested search procedure where an outer-loop systematically decreases δ while an inner-loop finds feasible solutions in the variables α and β for fixed δ. The procedure terminates when the smallest γ is found which admits a feasible solution. This algorithm is denoted as the *LP-Spectral Overbounding and Factorization (LPSOF) Algorithm*[23, 24]. By fundamental properties of linear programming, it can be shown to be globally convergent to the *globally optimal solution* of the nonlinear constrained minimax problem defined by (4.23, 4.24).

For use in overbounding the additive uncertainty set, the LPSOF algorithm is applied to overbounding $\ell = \ell_A^{1-\kappa}$ on the grid $\omega_i, \ i = 1, ..., n_s$. In principle, ensuring spectral factorizability requires that the inequality (4.22) be satisfied in-between grid points as well as on the grid points. Several modifications of the LPSOF algorithm are discussed in [23] which ensure proper behavior in-between grid points. However, for the present paper, these additional complications are not required. Due to Assumption 2', proper behavior in-between grid points can be be ensured by adding a small quantity to $\ell_A^{1-\kappa}$ at each grid point before overbounding.

Once the solution β/α is found, it can be spectrally factored by factoring $\beta = b^*b$ and $\alpha = a^*a$ separately as polynomials. We then choose $W_A = b/a$. An excellent algorithm for polynomial factorization which avoids solving for roots, is given in Kucera [19].

5 Robust Control Design

Once the nominal plant $\hat{P}(z)$ and additive uncertainty $\Delta_A = \Delta W_A(z)$ have been characterized using the methods above, it becomes a well-known problem in the robust control literature to find a controller with desired robustness properties. For example, there is a general framework for robust control synthesis based on the μ measure [13, 14] which can be invoked at this point. Alternatively, the discussion here will concentrate on an H_∞ approach based on the weighted mixed-sensitivity H_∞ problem. General references for this section are [21] and [10].

For use with software packages such as [2, 11] which are applicable to s-domain control design, it is necessary to use a Tustin transformation (cf., [11]) to convert the "z" plane quantities $\hat{P}(z)$ and $W_A(z)$ to the "w" plane for control design. As indicated in Fig. 2 an inverse Tustin transformation is then used to convert the control design back to sampled-data form for implementation. Since the H_∞ norm is invariant under the Tustin transformation, robustness and performance bounds satisfied by the controller in the w plane will carry over when implemented in the z domain.

It is worth mentioning that a controller designed for $\hat{P}$ with robustness properties over a plant set defined by W_A will not in general provide the *best* robust performance of all controllers robust over the set of plants consistent with the given data set. Rather, the best design must come out of a joint optimization over the identification and control design stages. This approach typically leads one to consider iterative refinements of the given control design (and hence the dashed feedback path in Fig. 2). Some preliminary methods for joint optimization and related iterative refinement algorithms are discussed in [7, 8, 30].

5.1 Robust Control Issues

It is useful to define the *sensitivity function:* $E(P,C) = (I + PC)^{-1}$ and the *Q-parameter:* $Q(P,C) = C(I + PC)^{-1}$.

A typical measure of control performance is to keep the infinity norm of the weighted sensitivity function small,

Nominal Performance Criteria:

$$\gamma \|W_1 E(\hat{P}, C)\|_\infty \leq 1 \ . \tag{5.1}$$

Here, γ is a scalar indicating the degree of performance which can be attained.

Define the *additive uncertainty set* associated with a particular weighting function W_A as,

$$\Omega_A = \{P : P - \hat{P} = \Delta W_A, \ \|\Delta\|_\infty \leq 1\} \ . \tag{5.2}$$

Many robustness bounds used in the literature require that the plants P in Ω_A are further restricted to have the same number of RHP unstable poles as

$\hat{P}$. However, in the present approach both the true plant $\mathcal{P}$ and estimate $\hat{P}$ are stable by assumption, so that this detail can be overlooked.

A criteria for all plants in Ω_A to be stabilized by compensator C can be written in terms of the Q parameter as,

Robust Stability Criteria:

$$\|W_A Q(\hat{P}, C)\|_\infty \leq 1 \ . \tag{5.3}$$

If performance is required for all plants P in an uncertainty set Ω_A the criteria becomes,

Robust Performance Criteria:

$$\gamma \|W_1 E(P, C)\|_\infty \leq 1 \ for \ all \ P \in \Omega_A \ . \tag{5.4}$$

5.2 Weighted Mixed Sensitivity problem

It can be shown that a necessary and sufficient condition for robust performance (5.4) (with implied robust stability (5.3)) can be written with respect to weighting W_A as,

$$\mu(T) < 1 \tag{5.5}$$

where,

$$T = \begin{bmatrix} \gamma W_1 E(\hat{P}, C) & -\gamma W_1 E(\hat{P}, C) \\ W_A Q(\hat{P}, C) & -W_A Q(\hat{P}, C) \end{bmatrix} \ . \tag{5.6}$$

Here, μ is the structured singular value introduced by Doyle [13, 14]. To find the controller which provides the best robust performance from (5.5) (i.e., μ-synthesis), it is required to maximize γ over choice of C. This is typically done in an iterative manner, by first fixing γ and finding a feasible solution to (5.5). If one exists, the value of γ is increased and a new feasible solution is sought. The largest value of γ which admits a feasible solution C is optimal. Methods for synthesizing robust controllers based on μ-synthesis are given in [2].

For the purposes of the numerical example in the next section, control design will be considered in terms of a related H_∞ criteria for robust performance. In particular, it can be shown that the following inequalities hold [10],

$$\|J\|_\infty \leq \mu(T) \leq \sqrt{2}\|J\|_\infty \tag{5.7}$$

where,

$$J = \begin{bmatrix} \gamma W_1 E(\hat{P}, C) \\ W_2 Q(\hat{P}, C) \end{bmatrix} \tag{5.8}$$

and $W_2 = W_A$ for the present application. This implies that designing a compensator to maximize γ subject to $\|J\|_\infty < 1$ will give a result to within 3db of the μ-synthesis approach. Finding such a C is called the *weighted mixed-sensitivity H_∞ problem* in the literature, and can be solved using available software [11].

6 Numerical Example

The overall approach to robust control design from raw data, is demonstrated on a flexible structures control example.

A 2-input 2-output multivariable plant $\mathcal{P} = \{\mathcal{P}^{n_2,n_1}\}$, $n_1 = 1,2$, $n_2 = 1,2$, is to be identified by performing two SIMO experiments. The plant is described by

$$\ddot{\eta} = -\Omega^2\eta - 2\zeta\Omega\dot{\eta} + \begin{bmatrix} 5.00 & -3.00 \\ -0.12 & 0.50 \\ 3.00 & 0.40 \end{bmatrix} u$$

$$y = \begin{bmatrix} 1.0 & -14.0 & 25.0 \\ 0.6 & -10.0 & 8.0 \end{bmatrix} \eta + \begin{bmatrix} v_1 \\ v_2 \end{bmatrix}$$

where $\Omega = \text{diag}[1.89\ 9.776\ 29.571]$, $\zeta = \text{diag}[0.156\ 0.0106\ 0.0068]$, and v_1, v_2 are uniformly distributed zero-mean white noises of standard deviations $\sigma_1 = 0.5$ and $\sigma_2 = 0.25$, respectively. This choice of $\mathcal{P}$ is motivated by a model identification experiment on the JPL Large Structure Control Laboratory [27].

Schroeder-Phased Sinusoidal Input Design. A single-channel input design $u_s(k, n_1)$ into the n_1th actuator is chosen as a sum of n_s sinusoids of the form (3.2). For the present simulation, $N_s = 256$, $n_s = 128$, $T = 0.05$ sec, $\alpha_i = \frac{1}{n_s}$, for $i = 1,...n_s$, and $T_p = 12.8sec$. A scaling factor of $\beta = 1.0607$ is chosen so that the input commands are within the actuator output limits of ± 1.5 nt. For each experiment, the deterministic part of the output reaches steady state within 2 periods, after which $m = 32$ periods of input/output data are collected.

DFT Frequency Domain Estimator. Denote the output data from the ℓth period at the n_2th sensor subjected to input $u_s(k, n_1)$ at the n_1th actuator as,

$$y_s^\ell(k, n_2, n_1) = y_s(k + (\ell - 1)N_s, n_2)$$

for $k = 0, ..., N_s - 1, \ell = 1, ..., m$ and $n_2 = 1,2$. A frequency domain estimate is constructed by taking DFT's on the time-domain data for the n_1th input/n_2th output channel,

$$P_s^{n_2 n_1}(\omega_i) = \frac{\frac{1}{m}\sum_{\ell=1}^m Y_s^\ell(\omega_i, n_2, n_1)}{U_s(\omega_i, n_1)}$$

where,

$$Y_s^\ell(\omega_i, n_2, n_1) = \frac{1}{N_s} \sum_{k=0}^{N_s-1} y_s^\ell(k, n_2, n_1)e^{-j\omega_i kT}$$

$$U_s(\omega_i, n_1) = \frac{1}{N_s} \sum_{k=0}^{N_s-1} u_s(k, n_1) e^{-j\omega_i kT} \ .$$

Two separate SIMO experiments (i.e., for $n_1 = 1, 2$) gives the total MIMO plant estimate, $P_s(\omega_i) = \{P_s^{n_2 n_1}(\omega_i)\}$, $n_1 = 1, 2$, $n_2 = 1, 2$, $i = 1, ..., n_s$. Figure 5 compares the magnitude plots of $\mathcal{P}^{11}$ and P_s^{11}.

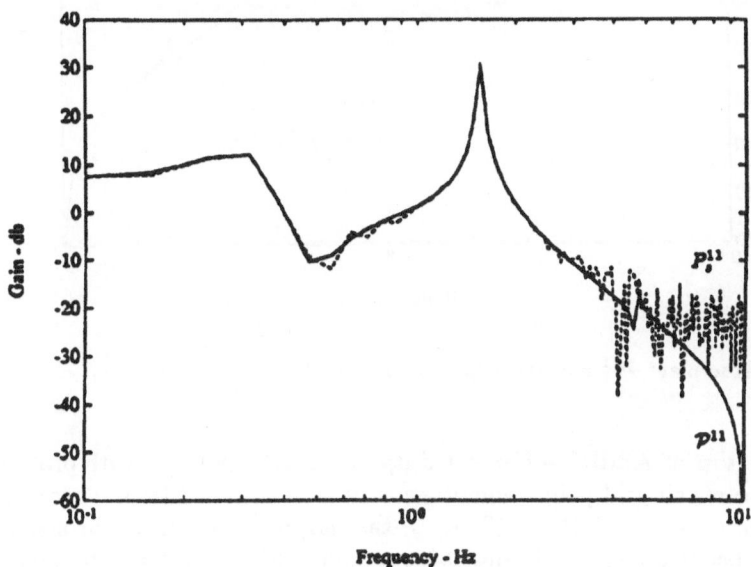

Fig. 5. True plant $\mathcal{P}^{11}$ and spectral estimate P_s^{11}

Curve Fitting/State-Space Realization. The MIMO data $P_s(\omega_i)$ is curve fitted using an SK iteration followed by a GN iteration as discussed in Sect. 3.4, giving rise to a rational transfer function matrix $P^o = B(z^{-1})/a(z^{-1})$ where B is a polynomial matrix and a is a polynomial. The present example fits a rational transfer function matrix P^o of 2 modes even though $\mathcal{P}$ has 3 modes. The mode not fitted represents unmodelled dynamics.

A state-space realization $\hat{P}$ is formed from $P^o = B/a$ by using block-controller form [18] and reduced further via balanced realization and truncation [15] (these last two steps could have been combined using the method of Sect. 3.5). The block-controller realization consists of 8 states. Performing balanced realization yields Grammian of [19.753 19.337 3.564 2.628 0.026 0.023 0.009 0.009]. Elimination of the last 4 states thus results in a 2-input/2-outputs, 2 mode state-space model $\hat{P}$ for $\mathcal{P}$. Figure 6 compares the magnitude plots of $\hat{P}^{11}$ and $\mathcal{P}^{11}$.

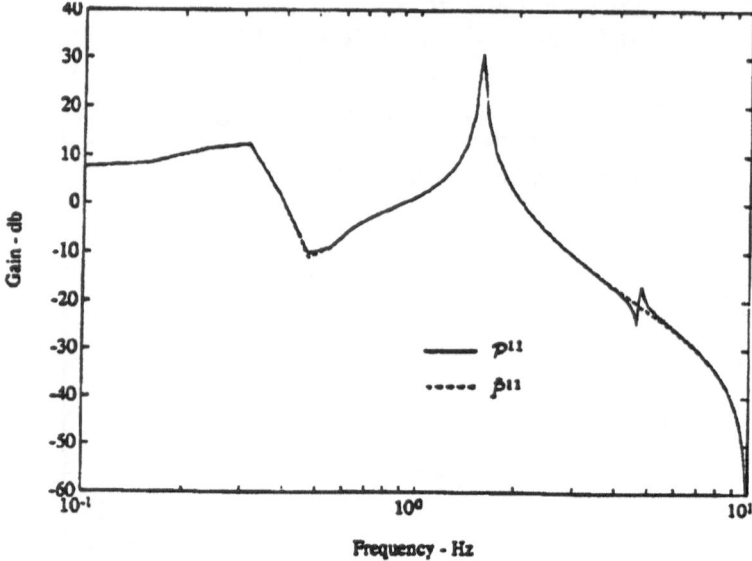

Fig. 6. True plant $\mathcal{P}^{11}$ and state-space estimate $\hat{P}^{11}$

Estimation of Additive Uncertainty. An overbound $\ell_A^{1-\kappa}$ with probability $1 - \kappa$ on all grid points simultaneously is determined from Theorem 7. In this case $1 - \kappa = (1 - \alpha)^{n_s}$ since the output covariances are assumed known. Letting $1 - \kappa = 0.5$ and 0.9999, values of $1 - \alpha$ are computed as $1 - \alpha = (1 - \kappa)^{1/n_2}$. Using Proposition 6 an upper bound on the additive uncertainty $\Delta_A(\omega_i) = \overline{\sigma}(\mathcal{P}(\omega_i) - \hat{P}(\omega_i))$ at each grid point ω_i $i = 1, ..., n_s$ is calculated to probability $1 - \alpha$. Figure 7 shows $\overline{\sigma}(P_s(\omega_i) - \hat{P}(\omega_i))$ and $\ell_A^{1-\kappa}(\omega_i)$ for $1 - \kappa = 0.5$ and 0.9999. For comparison, the true additive uncertainty $\overline{\sigma}(\mathcal{P}(\omega_i) - \hat{P}(\omega_i))$ is also computed and shown. Interestingly, it is seen that using the curve fit error $\overline{\sigma}(P_s(\omega_i) - \hat{P}(\omega_i))$ alone (as is sometimes done in practice) would not overbound the true additive uncertainty for this problem. It is also noted that the true additive uncertainty is overbounded in this case by $\ell_A^{1-\kappa}$ for either confidence levels of $1 - \kappa = 0.5$ or 0.9999. The added conservatism in the $1 - \kappa = .9999$ bound can be seen in Fig. 7 as the price to be paid for the greater statistical confidence. Finally, it can appreciated from the smoothness in the true additive uncertainty of Fig. 7 that there is no problem satisfying Assumption 2' in practical problems, even for lightly damped systems.

Spectrally Factorizable Overbound. The LPSOF algorithm outlined in Sect. 4.2 is used to compute a minimal-phase transfer function W_A of specified order $m_A = 2$ such that $|W_A|^2$ is a tight overbound on $\ell_A^{1-\kappa}(\omega)$. With this result, the additive uncertainty can be written in standard form for robust

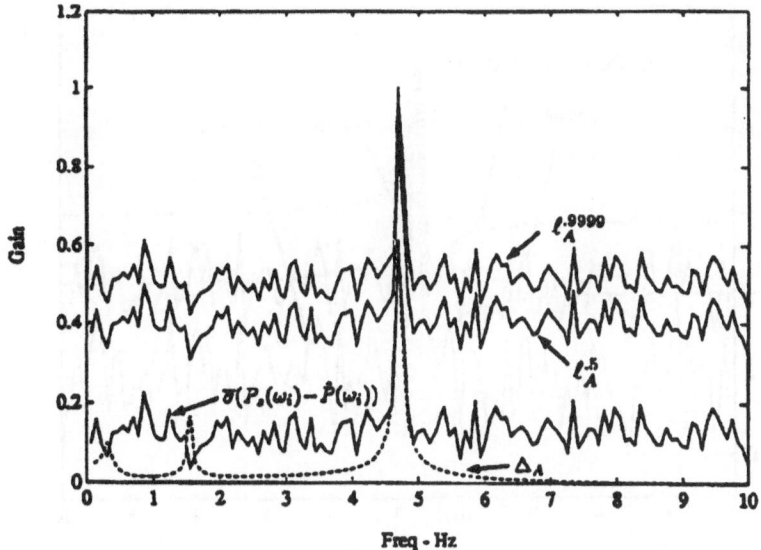

Fig. 7. $\Delta_A, \overline{\sigma}(P_s(\omega_i) - \hat{P}(\omega_i))$, and $\ell_A^{1-\kappa}$ for $1 - \kappa = 0.5$ and 0.9999

control design, i.e., $\Delta_A = \Delta W_A$ where $\|\Delta\|_\infty < 1$. Figure 8 shows W_A and $\ell_A^{1-\kappa}$ for $1 - \kappa = 0.5$ and 0.9999.

Robust Control Design. The plant estimate $\hat{P}$ and weighting function W_A are converted to the w plane using a Tustin transform. A controller is designed in the w plane by solving a weighted mixed-sensitivity H_∞ problem as outlined in Sect. 5.2 using Matlab Robust Control Toolbox of Chiang and Safonov [11]. The performance weighting W_1 is specified by $W_1(w) = .01 \cdot I_2(w+40)^2/(w+4)^2$ where I_2 is the 2x2 Identity matrix. The robustness weighting W_2 is specified as $W_A \cdot I_2$. Both W_1 and W_2 are converted into block controller state-space realizations for implementation in the Matlab software. The controller is of order 12=4 (for $\hat{P}$) + 4 (for W_1) + 4 (for W_2). The design is optimized by a line search over the performance weight γ, converging to an optimal design of $\gamma = 1.016$ for the low confidence case of $1 - \kappa = 0.5$ and $\gamma = 1.011$ for the high confidence case of $1 - \kappa = 0.9999$.

Results for the low confidence case of $1 - \kappa = 0.5$ are summarized in Figs. 9 and 10. Specifically, the low confidence controller is applied to the estimated plant, and Fig. 9 compares the open and closed-loop Bode plots corresponding to $\hat{P}^{11}$. The controller is then applied to the true system and Fig. 10 compares the open and closed-loop Bode plots corresponding to $\mathcal{P}^{11}$ As desired, the low confidence control design stabilizes the true plant and works as designed (i.e., provides performance at its designed level).

Results for high confidence case of $1 - \kappa = .9999$ are summarized in

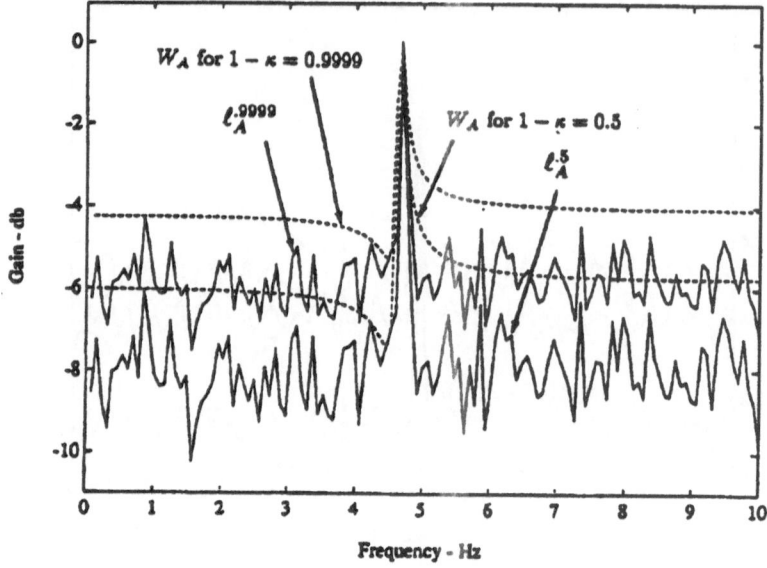

Fig. 8. W_A and $\ell_A^{1-\kappa}$ for $1-\kappa = 0.5$ and 0.9999

Figs. 11 and 12. In particular, the high confidence controller is applied to the estimated plant, and Fig. 11 compares the open and closed-loop Bode plots corresponding to $\hat{P}^{11}$. The high confidence controller is then applied to the true system and Fig. 12 compares the open and closed-loop Bode plots corresponding to $\mathcal{P}^{\infty\infty}$. As desired, the high confidence control design stabilizes the true plant, and works as designed (i.e., provides performance at its designed level).

In this example, it must be considered fortuitous that the low confidence controller worked as designed on the true system. Since the predicted performance of both the 50% and 99.99% confidence controllers was comparable, it would be prudent in this example to implement the higher confidence design.

7 Conclusions

This paper demonstrates a methodology for multivariable frequency domain identification which estimates quantities necessary for designing modern robust controllers. Specifically, a nominal plant estimate $\hat{P}$ and an additive uncertainty weighting W_A are estimated, where the true plant lies within the additive uncertainty set to a specified statistical confidence. Knowledge of $\hat{P}$ and W_A are then suitable for designing synthesizing μ-Synthesis or H_∞ controllers using modern robust control design software.

Since the additive uncertainty set is characterized to a specified statistical confidence level, a controller designed to have robustness properties for this

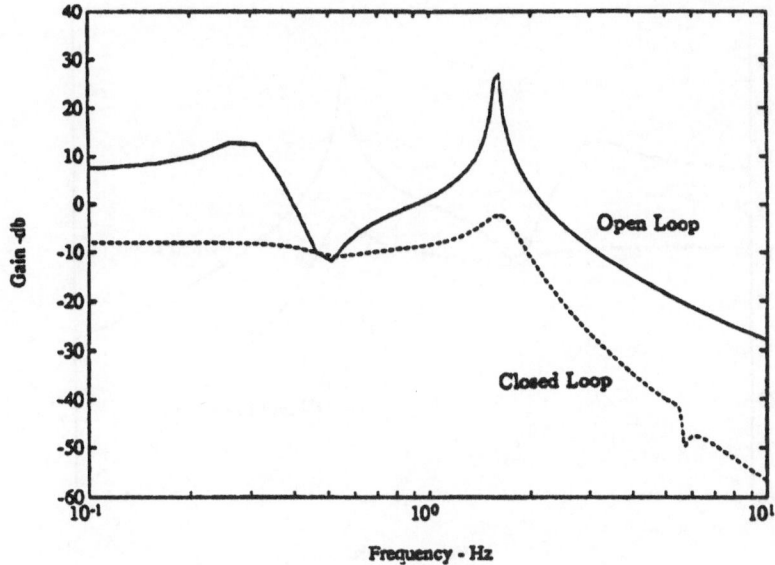

Fig. 9. Open and closed-loop gain (in w-plane) for $\hat{P}^{11}$, $1 - \kappa = 0.5$

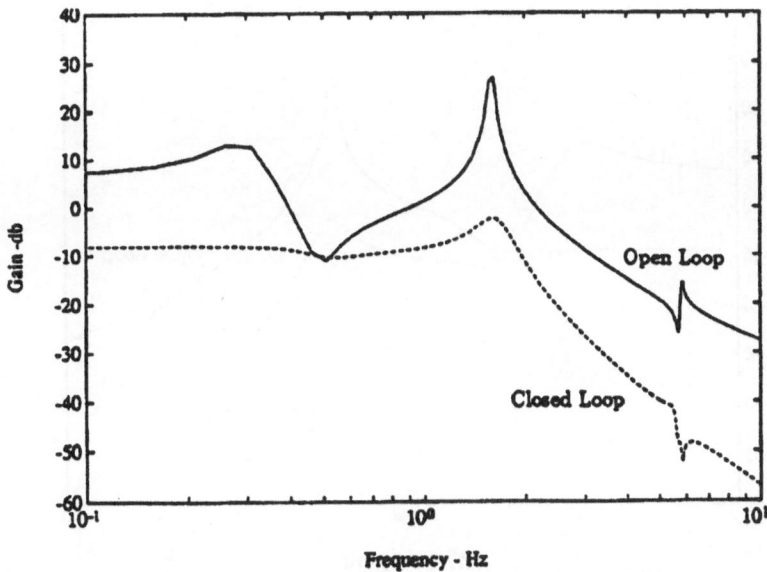

Fig. 10. Open and closed-loop gain (in w-plane) for $\mathcal{P}^{11}$, $1 - \kappa = 0.5$

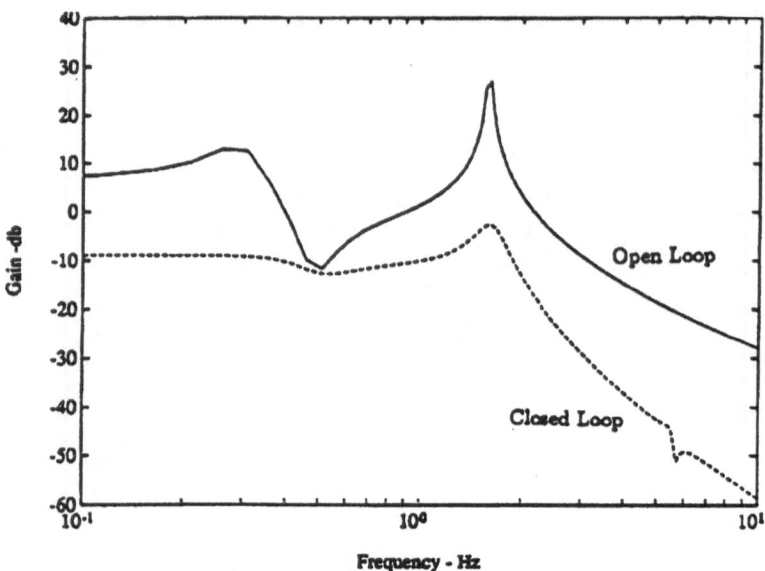

Fig. 11. Open and closed-loop gain (in w-plane) for $\hat{P}^{11}$, $1 - \kappa = 0.9999$

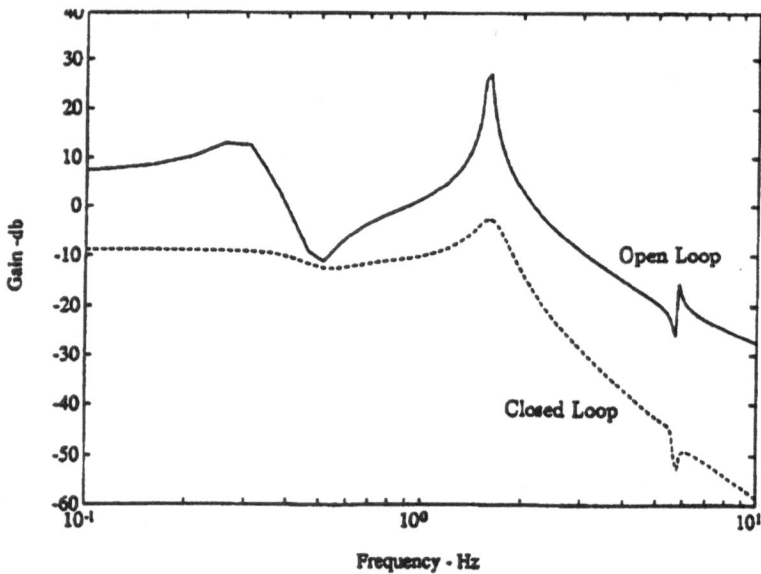

Fig. 12. Open and closed-loop gain (in w-plane) for $\mathcal{P}^{11}$, $1 - \kappa = 0.9999$

uncertainty set will work as designed on the true plant to the same confidence level. To demonstrate this concept, a numerical example was given in Sect. 6 where controllers were designed from raw data to work with 50% and 99.99% confidence, respectively. As desired, both controllers worked as designed when applied to the true system. In general applications, this methodology can provide a tool for performance/risk trade-off studies.

The present approach has dealt with the interpolation error "in-between" grid points in a practical manner i.e., by assuming that the additive uncertainty is overbounded everywhere if it is overbounded at the grid points with some small margin. It is emphasized that there is no obstacle to creating an statistical additive uncertainty set which includes the interpolation error in an explicit manner (this in fact was done in [4]) and overbounding it using the LPSOF algorithm (appropriate extensions of the LPSOF algorithm are available [23]). However, the practical usefulness of such an elaborate approach is questionable due to the huge conservatism in interpolation error bounds obtainable from M, ρ information. The main problem is due to the $1/(1-\rho^2)$ term which blows up very quickly for lightly damped modes.

The present approach has assumed that the noise transfer function matrix W_d is diagonal. This simplifies the analysis be ensuring that the errors in the elements of any column of the plant transfer function estimate (assuming SIMO experiments) are statistically independent. It is expected that the full W_d case can be handled by developing more general expressions based on the Bonferroni inequality which allows one to bound the probability of statistically dependent simultaneous events.

In this paper, a scalar matrix estimate $W_A \cdot I$ is computed to characterize the additive uncertainty set. However, more structure to the uncertainty can be retained by statistically overbounding submatrices of the additive uncertainty, and using the LPSOF algorithm on each overbound. These methods remain to be explored in the future.

8 Acknowledgements

This research was performed at the Jet Propulsion Laboratory, California Institute of Technology, under contract with the National Aeronautics and Space Administration.

References

[1] K.J. Astrom, *Introduction to Stochastic Control Theory*. Academic Press, New York, 1970.

[2] G.J. Balas, J.C. Doyle, K. Glover, A.K. Packard, R. Smith, *H-Infinity and Mu Control Analysis : Mu-Tools Manual*. The MathWorks Inc. 1990.

[3] D.S. Bayard, "Multivariable frequency domain identification via 2-Norm minimization," Proc. American Control Conference, Chicago, Illinois, pp. 1253-1257, June 1992.

[4] D.S. Bayard, "Statistical plant set estimation using Schroeder-phased multisinu-soidal input designs," J. Applied Mathematics and Computation (forthcoming); also, Proc. American Control Conference, Chicago, Illinois, pp. 2988-2995, June 1992; also JPL Internal Document D-8146, Jan 1991.

[5] D.S. Bayard, "High-order wide-band frequency domain identification using com-posite curve fitting," Proc. American Control Conference, Chicago, Illinois, pp. 3181-3185, June 1992.

[6] D.S. Bayard, "An algorithm for state-space frequency domain identification without windowing distortions," Proc. 31st IEEE Conference on Decision and Control, Tucson, Arizona, December 1992.

[7] D.S. Bayard, Y. Yam, and E. Mettler, "On the integration of on-orbit sys-tem identification with modern robust control tuning," Proc. 2nd USAF/NASA Workshop on System ID and Health Monitoring of Precision Space Structures, Pasadena, CA, March 1990.

[8] D.S. Bayard, Y. Yam, E. Mettler, "A criterion for joint optimization of identi-fication and robust control," IEEE Transactions on Automatic Control, *Special Mini-Issue on System Identification for Control Design*, vol. 37, no. 7, pp. 986-991, July 1992.

[9] D.S. Bayard, F.Y. Hadeagh, Y. Yam, R.E. Scheid, E. Mettler, M.H. Milman, "Automated on-orbit frequency domain identification for large space struc-tures," Automatica, vol. 27, no. 6, pp. 931-946, November 1991.

[10] R.Y. Chiang, *Modern Robust Control Theory.* Ph.D. Dissertation, Electrical Engineering Department, University of Southern California, December 1988.

[11] R.Y. Chiang and M.G. Safonov, *Robust-Control Toolbox.* The MathWorks, Inc. 1988.

[12] D.G. Childers, Ed. *Modern Spectrum Analysis.* IEEE Press, New York, 1978.

[13] J.C. Doyle, "Analysis of control systems with structured uncertainty," IEE Proc., Part D, vol. 129, pp.242,

[14] J.C. Doyle, J.E. Wall and G. Stein, "Performance robustness analysis for struc-tured uncertainty," Proc. 21st IEEE Conf. Decision and Control, Orlando, FL, pp. 629-636, December 1982.

[15] D.F. Enns, "Model reduction with balanced realization: An error bound and frequency-weighted generalization," Proc. 23rd IEEE Conf. on Decision and Control, Las Vegas, NV, Dec. 1984.

[16] W. Feller, *An Introduction to Probability Theory and its Applications.* 3rd ed., Wiley, New York, 1968.

[17] J.N. Juang, and R.S. Pappa, "An eigensystem realization algorithm for modal parameter identification and model reduction," J. Guidance, Control and Dy-namics, vol. 8, no. 5, pp. 620-627, Sept-Oct. 1985.

[18] T. Kailath, *Linear Systems,* Prentice-Hall, Englewood Cliffs, New Jersey, 1980.

[19] V. Kucera, *Discrete Linear Control: The Polynomial Equation Approach.* John Wiley and Sons, Chichester, 1979.

[20] S.Y. Kung, "A new identification and model reduction algorithm via singu-lar value decomposition," Proc. 12th Asilomar Conf. on Circuits, Systems and Computers, pp. 705-714, Pacific Grove, CA, November 1978.

[21] M. Morari and E. Zafiriou, *Robust Process Control.* Prentice Hall, Englewood Cliffs, New Jersey, 1989.

[22] C.K. Sanathanan and J. Koerner, "Transfer function synthesis as a ratio of two complex polynomials," *IEEE Trans. Auto. Contr.*, vol. 8, pp.56-58, 1963.

[23] R.E. Scheid, D.S. Bayard, "A linear programming algorithm for determining norm bounded uncertainty," JPL Internal Document JPL D-8145, December 19, 1990.

[24] R.E. Scheid, D.S. Bayard, Y. Yam, "A linear programming approach to characterizing norm bounded uncertainty from experimental data," Proc. American Control Conference, Boston, MA, pp. 1956-1958, June 1991.

[25] M.R. Schroeder, "Synthesis of low peak-factor signals and binary sequences of low auto-correlation," IEEE Trans. Information Theory, Jan. 1970.

[26] A.H. Whitfield, "Asymptotic behaviour of transfer function synthesis methods," International Journal of Control, vol. 45, no.3, pp. 1083-1092, March 1987.

[27] Y. Yam, "Frequency domain identification experiment phase II: Full system excitation," JPL Internal Document EM-343-1156, November 1989.

[28] Y. Yam, D.S. Bayard, F.Y. Hadeagh, E. Mettler, M.H. Milman, R.E. Scheid, *Autonomous frequency domain identification: Theory and experiment*. JPL Publication 89-8, April 1989.

[29] Y. Yam, D.S. Bayard, R.E. Scheid, "Frequency domain identification for robust large space structure control design," Proc. American Control Conference, Boston, MA, pp. 3021-3023, June 1991.

[30] Y. Yam, D.S. Bayard, R.E. Scheid, "Integrated identification and robust control tuning for large space structures," Proc. American Control Conference, San Diego, CA, 1990.

[31] P. Young, and Patton, R.J., "Comparison of test signals for aircraft frequency domain identification," AIAA J. Guidance, Dynamics, and Control, May-June 1990.

Time Domain Approach to the Design of Integrated Control and Diagnosis Systems

Hamid Ajbar and Jeffrey C. Kantor*

Dept. of Chemical Eng., Univ. of Notre Dame, Notre Dame IN 46556

Abstract:

The paper describes a time domain approach to model faults in an integrated control and diagnosis system. The design of the integrated system is done in the framework of linear optimal control for persistent inputs. The tradeoffs that govern the control and detection objectives are delineated. The effect of modeling errors is incorporated in the analysis by providing simple analytical expressions for the false alarm threshold and the size of the smallest detectable fault.

1 Introduction

Quantitative approaches based on analytical redundancy are recognized as being effective methods for fault detection isolation and accommodation (FDIA), of sensor, actuator and component faults [7, 8]. The appeal of these methods lies in the fact that it is possible to exploit the analytical redundancy contained in the static and dynamic relations among the system inputs and measured outputs without need for much physical redundancy. A problem connected with the use of these model-based approaches is that significant interactions exist between the operation of a control system and the diagnostic module in charge of faults detection. In fact, the objectives of disturbance regulation and fault detection are frequently in conflict. These conflicts have been carefully delineated by Nett and coworkers [13] in the general context of linear multivariable control. The main source of these tradeoffs is the presence of model-plant mismatch. The mismatch can cause the performance of the control system to deteriorate and trigger false alarms by the detection system. On the other hand raising the threshold can result in missing serious faults.

Nett in [13] proposed a four parameter controller for the integration of fault detection and control. The controller has two outputs, one used to manipulate process parameters, the other designed to track the fault signal, in the same way as the innovations (residuals) used in the FDIA algorithms.

---

* Corresponding author. Email: Jeffrey.Kantor@nd.edu

In this paper we propose a time domain approach to model faults. Faults are modeled as exogenous inputs and described by bounds on their magnitude or other linear correlations. The use of ℓ_∞ norm to describe the set of possible faults is very practical. Faults are detected by comparing the magnitude of the alarm signal to a threshold.

The specifications of the inputs-outputs signals in time domain enable us to use the techniques of linear optimal control developed in [1, 2, 9, 10] to the design and analysis of the integrated control and diaganosis system.

Solutions to the design problem can be set up as a linear program where the unknowns are the impulse response coefficients of each of the parameters of the controller. Using this framework, simple expressions for the false alarm threshold and the minimum detectable faults are derived.

2 Nomenclature

ℓ_1 Space of absolutely summable sequences. If $x = x(k)_{k=0}^\infty \in \ell_1$ then $||x||_1 = \sum_{k=0}^\infty |x(k)| < \infty$.

ℓ_∞ Space of all bounded sequences of real numbers. If $x = x(k)_{k=0}^\infty \in \ell_\infty$ then $||x||_\infty = sup_k|x(k)|$.

ℓ_∞^n Space of n-tuples of elements of ℓ_∞. If $x = (x_1, \cdots, x_n) \in \ell_\infty^n$ then $||x||_\infty = max_k||x_k||_\infty$.

$\ell_\infty^{n_s}$ Subspace of ℓ_∞^n of dimension $s < n$.

$\mathcal{L}_{TV}^{n \times m}$ Space of all bounded linear causal operators mapping ℓ_∞^m to ℓ_∞^n. If $H \in \mathcal{L}_{TV}^{n \times m}$ then $||H|| := sup_{x \neq 0} \frac{||Hx||_\infty}{||x||_\infty}$

$\mathcal{L}_{TI}^{n \times m}$ Subspace of $\mathcal{L}_{TV}^{n \times m}$ consisting of time-invariant operators. For each $H \in \mathcal{L}_{TI}^{n \times m}$ corresponds a unique $h \in \ell_1^{n \times m}$ where h_{ij} is the pulse response of H_{ij}, the component of H mapping the jth input to the ith output.

$||H||_\mathcal{A}$ The induced operator norm on $\mathcal{L}_{TI}^{n \times m}$. $||H||_\mathcal{A} = ||h||_1$, where $h = (h_{ij})$ is the pulse response matrix of H.

$||.||_{\infty,\tau}$ The truncated norm. If $x \in \ell_{\infty,\tau}^n$ then $||x||_{\infty,\tau} = max_{i=1,\cdots,n} ||x_i||_{\infty,\tau}$, where $||x_i||_{\infty,\tau} = sup_{k=0,\tau}|x_i(k)|$.

3 A Motivating Example

The purpose of this section is to provide a simple illustration of the coupling that exists between the control system and the fault diagnosis module in a functioning reliable feedback system.

Consider a minimal realization of a linear shift-invariant system described by the nominal state space representation

$$x(k+1) = \Phi x(k) + \Gamma u(k) \tag{1}$$
$$y(k) = C x(k) \tag{2}$$

The plant is assumed to be under multiplicative output uncertainties

$$\tilde{G}_p = (I + \Delta)G_p$$

where Δ is a linear bounded strictly casual operator such that $\|\Delta\| \le \delta < 1$. We will investigate the detection of faults occurring in the sensor. This task can be achieved by a simple estimator constructed as follows [3, 7]

$$\hat{x}(k + 1/k) = \Phi\hat{x}(k/k - 1) + \Gamma u(k) + K(y(k) - C\hat{x}(k/k - 1)) \quad (3)$$
$$\hat{y}(k) = C\hat{x}(k) \quad (4)$$

Where $\hat{x}(k + 1/k)$ is used to indicate an estimate of $x(k + 1)$ based on measurements available at time k. The estimation error $\tilde{x} = x - \hat{x}$ is such that

$$\tilde{x}(k + 1/k) = (\Phi - KC)\tilde{x}(k/k - 1) \quad (5)$$

where K is chosen such that system is asymptotically stable.

The sensor output is $y_s = y + b_s$ where y is the actual plant output and b_s represents the sensor bias. The faults occurring in the system is modeled by considering the signal b_s as representing deviations from the sensor ideal behavior $b_s = \eta_s + f_s$ where η_s is a zero mean sensor noise and f_s is the fault. The system input/output pulse transfer is then

$$\hat{y} = G_p u + Fa \quad (6)$$

where

$$a = y_s - C\hat{x} \quad (7)$$
$$u = G_c e \quad (8)$$
$$e = r - \hat{y} \quad (9)$$

G_c is the controller and $F = C(zI - \Phi)K$ is the filter transfer function. The signal $a(k) = y(k) - C\hat{x}(k)$ will be referred to as the filter residual.

Under the uncertainties the feedback system is described by the equations

$$e = \tilde{H}_{er}r + \tilde{H}_{eb}(\eta_s + f_s) \quad (10)$$
$$a = \tilde{H}_{ar}r + \tilde{H}_{ab}(\eta_s + f_s) \quad (11)$$
$$v = \Delta w \quad (12)$$

The $\tilde{H}_{ji}$ are transfer function matrices dependant upon the model error, and (w, v) is the input-output pair associated with Δ. The $\tilde{H}_{ji}$ can be obtained from the general equation [12, 11]

$$\tilde{H}_{ji} = H_{ji} + H_{jv}(I - \Delta H_{wv})^{-1}\Delta H_{wi}, \quad j = e, a \quad i = r, b$$

where the H_{ji} are the nominal transfer functions.

In the case of multiplicative output uncertainties these equations become

$$\tilde{H}_{er} = S_c - S_c(I + G_pG_cS_f)(I + \Delta E_c E_f)^{-1}\Delta E_c \qquad (13)$$

$$\tilde{H}_{eb} = E_c E_f + S_c(I + G_pG_cS_f)(I + \Delta E_c E_f)^{-1}\Delta E_c E_f \qquad (14)$$

$$\tilde{H}_{ar} = S_f(I + \Delta E_c E_f)^{-1}\Delta E_c \qquad (15)$$

$$\tilde{H}_{ab} = S_f(I + \Delta E_c E_f)^{-1} \qquad (16)$$

where

$$S_c = (I + G_pG_c)^{-1}, \quad E_c = I - S_c$$
$$S_f = (I + F)^{-1}, \quad E_f = I - S_f$$

S_c and E_c are respectively the sensitivity and complementary sensitivity of the control system. In the same way S_f and E_f can be considered as being the sensitivity and complementary sensitivity of the diagnosis system.

It is known that the system is stable if

- S_c, S_f are stable.
- $\delta\|E_c E_f\|_A < 1$,
- $\delta\|E_c\|_A < 1$,

$\|.\|_A$, as defined in later sections, is the induced norm on the set of linear time-invariant bounded causal operators on ℓ_∞.

3.1 Coupling in the Nominal Case

In the absence of uncertainties, $\Delta = 0$, the equations above become

$$e = S_c r + (I - S_c)(I - S_f)(f_s + \eta_s) \qquad (17)$$

$$a = S_f(f_s + \eta_s) \qquad (18)$$

The objectives of the control system is to achieve set point tracking and to reject sensor noise, that is to make $\|S_c\|_A$ and $\|(I - S_c)(I - S_f)\|_A$ small. On the other hand fault detection module should reject fault and noise at the residual: make $\|S_f\|_A$ 'small'. Since $H_{e\eta_s} = (I - S_c)(I - H_{af_s})$, the control and the diagnosis objectives are clearly linked: A filter F, and hence S_f, designed for the open-loop to reject satisfactorily f_s at the residual may perform poorly with the controlled system.

One is interested in the false alarm threshold and the size of the minimum detectable fault. The derivation of these expressions is given later in Sect. 6. An upper bound on false alarm threshold is given by

$$\hat{T}h = \|H_{ab}\|_A\|\eta_s\|_{\infty,\tau} = \|S_f\|_A\|\eta_s\|_{\infty,\tau} \qquad (19)$$

and the correspondent size of the minimum detectable fault is such

$$(\|f\|_{\infty,\tau})_{min} \geq 2\hat{T}h\|S_f^{-1}\|_A \qquad (20)$$

where τ is the detection window $[0, \tau]$.

3.2 Effect of Modeling Errors

The presence of uncertainties in the process will couple in a more severe way the objectives of the control subsystem and those of the diagnosis subsystem. To see that, consider the expression of the residual

$$a = \tilde{H}_{ar}r + \tilde{H}_{ab_*}(\eta_* + f_*) \tag{21}$$

The term $\tilde{H}_{ar}$ is no longer zero. If we assume for the purpose of clarity that $(I + \Delta E_c E_f) \simeq I$, then

$$\tilde{H}_{ar} = H_{a\eta_*}\Delta(I - H_{er}) \tag{22}$$

Clearly the effect of the set point on the detection has to be minimized and we see that this will involve trading-off between the detection and the control objectives. The presence of model uncertainties will lead to a degradation of the control performances and in turn will lead to false alarms. On the other hand raising the threshold will lead to missing serious faults. An upper bound on the false alarm threshold, taken from the development that will follow in Sect. 6, is given by

$$\hat{Th} = \frac{||S_f||_A||E_c||_A}{1 - \delta||E_c E_f||_A}||r||_{\infty,\tau} + \frac{||S_f||_A}{1 - \delta||E_c E_f||_A}||\eta_*||_{\infty,\tau} \tag{23}$$

and the minimum detectable fault is such that:

$$(||f||_{\infty,\tau})_{min} \geq 2\hat{Th}||(I + \Delta E_c E_f)S_f^{-1}||_A \tag{24}$$

3.3 Example

Consider the following SISO example

$$G_p(s) = \frac{5e^{-2s}}{10s + 1}$$

The plant features uncertainties in the steady-state gain. The perturbed plant is

$$\tilde{G}p(s) = \frac{5(1 + \Delta)e^{-2s}}{10s + 1}$$

with $|\Delta| \leq 0.5$. Sampling with a period of 2 yields

$$Gp(z) = \frac{0.5}{z - 0.8187}$$

The reference point and the noise are assumed to be bounded in magnitude: $||r||_\infty \leq 0.1$ and $||\eta_*||_\infty \leq 0.01$. The nominal design leads to the controller $G_c = \frac{z-0.8187}{0.5(z-1)}$ and a filter $F(z) = \frac{0.5k}{z-0.8187}$. The gain k is restricted to be $-0.36 \leq k \leq 3.63$ in order to assure that the estimation error goes to zero asymptotically.

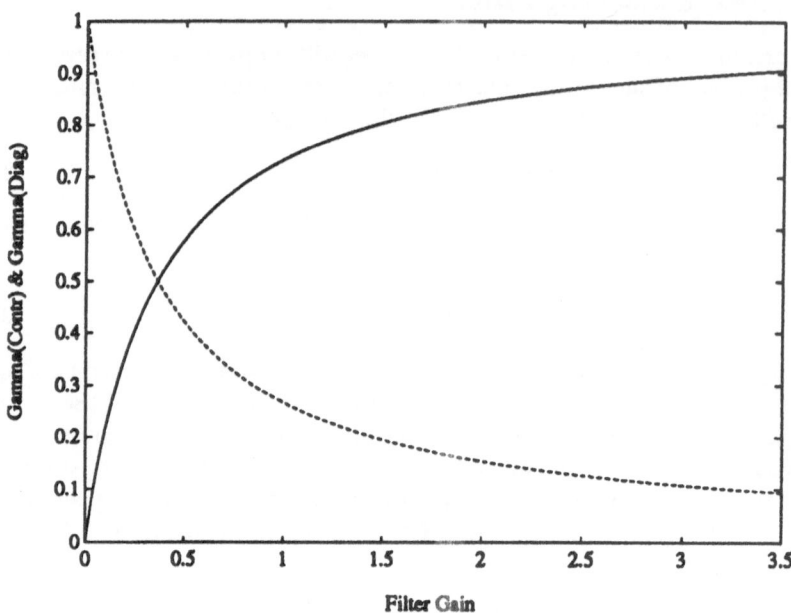

Fig. 1. Trade-offs between control and diagnosis objectives. Plot legend: gain of $H_{e\eta_s}$ - solid, gain of H_{af_s} - dashed

We will denote by $\gamma^{control}$ and γ^{diag} respectively the gains of $H_{e\eta_s}$ and H_{af_s}. The two indices represent in a way the performance of control subsystem and the diagnosis subsystem. Figure 1 shows the plot for various values of the filter gain. Increasing the filter gain will lead to a good fault detection but poor noise rejection at the output. A compromise can be found in the intersection of the two plots. For the value of the gain $k = 0.37$ the threshold and the minimum detectable faults are $\hat{T}h = 0.0150$ and $(||f|_\infty)_{min} \geq 0.061$

In the presence of modeling errors, a good choice for control performance measure would be the gain of $H_{a\eta_s}$, which will be denoted by $\gamma^{control}$. The level of set point rejection at the detection signal, the gain of H_{ar}, will be chosen as the diagnosis performance measure. It will be denoted by γ^{diag}. In this case, the controller is redesigned to achieve robust stability, An ℓ_∞ design leads to a controller

$$G_c = \frac{2.441z^5 - 3.9974z^4 + 1.6366z^3 - 0.4472z^2 + 0.7216z - 0.2954}{z^5 - 2.0392z^4 + 0.9995z^3 + 0.2204z - 0.1804}$$

This guarantee that $\delta||E_c||_{\mathcal{A}} = 0.721 < 1$. In order to ensure that $\delta||E_cE_f||_{\mathcal{A}} < 1$, the filter speed has to be detuned reducing the gain range to $-0.24 \leq k \leq 2.11$. Figure 2 shows the closed loop response to a step. Figures 3 and 4 show how γ^{diag} and γ^{contr} are accentuated by uncertainties, hence increasing the

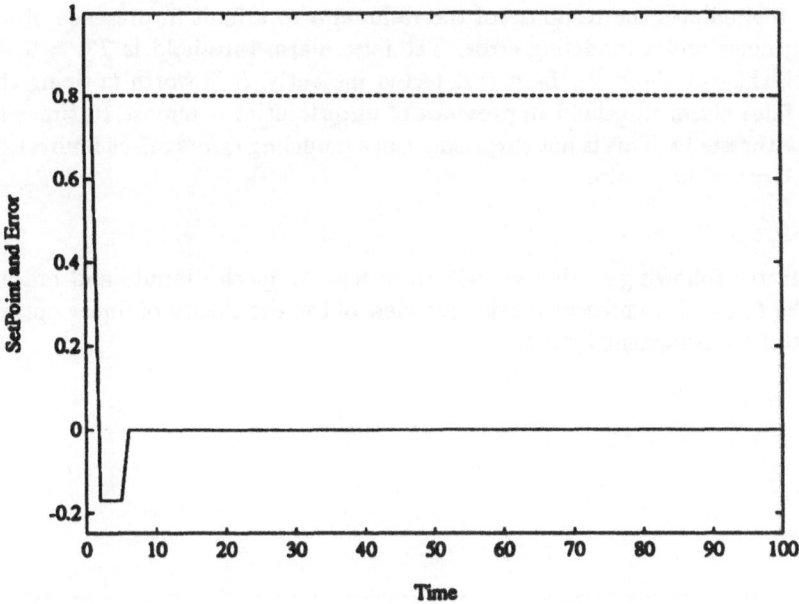

Fig. 2. Closed loop response. Legend: set point - dashed, error - solid

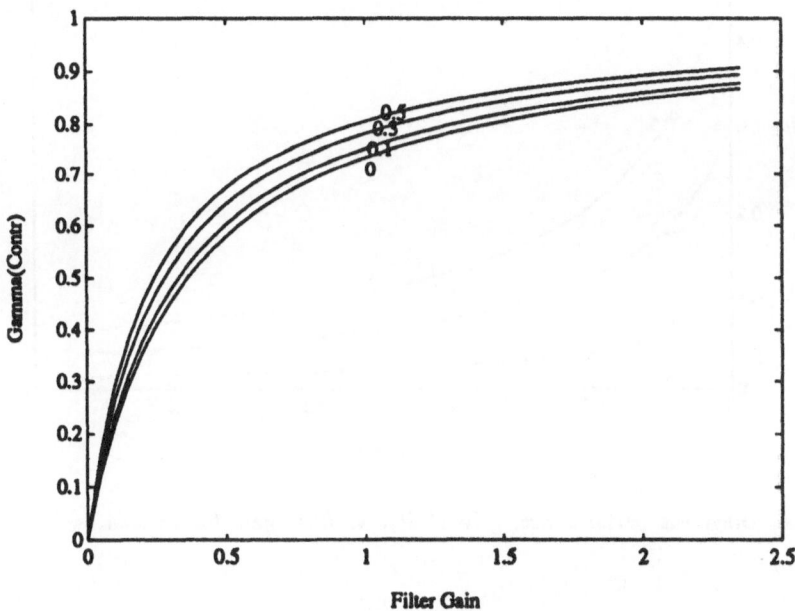

Fig. 3. Control performance: gain of $H_{a\eta_s}$ vs filter gain for various uncertainty sizes

tradeoffs between control and diagnosis objectives. For a filter gain of 0.37 Fig. 5 simulates the response of the residual a to a fault in presence of uniform noise and a modeling error. The false alarm threshold is $\hat{T}h = 0.145$. With this threshold the fault is detected instantly. It is worth noticing that the false alarm threshold in presence of uncertainties is almost 10 times the noise threshold. This is not surprising since modeling errors affect innovations much more than noise.

In the following section we will show how to specify inputs and outputs in the ℓ_∞, and we present a brief overview of the the theory of linear optimal control for persistent inputs.

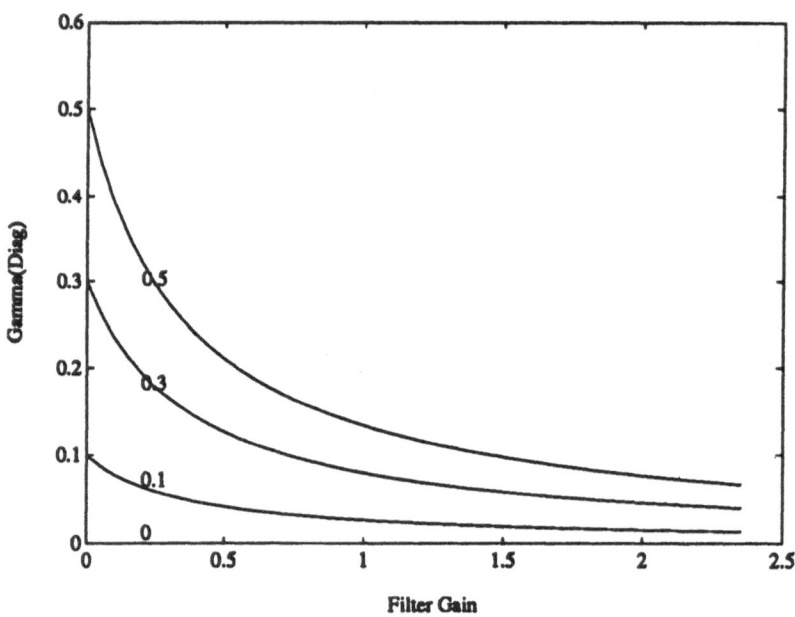

Fig. 4. Diagnosis performance: gain of H_{ar} vs filter gain for various uncertainty sizes

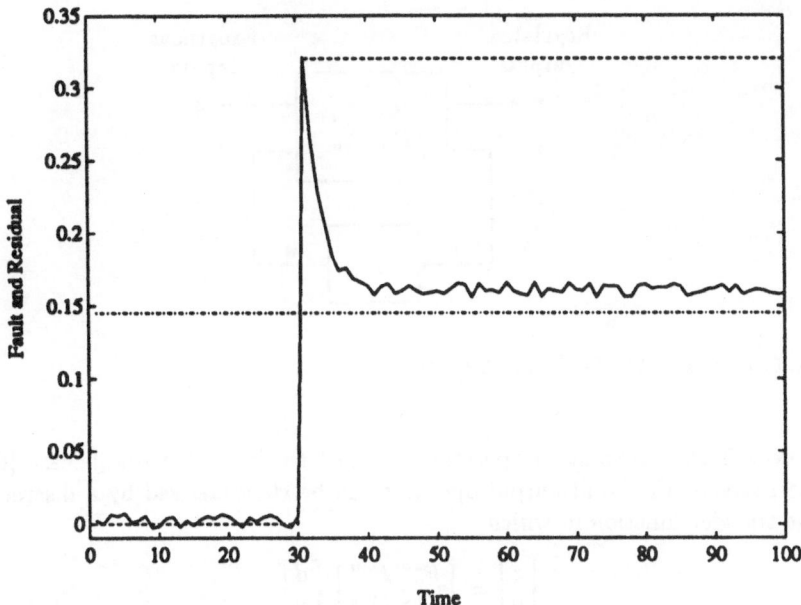

Fig. 5. Residual response to a fault. Plot legend: residual-solid, fault - dashed. Uniform noise of magnitude 0.01. Set point is a step of 0.1. Uncertainty size $\Delta = 0.3$. Threshold=0.145

4 ℓ_∞ Linear Optimal Control

Consider the time-invariant multivariable linear system shown in Fig. 6. The variables are defined as follows:

> z regulated outputs
> y measured outputs
> d unmeasured disturbances
> u manipulated inputs

Each element of these discrete-time signal vectors is a semi-infinite sequence. For a signal vector x, $x_i(k)$ denotes the ith element of x at time k, $k \geq 0$. The notation x_i denotes the i_{th} component sequence, and $x(k)$ denotes the value of the vector signal at time k. The ∞-norm is defined as

$$\|x\|_\infty = \max_{1 \leq i \leq n} \sup_{k > 0} |x_i(k)|$$

The set l_∞^n consists of all signals such that $\|x\|_\infty \leq \infty$.

The operator mapping $[d, u]^T \to [z, y]^T$ can always be represented as

$$z = \tilde{P}^{zd} * d + \tilde{P}^{zu} * u \tag{25}$$

$$y = \tilde{P}^{yd} * d + \tilde{P}^{yu} * u \tag{26}$$

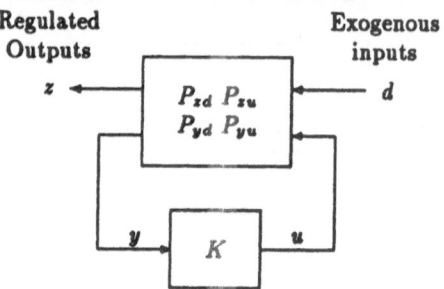

Fig. 6. Generic setup for feedback control

where $*$ is the convolution operator extended to the multivariable case [6]. Alternatively, the input-output operator can be characterized by a discrete-time transfer function in which

$$\begin{bmatrix} z \\ y \end{bmatrix} = \begin{bmatrix} P^{zd} & P^{zu} \\ P^{yd} & P^{yu} \end{bmatrix} \begin{bmatrix} d \\ u \end{bmatrix}$$

With control in place, the output of the closed-loop system is given by

$$z = H^{zd}d$$

In order to specify control system performance it is first necessary to specify classes of signals. Given classes of disturbances and acceptable responses, the nominal design objective will then be to satisfy the regulatory constraints for all disturbances in a specified set.

Let W^d be a multivariable transfer function. The *disturbance set* $D \subset l_\infty^n$ is defined as

$$D \equiv \left\{ d \in l_\infty^n : \|W^d d\|_\infty \leq 1 \right\}$$

This technique can be used define a set of disturbances that are bounded in magnitude or rates of change. These sets can include or exclude persistent disturbances.

Similarly, a *regulated set* Z is defined as the set of all bounded signal z such that $\|W^z z\|_\infty \leq 1$ where W^z is an appropriately chosen weighting operator.

The nominal objective for regulatory design is then to find a control K such that $z \in Z$ for all $d \in D$. This construction encompasses frequently used regulatory criteria such as the rejection of persistent disturbances, overshoot with respect to step inputs, and settling times. Significantly, this specification can also include hard time-domain constraints on the elements of z.

The following theorem is the key result that is the basis for the proposed research.

Theorem 1. *Let the disturbance set D and the regulation set Z be defined as above. Assume that D and Z are bounded. Then the stable closed-loop operator H^{zd} maps $D \to Z$ if and only if there exists a conformable operator Λ such that*

$$W^z H^{zd} = \Lambda W^d \tag{27}$$

where $\|\Lambda\|_A \leq 1$.

The set of all ΛW^d with $\|\Lambda\|_A \leq 1$ constitutes what is called the 'polar' in convex analysis [15]. The 'polar' of D is in the space dual to l_∞^{nd} consisting of those operators which map D to signals with norm less than one. The above theorem says that the weighted closed-loop transfer function must lie in the polar of D. A sketch of this proof is available in [9, 10].

This characterization of closed-loop performance is significant for several reasons. First, it characterizes performance in the time-domain and can account for various kinds of hard constraints. Traditional frequency domain performance criteria, such as steady-state disturbance rejection, can also be cast in this format. This is described in more detail below.

Second, this technique is *not equivalent* to the l_1 norm minimization techniques that have been proposed by Vidaysagar, Dahleh and Pearson, and others [5, 16]. In those techniques, the closed-loop performance is measured by the weighted norm $\|W^z H^{zd}\hat{W}^d\|_A$ where $\hat{W}^d$ is a weighting matrix. The selection of $\|\hat{W}^d\|_A$ is problematical. For the l_∞ technique proposed here to be equivalent to l_1 minimization it would be necessary for $W^d \hat{W}^d = I$. This is not generally possible since W^d is normally 'long and skinny' and not of full row rank. Thus l_1 optimal solutions would typically unnecessarily conservative.

Third, the formula $W^z H^{zd} = \Lambda W^d$ can be recast as convolutions involving $\tilde{H}^{zd}$ and $\tilde{\Lambda}$. Using a linear-fractional transformation $H^{zd} = T_1 + T_2 Q T_3$, this yields a system of linear constraints involving a stable control parameter, Q, and a bounded parameter Λ that characterizes closed-loop performance. This provides a computationally convenient parameterization of all stabilizing controls that meet a set of performance constraints.

Finally, what we term an l_∞ optimal control can be found by computing a control parameter Q which minimizes $\|\Lambda\|_A$. This effectively rescales the disturbances to find the largest class which can be mapped into the regulated set.

The nominal design problem is stated as follows.

$$\min_{Q, \Lambda \text{stable}} \gamma$$

$$T_1 + T_2 Q T_3 = \Lambda W^d \tag{28}$$

$$\|\Lambda\|_A \leq \gamma$$

To obtain a solution , a finite time horizon N, is chosen for Q. The unknowns are then the impulse response coefficients of Q and Λ where

$$Q = \sum_{k=0}^{N} Q(k)q^k, \qquad\qquad \Lambda_{ij} = \sum_{k=0}^{M} \Lambda_{ij}(k)q^k$$

The solution procedure is to match the coefficients of the forward shift operator q in Equation (28). Given a fixed horizon for Q, the horizon M for Λ is chosen so the right and left hand sides have the same order.

5 Input Sets Specification in ℓ_∞

All external signals that enter the system will be referred to as inputs. These will include measured and unmeasured disturbances, reference inputs noise sources and possible faults. The idea of specifying the set of inputs is that if the controller is designed to perform adequately for all inputs that can be reasonably expected, then it will perform well in actual practice. There are two distinct dangers to this approach. First, the controller may be too conservative if the model for the input class includes inputs which are not physically reasonable. Second, the opposite may happen if the class of inputs is not made large enough. The goal of the designer is to include an input model which is as physically accurate as possible. In this section we will show how to represent sets for inputs that are ℓ_∞ bounded.

5.1 Disturbances and Set Points Sets Representation

Unmeasured disturbances can be generally represented by appropriate bounds on magnitude, rate of change or possible correlations among them. The set of disturbances can be written as

$$D_d = \{d \in \ell_\infty^{n_d} : ||W^d d||_\infty \leq 1\} \tag{29}$$

where W^d is a stable causal weighting matrix. W^d contains typically bounds on the magnitude and the rate of change of the elements in the disturbance vector. For a scalar signal these bounds would be written as

$$\begin{cases} |d(k)| & < \text{a} \\ |d(k) - d(k-1)| < \text{b} \end{cases} \tag{30}$$

which leads to

$$W^d = \left[\begin{matrix} \frac{1}{a} \\ \frac{1-\frac{1}{q}}{b} \end{matrix} \right] \tag{31}$$

where q is the forward shift operator. In other situations W^d might also contain ratios of polynomials in q that represent 'frequency' domain weighting. To illustrate this, consider the following disturbance set

$$D_d = \{d \in \ell_\infty \ : \ ||\hat{W}^{-1}d||_\infty \le 1\} \tag{32}$$

Let d' be any input such

$$d' = \hat{W}^{-1}d \tag{33}$$

Following the reasoning used by Morari and Zafiriou [12] to the construction of weighting functions for L_2 bounded signals, consider the following weighting function

$$\hat{W}(q) = \frac{\beta q - 1}{\beta(q - 1)} \tag{34}$$

where $\beta > 1$. Any input $d'(k) = \beta^{-k}$ satisfies

$$||d'||_\infty \le 1 \tag{35}$$

and when passing through $\hat{W}$ it gives rise to a unit step $d(k) = 1$. On the other hand any input

$$d'(k) = \alpha^{-k} \quad \alpha > 1, \quad \alpha \ne \beta \tag{36}$$

satisfies $||d'||_\infty \le 1$, and after passing through the weight $\hat{W}$ the signal becomes

$$d(k) = \frac{\beta q - 1)}{\beta(q - 1)}\alpha^{-k} \tag{37}$$

which is a step modified by a lead if $\alpha > \beta$ or a lag if $\alpha < \beta$. Thus the set D_d characterizes not only steps but a lager class of inputs. The controller designed to reject step disturbances in D_d will also work for all class of disturbances 'similar' to steps. Similar reasoning can be applied to define weighted sets for noises and reference inputs.

5.2 Constraints on Manipulative Variables

Because the manipulative variables are usually subjected to constraints in magnitude and in rate of change, control input set can be described as

$$D_u = \{u \in \ell_\infty^{m_u} : ||W^u||_\infty \le 1\} \tag{38}$$

with $W^u = \left[\begin{array}{c} \frac{1}{a} \\ 1 - \frac{q^{-1}}{b} \end{array}\right]$, where a and b represent respectively bounds on magnitude and rate of change of the control input.

5.3 Faults Modes Representation

In general two typical faults modes can arise: abrupt faults, i.e. step-like changes and slowly developing faults, like bias. The faults will be bounded either in magnitude or in rate or in both. The purpose of the alarm is to detect significant deviations in the signals b_s and b_a which represent degradation in the sensor and actuator behavior. The detection is based on the transient behavior of the system after the occurrence of the fault. So in contrast to the control problem, the performance of the alarm is measured over a finite period of time. Consequently the detection scheme is based on a measure of the alarm signal over a detection window $[0, \tau]$.

Thus a fault occurring either in the sensors or in the actuators can be represented as follows

$$D_f = \{ f \in \ell_\infty^{n_f} : ||W^f f||_{\infty, \tau} \le 1 \} \tag{39}$$

where τ is the detection window. W^f will have typically the form, $W^f = \begin{bmatrix} \frac{1}{a_f} \\ \frac{1-q^{-1}}{b_f} \end{bmatrix}$, where a_f and b_f are, respectively, bounds on the magnitude and the rate of change. The specification of the faults using this time-domain approach is intuitive and practical. It is worth noticing that some knowledge of the nature of the faults, taken for instance from the process history, will be of a big help to specify tightly the set, and to avoid a conservative design of the controller.

5.4 Performance Specifications

Control Objectives Specifications. The control performances for closed loop feedback system is usually measured in terms of its disturbance rejection, set point tracking or some other desired performances. To express this performance quantitatively, bounds may be imposed on the absolute magnitude of deviations of regulated variables, their rates of change, on moving averages, or on possible correlations among the regulated variables. Constraints such as these can be lumped into a single weighting matrix which describes regulatory performance. The set of acceptable process outputs y_p to an unmeasured disturbances can be defined as

$$D_{y,d} = \{ y_p \in \ell_\infty^{n_y} : ||W^{y,d} y_p||_\infty \le \gamma_d^{perf} \le 1 \} \tag{40}$$

The parameter γ_d^{perf} is a measure of this aspect of regulation performance. Smaller γ_d^{perf} implies better performance. As example of performance requirement, bounds a_1, a_2 can be imposed respectively on the magnitude and on the rate of change of the response to a disturbance. A typical $W^{y_p d}$ will be,

$$W^{y_p d} = \left[\begin{array}{c} \frac{1}{a_1} \\ \frac{1-q^{-1}}{a_2} \end{array} \right] \tag{41}$$

If the disturbance is to be completely suppressed at steady state one can impose an integral action

$$W^{y_p d} = \frac{q^{-1}}{q^{-1} - 1} \tag{42}$$

Similarly for the suppression of noise, the set of desired responses to various types of noise can be defined by

$$D_{y_p \eta} = \{y_p \in \ell_\infty : \|W^{y_p \eta} y_p\|_\infty \le \gamma_\eta^{perf} \le 1\} \tag{43}$$

where γ_η^{perf} is a measure of the desired level of noise suppression. The response to set point tracking can also be implemented in the same way. The appropriate setpoint-output set is given in terms of the error, $e = r - y_p$

$$D_{y_p r} = \{y_p \in \ell_\infty : \|W^{y_p r}(r - y_p)\|_\infty \le \gamma_r^{perf} \le 1\} \tag{44}$$

In multivariable case, additional off-diagonal elements in the weighting matrices described above can be added to express bounds involving correlated responses among elements in y_p.

Diagnosis Objectives Specifications. We specify a set for the allowed responses of the alarm, specifying how the alarm will diagnose the eventual faults and reject the set point changes , the disturbances and the various types of noises.

$$D_a = \{a \in \ell_\infty^{n_a} : \|W^a a\|_{\infty, \tau} \le \gamma_{alarm}\} \tag{45}$$

where W^a is a weighting matrix that represents the desired diagnosis performance: As an example W^a can represent appropriate bounds on the magnitude or the rate of change of the alarm signal.

The alarm response can also be implemented in such a way that the alarm 'tracks' faults in the system. Then the appropriate alarm-fault set is given in terms of the error between the actual fault and the alarm.

$$D_{af} = \{a \in \ell_\infty^{n_a} : \|W^{af}(a - f)\|_{\infty, \tau} \le \gamma_{af}\} \tag{46}$$

where W^{af} is an appropriate weighting matrix

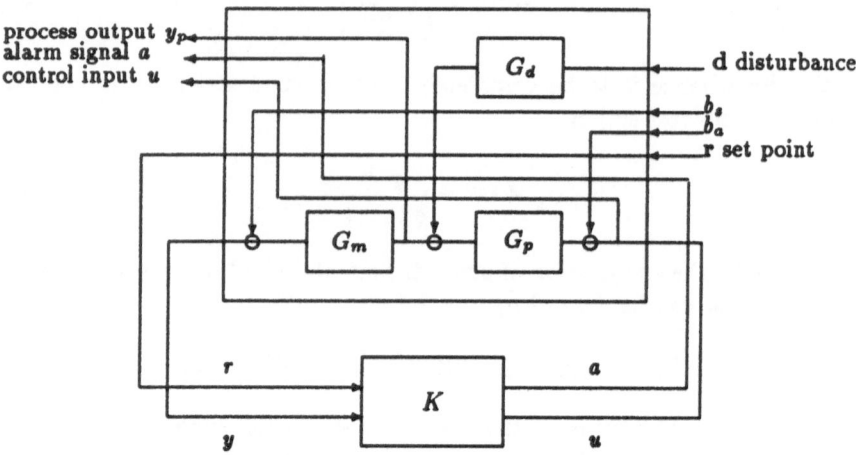

Fig. 7. Feedback scheme for integrated control and diagnosis

5.5 Nominal Design for Integrated Control and Diagnosis

We consider the general feedback scheme of Fig. 7, where the inputs and the outputs are structured as follows:

- $w = [d, b_s, b_a, r]^T$: n_w-dimensional exogenous input vector.
- $v = [a, u]^T$: n_v-dimensional control input vector.
- $y = [r, -y_m - b_s]^T$: n_y-dimensional measurement vector.
- $z = [y_p, a, u]^T$: n_z-dimensional vector of regulated outputs.

The faults occurring in the system are modeled by considering the signals b_s and b_a as representing deviations from the sensors and actuators ideal behaviors

$$b_s = \eta_s + f_s \text{ and } b_a = \eta_a + f_a$$

where η_s, η_a are zero mean sensor and actuator noise, respectively; f_s and f_a are faults in the sensors and actuators

Control and Diagnosis Objectives. To simplify the developments, the nominal plant is assumed to be stable. Otherwise a double coprime factorization is used. The closed loop transfer function matrix, H_{zw} for the control/diagnostic system is parameterized and its elements are shown in Table

1, where

$$Q_{ar} = K_{ar} - K_{ay}S_oG_mG_pK_{ur}$$
$$Q_{ay} = K_{ay}S_o$$
$$Q_{ur} = K_{ur} - K_{uy}S_oG_mG_pK_{ur}$$
$$Q_{uy} = K_{uy}S_o$$

and $S_o = (I + G_mG_pK_{uy})^{-1}$ is the output sensitivity matrix.

	d	b_s	b_a	r
y_p	$(I - G_pQ_{uy}G_m)G_d$	$-G_pQ_{uy}$	$(I - G_pQ_{uy}G_m)G_p$	G_pQ_{ur}
a	$-Q_{ay}G_mG_d$	$-Q_{ay}$	$-Q_{ay}G_mG_p$	Q_{ar}
u	$-Q_{uy}G_mG_d$	$-Q_{uy}$	$I - Q_{uy}G_mG_p$	Q_{ur}

Table 1. Parameterized closed-loop transfer function H_{zw}

The nominal control objectives can be stated as follows:

1. Reject unmeasured disturbances d at the output y_p $\Rightarrow$ Make H_{y_pd} small.
2. Achieve set point tracking $\Rightarrow$ Make $H_{y_pr} - I$ small.
3. Reject actuators and sensors noises and faults at y_p $\Rightarrow$ Make $H_{y_p\eta_a}$, $H_{y_p\eta_s}$, $H_{y_pf_a}$ and $H_{y_pf_s}$ small.
4. Achieve previous objectives with limited control action u $\Rightarrow$ H_{ud}, H_{ub_a}, H_{ub_s} and H_{ur} are to be 'small' in a certain way.

The Nominal diagnosis objectives can be summarized as follow:

1. Diagnostic faults in sensors and actuators $\Rightarrow$ Make $H_{af_s} - I$ and $H_{af_a} - I$ small.
2. Reject noises at the alarm a $\Rightarrow$ Make $H_{a\eta_s}$ and $H_{a\eta_a}$ small.
3. Reject disturbance at the alarm a $\Rightarrow$ H_{ad} small.
4. Reject set point r at the alarm a $\Rightarrow$ Make H_{ar} small.

The examination of these design objectives reveals the presence of certain tradeoffs. Some of them are familiar, such as the tradeoffs among disturbance rejection, setpoint tracking, and limited control action. These tradeoffs constrain the design of the control subsystem parameterized by Q_{ur} and Q_{uy}. Design tradeoffs also appear in the diagnostic subsystem parameterized by Q_{ar} and Q_{ay}. The conflict is for instance between noise rejection and fault detection which have the same transfer function. In the nominal case where no model error is present, there appears to be no tradeoffs between the design of the control and the diagnostic subsystems. This, however, changes when model error is introduced.

6 Quantification of the Effect of Modeling Errors on Diagnosis

The guaranteed levels of control performance are degraded by model uncertainty. Moreover, model uncertainty also couples the subsystems design such that there are tradeoffs between diagnosis and control performances.

To show how model uncertainty imposes additional restrictions on the integrated control/diagnostic design and to emphasize the conflicts among different design objectives, we will obtain the perturbed closed-loop transfer function in presence of modeling error.

Since our design relies on the input-output properties of the system, a suitable approach to quantify the effect of modelling errors is based on the ideas of threshold selector proposed in [7, 11]. The effect of modeling errors is incorporated in the analysis by providing analytical expressions for the optimal threshold in presence of uncertainties and noise. Also computed is the set of the minimum detectable faults.

6.1 False Alarm Threshold

The alarm is implemented by monitoring the magnitude $||a||_{\infty,\tau}$ of the signal a. We consider now the selection of appropriate thresholds for $||a||_{\infty,\tau}$ in the presence of noise and uncertainties. A threshold must compromise between erroneously raising an alarm (a false alarm) and erroneously not raising an alarm (a missed fault). It is possible to compute thresholds that are optimum in some sense.

The expression for the alarm signal a is

$$a = \tilde{H}_{ad}d + \tilde{H}_{a\eta}\eta + \tilde{H}_{ar}r + \tilde{H}_{af}f$$

where the $\tilde{H}_{aj}$ stands for the transfer function dependent on the model error Δ between the alarm signal and the input j, and d, r, η and f are respectively disturbance, set point, noise and fault either in the sensor or in the actuator. A false alarm threshold, adapted from definitions given in [11], is defined as

$$\overline{Th}(\tau) = \sup_{d,\eta,r,\Delta} \{||a||_{\infty,\tau} \ni ||f||_{\infty,\tau} = 0\}$$

where τ is the detection window $[0,\tau]$. Thus

$$\overline{Th}(\tau) = \sup_{d,\eta,r,\Delta} \{||\tilde{H}_{ad}d + \tilde{H}_{a\eta}\eta + \tilde{H}_{ar}r||_{\infty,\tau}\} \qquad (47)$$

Note that because of the fact that the detection is concerned with the transient behavior following the occurrence of a fault, the appropriate norm is defined on a finite interval. This will induce an operator norm dependant on τ, that is a transient gain. In order to get a bound on the false alarm threshold, we will use the fact that any $u \in \ell_{\infty,\tau}$ can be extended to ℓ_∞ by setting $u_k = 0 \ \forall \ k > \tau$. Hence for any operator $H \in \mathcal{L}_{TI}$ acting on u, we have

$$\|Hu\|_{\infty,\tau} \leq \|Hu\|_{\infty} \leq \|H\|_A \|u\|_{\infty,\tau}$$

Recasting the expression of the false alarm threshold we can write

$$\overline{Th}(\tau) \leq \|\tilde{H}_{ad}d\| + \|\tilde{H}_{a\eta}\eta\| + \|\tilde{H}_{ar}r\|$$

This expression can be further simplified by taking into account the precise kind of uncertainties that occur in the plant, as we will see later.

6.2 Minimum Detectable Fault

The alarm test $\|a\|_{\infty,\tau} \geq \hat{T}h$ unambiguously establishes that a fault has occurred. However, when $\|a\|_{\infty,\tau} < \hat{T}h$, the fault might still have occurred but the test is not sufficiently sensitive to distinguish the fault from the effects of noise and model error. It is useful, therefore, to find the minimum fault that will raise the alarm. The minimum detectable fault is the smallest fault f such

$$\inf_{f,d,r,\eta,\Delta} \|a\|_{\infty,\tau} > \overline{Th}(\tau)$$

which is equivalent to

$$\inf_{f,d,r,\eta,\Delta} \|\tilde{H}_{af}f + \tilde{H}_{ad}d + \tilde{H}_{ar}r + \tilde{H}_{a\eta}\eta\|_{\infty,\tau} > \sup_{d,r,\eta,\Delta} \|\tilde{H}_{ad}d + \tilde{H}_{ar}r + \tilde{H}_{a\eta}\eta\|_{\infty,\tau}$$

Following [7], A sufficient condition to ensure this is that

$$\inf_{f,\Delta} \|\tilde{H}_{af}f\|_{\infty,\tau} > 2 \sup_{d,r,\eta,\Delta} \|\tilde{H}_{ad}d + \tilde{H}_{ar}r + \tilde{H}_{a\eta}\eta\|_{\infty,\tau} \qquad (48)$$

For the right hand side we have the estimate

$$\|\tilde{H}_{ad}d + \tilde{H}_{ar}r + \tilde{H}_{a\eta}\eta\|_{\infty,\tau} < \hat{T}h$$

If we extend f to ℓ_∞ by letting $f(k) = 0 \ \forall \ k > \tau$ then $\|f\|_{\infty,\tau} = \|f\|_\infty$. Recalling that

$$\tilde{H}_{af} = H_{af}\tilde{H}$$

where H_{af} is the nominal transfer function and $\tilde{H}$ is a transfer function dependant on Δ, we can write that

$$\|f\|_\infty = \|\tilde{H}H_{af}^{-1}H_{af}\tilde{H}^{-1}f\|_\infty$$

which implies that

$$\|H_{af}\tilde{H}^{-1}f\|_\infty \geq \frac{\|f\|_{\infty,\tau}}{\|\tilde{H}H_{af}^{-1}\|}$$

This equation combined with Equation (48) provides the following sufficient condition for the fault to raise the alarm

$$\|f\|_{\infty,\tau} \geq 2\xi\hat{T}h$$

where $\xi = \|\tilde{H}H_{af}^{-1}\|$.

6.3 Effect of Uncertainties on Sensor Faults Detection

Consider the nominal plant subject to multiplicative output uncertainties: $\tilde{G}_p = (I + W\Delta)G_p$ where G_p is the stable nominal plant, W is a stable causal weighting matrix and $\Delta \in \mathcal{L}_{TV}$ such $\|\Delta\| < 1$. Following [4], $\tilde{G}_p$ is BIBO stable if and only if

$$M = -G_p Q_{uy} G_m W \text{ is such that } \|M\|_{\mathcal{A}} \leq 1$$

As remarked by Nett et al. [13], the transfer function $\tilde{H}_{ar}$ is the key to understand the interactions between the control and diagnostic objectives. The perturbed transfer function $\tilde{H}_{ar}$ is given by

$$\tilde{H}_{ar} = H_{ar} + H_{a\eta_s} G_m W \Delta (I + G_p K_{uy} S_o G_m W \Delta)^{-1} H_{y_s r}$$

Since $H_{ar} = Q_{ar}$, we can always set in the nominal design $Q_{ar} = 0$. Therefore we have

$$\tilde{H}_{ar} = H_{a\eta_s} G_m W \Delta (I - M\Delta)^{-1} H_{y_s r} \tag{49}$$

$H_{a\eta_s}$ and $H_{y_s r}$, the first and the last factors of this expression, appear in the nominal design. Based on the nominal design,

- $H_{y_s r} = (G_p Q_{ur})$ should be designed to be close to identity,
- $H_{a\eta_s} = (-Q_{ay})$ should be designed to be 'small',
- $M = -G_p Q_{uy} G_m W$ should be 'small' for robust stability.

Following Sect. 6, We can get the following expressions for false alarm threshold and the minimum detectable faults. Using the inequality

$$\|H(I - \Delta M)^{-1}\| \leq \frac{\|H\|_{\mathcal{A}}}{1 - \|M\|_{\mathcal{A}}} \quad \forall \Delta \ni \|\Delta\| < 1 \text{ and } M \ni \|M\|_{\mathcal{A}} < 1$$

a bound on the false alarm threshold is given by:

$$\overline{Th} \leq \hat{Th} = \frac{\|H_{ad}d\|_{\mathcal{A}}}{1 - \|M\|_{\mathcal{A}}} + \frac{\|H_{a\eta_s}\|_{\mathcal{A}}}{1 - \|M\|_{\mathcal{A}}} + \frac{\|H_{a\eta_s} G_m W\|_{\mathcal{A}} \|H_{y_s r}\|_{\mathcal{A}}}{1 - \|M\|_{\mathcal{A}}}$$

The different expressions in the threshold represent the effect of the disturbance d, noise, η_s and the set point r, on the threshold. The quantity $\|M\|_{\mathcal{A}}$ represents the effect of stability robustness.

The size of the minimum detectable fault is given by

$$\|f_s\|_{\infty, r} \geq 2\xi \hat{Th}$$

where

$$\xi = \|(I - \Delta M) H_{af_s}^{-1}\|$$

6.4 Example: Sensor Fault Detection

The example is the loop corresponding to the side draw of the heavy oil fractionar shell benchmark control problem [14]. The open loop discrete-time model is

$$y = 5.72z^{-4}\frac{(0.0328z + 0.0317)}{z - 0.9355}u + 1.52z^{-4}\frac{(0.0392z + 0.1087)}{z - 0.8521}d$$

The plant features uncertain steady-state gains which are represented as $K = K_0(1 + W\Delta)$ where $K_0 = 5.72$ is the nominal gain, $W = 0.0996$ and Δ is such that $|\Delta| \leq 1$. The plant is then $\tilde{G}_p = (I + W\Delta)G_p$. The measurement device is a pure delay, $G_m(z) = z^{-1}$.

The following sets are assumed for the integrated system inputs:

- Disturbances are bounded in magnitude and in rate

$$D_d = \{d \in \ell_\infty : \| \begin{bmatrix} 1 \\ \frac{1-q^{-1}}{0.5} \end{bmatrix} d\|_\infty \leq 1\}$$

- Reference points are also bounded in magnitude and rate

$$D_r = \{r \in \ell_\infty : \| \begin{bmatrix} 1 \\ \frac{1-q^{-1}}{0.5} \end{bmatrix} r\|_\infty \leq 1\}$$

- Manipulative variables are constrained in both magnitude and rate of change.

$$D_u = \{u \in \ell_\infty : \| \begin{bmatrix} 1 \\ \frac{1-q^{-1}}{0.3} \end{bmatrix} u\|_\infty \leq 1\}$$

- All noise sources are assumed to be bounded in magnitude

$$D_\eta = \{\eta \in \ell_\infty : \|\eta\|_\infty \leq 1\}$$

- Eventual faults in sensors or actuators are assumed to be bounded in magnitude and rate of change.

$$D_f = \{f \in \ell_\infty : \| \begin{bmatrix} 1 \\ \frac{1-q^{-1}}{0.4} \end{bmatrix} f\|_{\infty,r} \leq 1\}$$

Performances objectives include:

- Integral action for disturbances rejection and set point tracking: $W^{y,d} = \frac{q^{-1}}{1-q^{-1}}$, $W^{y,r} = \frac{q^{-1}}{1-q^{-1}}$.
- Tracking faults at the alarm: $W^{a,f_*} = \frac{q^{-1}}{1-q^{-1}}$.
- Attenuating sensor noise at the output y_p and at the alarm: $W^{y,\eta_*} = \frac{1}{0.01}$, $W^{a\eta_*} = \frac{1}{0.01}$.
- Rejecting unmeasured disturbances d at the alarm: $W^{ad} = \frac{1}{0.01}$.

The design objectives of nominal performance with robust stability can be cast in the ℓ_∞ framework as follows

- Control Objectives
  1. Reject disturbance d at y_p

  $$||W^{y_pd}y_p||_\infty \le \gamma_{y_p d} \le 1 \; \forall \, d \; \ni \; ||W^d d||_\infty \le 1$$

  2. Achieve set point tracking

  $$||W^{y_pr}(r - y_p)|_\infty \le \gamma_{y_p r} \le 1 \forall \, r \; \ni \; ||W^r r||_\infty \le 1$$

  3. Reject noise η_s at y_p

  $$||W^{y_p\eta_s}y_p||_\infty \le \gamma_{y_p \eta_s} \le 1 \; \forall \, \eta_s \; \ni \; ||W^{\eta_s}\eta_s||_\infty \le 1$$

  4. Achieve the three objectives with limited control action
  $$||W^u u||_\infty \le 1 \;\; \forall \, d \; \ni \; ||W^d d||_\infty \le 1, \forall \, r \; \ni \; |||W^r r||_\infty \le 1 \text{ and}$$
  $$\forall \, \eta_s \; \ni \; ||W^{\eta_s}\eta_s||_\infty \le 1$$

- Diagnosis Objectives
  1. Diagnose f_s, i.e. make $a - f_s$ 'small'

  $$||W^{af_s}(a - f_s)||_\infty \le \gamma_{af_s} \le 1 \;\; \forall \, f_s \; \ni \; ||W^{f_s}f_s||_\infty \le 1$$

  2. Reject noise η_s at alarm a

  $$||W^{a\eta_s}a||_\infty \le \gamma_{a\eta_s} \le 1 \;\; \forall \, \eta_s \; \ni \; ||W^{\eta_s}\eta_s||_\infty \le 1$$

  3. Reject disturbance d at alarm a

  $$||W^{ad}a||_\infty \le \gamma_{ad} \le 1 \; \forall \, d \; \ni \; ||W^d d||_\infty \le 1$$

  4. Reject set point r at alarm a

  $$||W^{ar}a||_\infty \le \gamma_{ar} \le 1 \; \forall \, r \; \ni \; ||W^r r||_\infty \le 1$$

- Robust stability

$$M = -G_p Q_{uy} G_m W \; \ni \; ||M||_A \le 1$$

where $\tilde{G}_p = (I + W\Delta)G_p$ and Δ is such that $||\Delta|| < 1$.

The nominal design for the control subsystem parameter Q_{uy} and Q_{ur}, and the alarm subsystem parameter Q_{ay} and Q_{ar} are decoupled. Using the techniques described before, we formulate a separate linear program solution for each subsystem. As mentioned before the parameter Q_{ar} can be taken equal to zero.

LP1: Nominal Design of Q_{uy} with Robust Stability. The design objective is to minimize weighted measures of the nominal control performance and the stability robustness. Three performance indices are introduced: $\gamma_{y_p d}$ measures disturbance rejection, $\gamma_{y_p \eta_\bullet}$ measures noise rejection and $\gamma_{r_\bullet}$ measures stability robustness. Each of these should be made small subject to the closed-loop constraints described above. A linear programming problem is then set as

$$\min_{Q_{\bullet y}} \alpha_{y_p d}\gamma_{y_p d} + \alpha_{y_p \eta_\bullet}\gamma_{y_p \eta_\bullet} + \alpha_{r_\bullet}\gamma_{r_\bullet}$$

subject to

$$\begin{bmatrix} W^{y_p d}G_d \\ 0 \end{bmatrix} - \begin{bmatrix} W^{y_p d}G_p \\ W^u \end{bmatrix} Q_{uy}G_m G_d = \Lambda_1 W^d$$

$$- \begin{bmatrix} W^{y_p \eta_\bullet}G_p \\ W^u \end{bmatrix} Q_{uy} = \Lambda_2 W^{\eta_\bullet}$$

$$-G_p Q_{uy}G_m W = \Lambda_3$$

$$\|\Lambda_1\|_A \leq \gamma_{y_p d}, \qquad \|\Lambda_2\|_A \leq \gamma_{y_p \eta_\bullet}, \qquad \|\Lambda_3\|_A \leq \gamma_{\gamma_{r_\bullet}}$$

$$\alpha_{y_p d} + \alpha_{y_p \eta_\bullet} + \alpha_{r_\bullet} = 1$$

The weighting coefficients α_i are used to tradeoff among the three performance indices. The tradeoffs can be shown by plotting $\gamma_{y_p d}$ or $\gamma_{y_p \eta_\bullet}$ as functions of $\gamma_{r_\bullet}$.

LP2: Nominal Design of Q_{ur}. The design objective is to minimize $\gamma_{y_p r}$ as a measure of set point tracking. A linear programming problem is then set as

$$\min_{Q_{\bullet y}} \gamma_{y_p r}$$

subject to

$$\begin{bmatrix} W^{y_p r} \\ 0 \end{bmatrix} - \begin{bmatrix} W^{y_p r}G_p \\ W^u \end{bmatrix} Q_{ur} = \Lambda W^r$$

$$\|\Lambda\|_A \leq \gamma_{y_p r}$$

LP3: Nominal design of Q_{ay}. Three performance indices are introduced to measure the performance of the diagnostic subsystem: $\gamma_{a f_\bullet}$ measures sensor fault diagnosis, $\gamma_{a\eta_\bullet}$ measures noise rejection, and γ_{ad} measures disturbance rejection at that alarm. The linear program is given by

$$\min_{Q_{\bullet y}} \alpha_{a f_\bullet}\gamma_{a f_\bullet} + \alpha_{a\eta_\bullet}\gamma_{a\eta_\bullet} + \alpha_{ad}\gamma_{ad}$$

subject to

$$W^{a f_\bullet}(I + Q_{ay}G_m G_d) = \Lambda_1 W^{f_\bullet}$$

$$-W^{a\eta_\bullet}Q_{ay} = \Lambda_2 W^{\eta_\bullet}$$

$$-W^{ad}Q_{ay}G_m G_d = \Lambda_3 W^d$$

$$\|\Lambda_1\|_A \leq \gamma_{a f_\bullet}, \qquad \|\Lambda_2\|_A \leq \gamma_{a\eta_\bullet}, \qquad \|\Lambda_3\|_A \leq \gamma_{ad}$$

$$\alpha_{a f_\bullet} + \alpha_{a\eta_\bullet} + \alpha_{ad} = 1$$

The weighting coefficients α_i provide the tradeoffs between the performance indices for this subsystem.

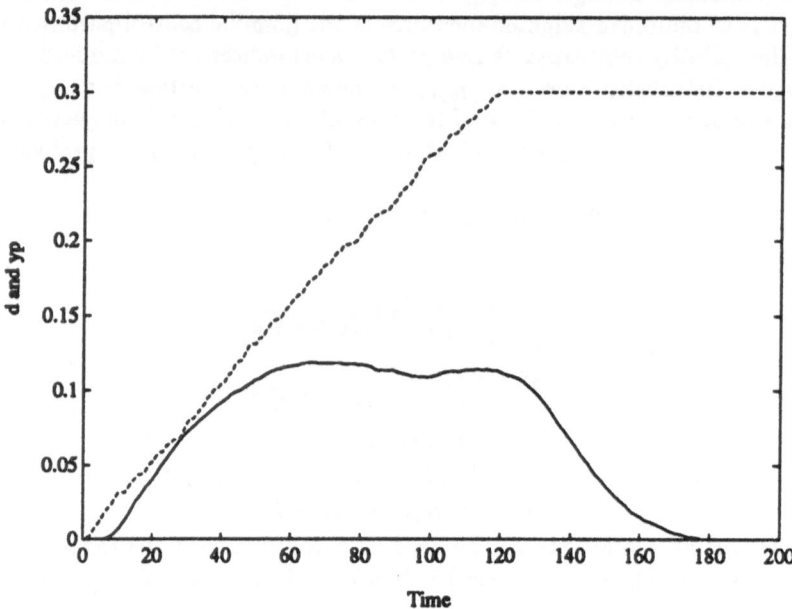

Fig. 8. Closed loop response to a ramp-type disturbance. Plot legend: disturbance
- dashed, output - solid. Uniform noise of magnitude 0.01

Results.

- LP1: For the values of $\alpha_{y,d} = 0.5$, $\alpha_{r_s} = 0.3$ and $\alpha_{y,\eta_s} = 0.2$, the controller Q_{uy} is

$$Q_{uy} = 0.491z^{-10} - 0.315z^{-9}$$

and the performance indices are

$$\gamma_{y,d} = 3.01 \ , \ \gamma_{r_s} = 0.116 \ , \gamma_{y,\eta_s} = 116.53$$

This indicates that the maximum bound on the disturbance that the process can tolerate without violating the constraints is $\frac{1}{3.10} = 0.32$. The system is robust stable, the value of $\gamma_{r_s} = 0.116$ indicates that the system will be stable for any $w < \frac{w}{0.116} = 1.164$ which allows a large variation in the gain. The maximum noise that the system will allow should be bounded in magnitude by $\frac{1}{116.53} = 0.086$. Figure 8 shows the closed loop response of the system to a ramp disturbance in presence of uniform noise of magnitude 0.01.

- LP2: The controller Q_{ur} is

$$Q_{ur} = 5.55 10^{-4}z^{-6} + 5.41 10^{-2}z^{-5} + 8.24 10^{-04}z^{-4} + 8.44 10^{-04}z^{-3}$$
$$- 1.37z^{-2} + 4.37 10^{-04}z^{-1} + 1.49$$

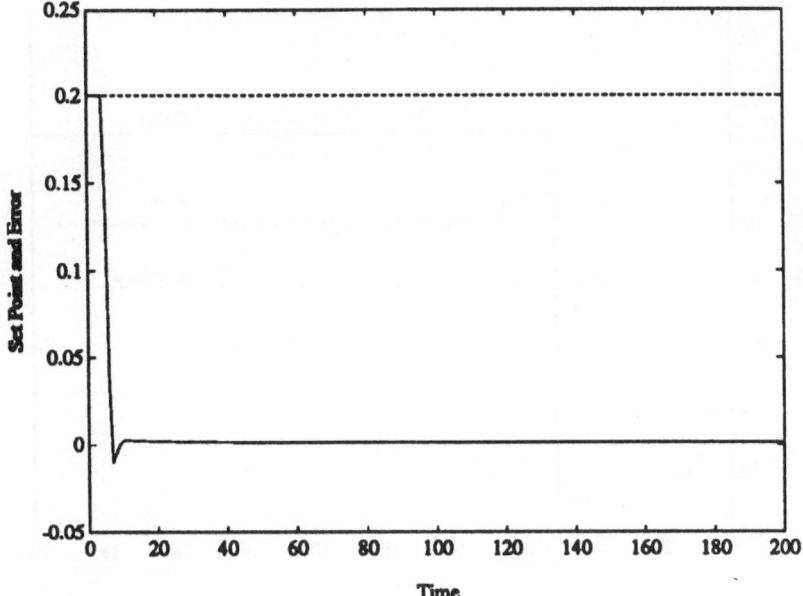

Fig. 9. Closed loop response to a step set point. Plot legend: set point - dashed, error - solid

The set point tracking performance measure is $\gamma_{y_p r} = 4.870$ which means that the set point magnitude and rate have to be reduced by $\frac{1}{4.87} = 0.205$. Figure 9 shows the closed loop response to a step in the set point.

- LP3: For $\alpha_{af} = 0.5$, $\alpha_{ad} = 0.3$ and $\alpha_{\eta_s} = 0.2$ the performances are:

$$\gamma_{af} = 1.535, \quad \gamma_{ad} = 15.29, \quad \gamma_{a\eta_s} = 119.0$$

This means that the maximum bounds on the magnitude and the rate of change of the fault should be $\frac{1}{1.535} = 0.651$ respectively. Also the bounds on the disturbance are to be reduced by $\frac{1}{15.29} = 0.065$, and the noise magnitude is $\frac{1}{119.0} = 0.0084$. The controller is order 15 and is

$$Q_{ay} = 0.0997z^{-15} - 1.0996$$

Figure 10 shows response of sensor alarm to a step like fault, in presence of a step disturbance, uniform noise and modeling error of $\Delta = 0.5$. The fault has been detected with a threshold of 0.20.

- Figure 11 shows the tradeoffs between nominal performance and stability robustness. The more system is robust stable (γ_{r_s} small) the more affected is the nominal performance ($\gamma_{y_p d}$ large).

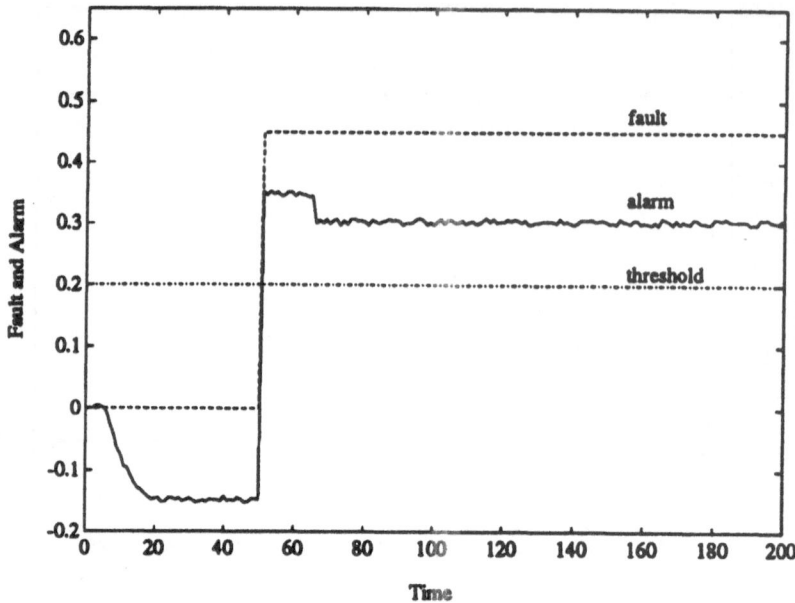

Fig. 10. Alarm signal response to sensor fault. Plot legend: fault - dashed, alarm signal - solid, threshold level - doted. Sensor noise magnitude is 0.01. Set point is a step of 0.1. Disturbance is a step of 0.1. Uncertainty size is $\Delta = 0.5$

- As shown from the Sect. 6.3, the false alarm threshold depends on the level of stability robustness $\|M\|_{\mathcal{A}} = \gamma_{rs}$ of the feedback system. Figure 12 shows this variation. The more robust stable the system is ,($\|M\|_{\mathcal{A}}$ small), the lower the threshold is.
  Naturally since the nominal performance and robust stability are in conflict the lower the threshold is the more the nominal performance deteriorates, as shown in Fig. 13.

7 Conclusions

The scheme for the integrating fault detection and control has some attractive features. Since it is cast in the framework of a multivariable control problem there is a highly developed body of theory and practice to support robust design of four parameters. On the other hand there are important criticism of the four parameter controller. Even if one accepts that faults can be modeled as exogenous inputs, there is still the question of determining appropriate norms. The H_∞ and ℓ_∞/ℓ_1 norms assumes statically stationary signals. Essentially by definition this is not the case for faults. To circumvent this problem Kosut [11] suggests an extended norm that measures a

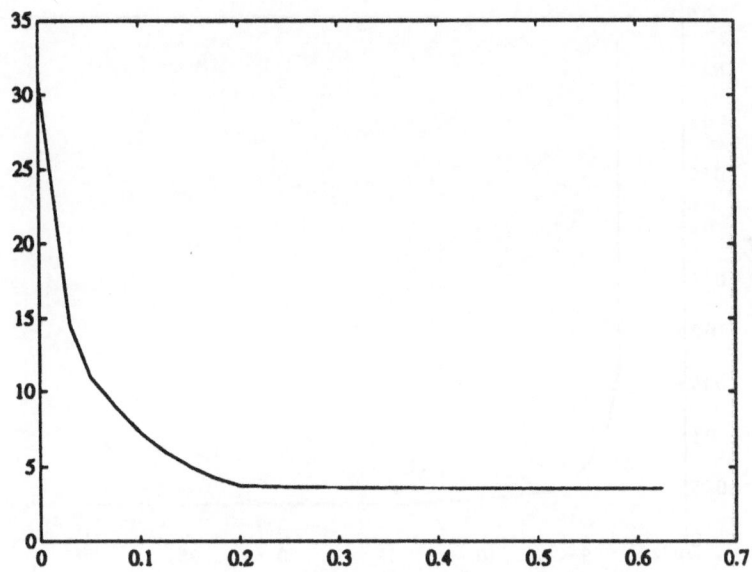

Fig. 11. Tradeoffs between nominal performance and stability robustness

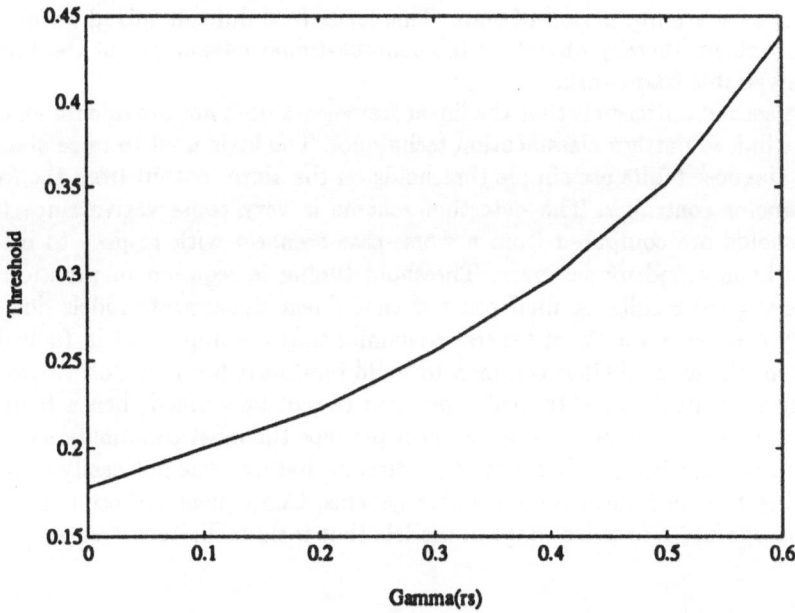

Fig. 12. Variations of sensor alarm threshold with the stability robustness

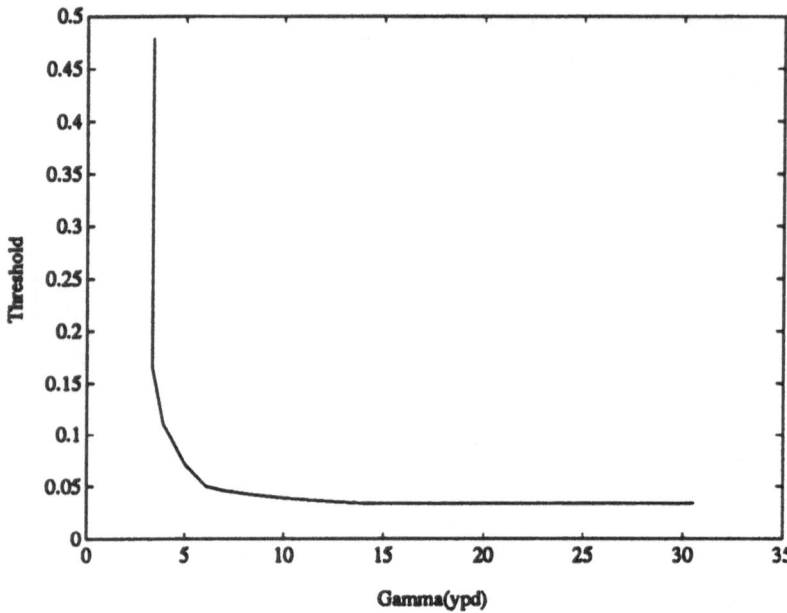

Fig. 13. Variations of sensor false alarm threshold with nominal performance

signal over a finite period of time. This leads to a difficult mixed norm design problem thereby obviating the computational advantages of the linear multivariable framework.

A second criticism is that the linear framework does not provide for an obvious link to pattern classification techniques. The logic used to raise alarms and diagnose faults use simple thresholds on the alarm output from the four parameter controller. The detection scheme is very conservative since the thresholds are computed from a worse-case scenario with respect to other disturbances and model error. Threshold tuning is required in practice to achieve good results. A final point is that linear dynamical models do not easily capture a variety of process dynamics that are important in fault detection. For example it is common to build limit switches in on/off valves to verify valve position. If the valve position cannot be verified then a fault is declared. This type of fault detection is perhaps the most commonly used in process industries, yet, because of its discrete nature, it is not easily formulated in terms of linear multivariable systems. Consequently there is a need to incorporate a broader range of models that include discrete events.

References

1. Ajbar H., Elias, A. and J.C. Kantor, "Multivariable Control System With

Integrated Diagnosis For Chemical Processes." *4th International Symposium on Process systems Engineering* PSE'91, Montebello, Canada, Aug. 5-9, 1991.

2. Ajbar H., Elias A., and J.C. Kantor, "Optimal Linear Control with Hard time Constraints," Submitted to *AIChE Journal*, 1992.

3. Astrom K. J., and B. Wittenmark, *"Computed Controlled Systems: Theory and Design."* Prentice-Hall, 1988.

4. Dahleh M. A., and Y. Otha, "A Necessary and Sufficient Condition for Robust BIBO Stability,' *Systems and Control Letters*, Vol. 11, pp. 271-275, 1988.

5. Dahleh M. A., and J.B. Pearson, Jr., "ℓ_1 optimal feedback controllers for MIMO discrete-time systems," *IEEE Trans. Automat. Contr.*, Vol. 32, pp. 314-322, 1987.

6. Desoer C. A., and M. Vidyasagar, *"Feedback systems: Input-Output Properties."* Academic Press. New York, 1975.

7. Emani-Naeini A. M., Akhter, M., and S. M. Rock, "Effect of Model Uncertainty on fault Detection: The Threshold Selector," *IEEE Trans. Automat. Control*, Vol. 33, pp. 1106-1115, 1988.

8. Frank P. M., "Fault Diagnosis in Dynamic Systems Using Analytical and Knowledge-based Redundancy-A Survey and Some New Results," *IFAC*, pp. 459-474, 1990.

9. Keenan M. R., and J.C. Kantor, "An ℓ_∞ optimal performance approach to linear feedback control," *The Second Shell Process Control Workshop*, 1988

10. Keenan M. R., and J.C. Kantor, "An ℓ_∞ Optimal performance approach to robust feedback control," *Proceedings of the ACC*. 1989.

11. Kosut R. L., and R.A Walker, "Robust fault detection: The effect of model error," *Proc. Amer. Contr. Conf*, 1094-1096, 1984.

12. Morari M., and E. Zafriou, *"Robust Process Control,"* Prentice-Hall, 1989

13. Nett C.N, Jacobson C. A., and A.T. Miller, "An Integrated Approach to Controls and Diagnostics: The 4-Parameter Controller," *Proceedings of the ACC*, pp. 824-835, 1988.

14. Prett D. M., and C. E.García, *"Fundamental Process Control,"* Butterworks Series in Chem. Eng., 1988.

15. Schrijver, A. *"Theory of Linear and Integer Programming,"* John Wiley, 1986.

16. Vidyasagar, M. "Optimal rejection of persistent bounded disturbances," *IEEE Trans. Automat. Control*, Vol. 31, pp. 527-534, 1986.

Identification of Ill-Conditioned Plants — A Benchmark Problem*

Elling W. Jacobsen[1] and Sigurd Skogestad[2]

[1] Centre for Process Systems Engineering, Imperial College, London SW7 2BY, United Kingdom.

[2] Chemical Engineering, University of Trondheim — NTH, N-7034 Trondheim, Norway.

Abstract:
This note provides a simple process example from chemical engineering which is proposed as a challenge problem for multivariable identification. The process considered is a simple heat-exchanger with two inputs and two outputs. It is strongly interactive and also ill-conditioned. A single slow pole, resulting from the interactions, is dominating all the individual open-loop responses. Attempting to identify a model based on fitting the individual transfer-matrix elements will usually result in a multivariable model which incorrectly has this dominant pole repeated. Such a model, although a reasonable model for the open-loop dynamics, yields a poor prediction of the process behavior under feedback control, in particular when considering partial control.

The note includes a description of the process, a file for generating open-loop "experimental" data and an example demonstrating that classical identification employing an ARMAX-type of model yields a model which is poor for feedback control studies of the process.

1 Introduction

Most published work on the identification of dynamic models from experimental data has been concentrated on the single-input-single-output (SISO) case. This is also reflected in the literature on process dynamics and control, where linear dynamic models usually are obtained by fitting input-output data from a plant or nonlinear simulation to a low-order transfer-function. In cases where the process is multivariable, the transfer-matrix is usually obtained by fitting the transfer-matrix elements *independently*. However, obtaining reasonable models for the individual transfer-function elements does not guarantee a reasonable multivariable model. This is in particular true for ill-conditioned processes which is the subject of this note. Ill-conditioned processes are commonplace in the chemical process industry and include, for example, high-purity distillation columns (Skogestad et al. [6]).

* Financial support from the Royal Norwegian Council for Scientific and Industrial Research (NTNF) is greatly acknowledged.

Skogestad and Morari [5] argue that fitting the transfer-matrix elements independently may easily lead to poor models for ill-conditioned processes unless one explicitly takes into account the coupling between the gains of the different elements. In particular, one is not able to obtain a good model of the low-gain direction of the plant, and the model will easily have the wrong sign of the determinant of the steady-state gain matrix, and the model will be useless for control studies. This problem may, however, usually be corrected as the sign of the determinant and its approximate value in many cases is known a priori (Kapoor and McAvoy [3], Jacobsen et al. [1]).

Another, and more fundamental problem in the identification of ill-conditioned processes, is the fact that such plants often have a *single* "slow" pole (large time-constant) which tends to dominate all responses of the plant (Jacobsen and Skogestad [2]). This dominating pole is a result of interactions in the process, and is thus shared by all the transfer-matrix elements. As shown by Jacobsen and Skogestad [2], fitting the transfer-matrix elements independently such that they all contain the dominating pole, will usually result in an inconsistent model with several poles equal to the dominating pole of the process. This inconsistency will result in a poor prediction of the process under partial feedback control, that is, with only some of the process outputs under feedback control.

The general literature on identification has so far not focused very much on multivariable issues, and the particular problems that may be encountered for ill-conditioned processes mentioned above, do not seem to have been discussed. In this note we therefore present data for an ill-conditioned process which we believe represents a "new" and difficult problem in multivariable identification.

We start the note by presenting a model and a set of input-output data of a heat-exchanger which is ill-conditioned. In addition to providing data for the process we also discuss briefly some specific process properties which are of interest for the identification problem. Having presented the problem we employ a fairly standard identification technique and show that it results in an inconsistent model which is poor for control studies of the plant. The objective of the example is to demonstrate that obtaining reasonable models for the individual transfer-matrix elements does not guarantee that the multivariable properties have been reasonably captured.

2 Process Description

The process we consider is a simple heat-exchanger where heat is transferred between a cold and a hot flow (see Fig. 1). Each side of the heat-exchanger is approximated as a single, perfectly mixed tank. Neglecting variations in liquid volume and heat accumulated in the walls yields a model with two states. The model derivation is given in Appendix 1. The linear model $y(s) = G(s)u(s)$

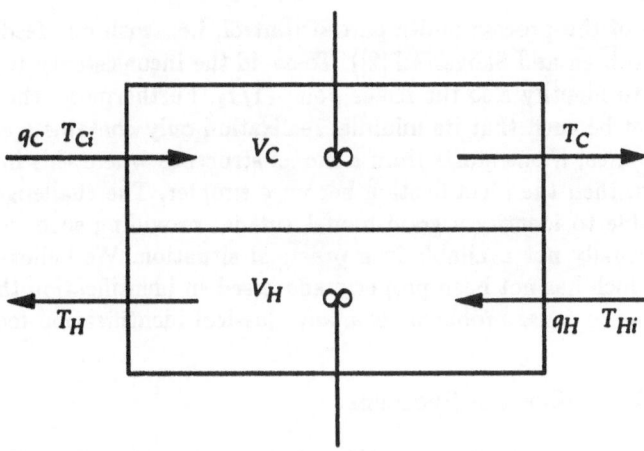

Fig. 1. Simple heat exchanger

is given by

$$G(s) = \frac{1}{(1 + \tau_1 s)(1 + \tau_2 s)} \begin{pmatrix} k_{11}(1 + z_1 s) & k_{12} \\ k_{21} & k_{22}(1 + z_1 s) \end{pmatrix}, \qquad (1)$$

where $\tau_1 = 100$; $\tau_2 = 2.44$; $z_1 = 4.76$; $k_{11} = -k_{22} = -1874$; and $k_{12} = -k_{21} = 1785$. Here $y = [T_C \quad T_H]$ is the cold and hot exit temperatures and $u = [q_C \quad q_H]$ are the cold and hot inlet flow rates. The first thing to note about the model is that there are two pole-zero cancellations such that the model contains only two and not four states. The model is also relatively ill-conditioned with a steady-state condition number of 41. The physical explanation for the ill-conditioning is simply that the two exit temperatures are almost the same ($T_C = 61.59\,^\circ C$ and $T_H = 63.41\,^\circ C$ in our case), and it is very difficult to change them independently. In particular, it is difficult to make them closer or further apart (this is the low-gain direction of the process) whereas we may easily make them *both* hotter or colder (this is the high-gain direction of the plant). An analysis of the model reveals that the slow pole $-1/\tau_1$ is related to the high-gain direction of the plant while $-1/\tau_2$ is related to the weak direction. The steady-state gain related to the slow pole is hence 41 times larger than the gain related to the fast pole.

The open-loop responses of the process model (1) are almost pure first-order responses with a time-constant equal to τ_1. Thus, a reasonably good fit of the individual transfer-matrix elements is obtained by first-order transfer-functions with time-constant $\tau_1 = 100$ min. However, the resulting model

$$G(s) = \frac{1}{1 + \tau_1 s} \begin{pmatrix} k_{11} & k_{12} \\ k_{21} & k_{22} \end{pmatrix} \qquad (2)$$

contains two poles at $-1/\tau_1$ and is thus inconsistent with the true process (1) which has only one pole at this location. The inconsistency results in a poor prediction of the process under partial control, i.e., with one feedback loop closed (Jacobsen and Skogestad [2]). To avoid the inconsistency it is at least necessary to identify also the faster pole $-1/\tau_2$. Furthermore, the identified model must be such that its minimal realization only contains a single slow pole. Of course, if one starts from a model structure where this information is included, then the identification becomes simpler. The challenge is to see if one is able to identify a good model *without* providing such information which is usually not available in a practical situation. We believe this is a problem which has not been properly addressed in identification theory, and which seems to cause problems for many classical identification methods.

2.1 The Identification Problem

In Appendix 2 we provide a Matlab file for generating open-loop "experimental" data using the linear model (1). The data are produced using a multivariable experiment, i.e., simultaneous perturbations in the two inputs. Noise is added to the inputs as well as the outputs.

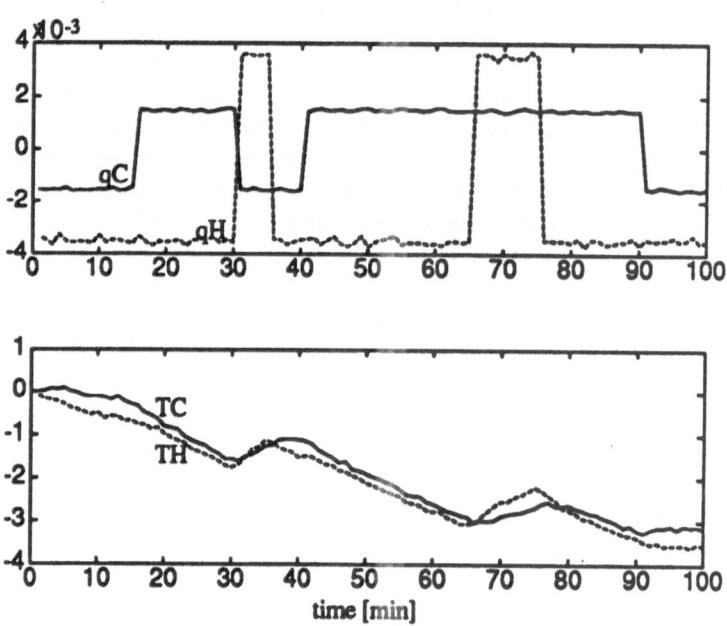

Fig. 2. Input and output data used for identification of heat-exchanger. The data were generated with the Matlab file given in Appendix 2

Figure 2 shows the 100 min. input sequence (including noise) and the resulting outputs generated using the Matlab file. The inputs to the process contain 3 % white noise, while the outputs have white noise with variance $0.03\,^\circ C$ (which is very small compared to practical situations).

The identification problem is to come up with a reasonable multivariable dynamic model based on these data alone, i.e., based on the noise-free inputs and the noisy measurements. One should not supply any knowledge about the special multivariable structure of the model as given by (1). The identified model is intended to be used for feedback control studies, and two different cases are of interest.

1. Partial control: Output y_1 is controlled using input u_1 while y_2 is left uncontrolled.
2. Multivariable control: Both y_1 and y_2 are controlled using both inputs.

In both cases the responses to set-point changes as well as disturbances in the inputs should be considered and compared with those of the correct model (1). The intention of the challenge problem is that one should identify the model based on open-loop data only. If one is allowed to use closed-loop data we believe the identification becomes simpler.

3 MISO-Identification using an ARMAX-type model

In this section we employ a fairly standard identification technique to the data generated using the Matlab file given in Appendix 2. We employ the Matlab System Identification Toolbox (Ljung [4]) and use MISO-identification with an ARMAX-type model structure. In the identification we fit each output with a strictly proper second order model which is the same structure as the true model (1). The model resulting from this identification is given by

$$
G(s) = \begin{pmatrix} \dfrac{-2025(5.218s+1)}{(2.027s+1)(110.7s+1)} & \dfrac{1871(0.0263s+1)}{(2.027s+1)(110.7s+1)} \\[4mm] \dfrac{-1795(-0.0933s+1)}{(1.404s+1)(110.5s+1)} & \dfrac{2049(3.947s+1)}{(1.404s+1)(110.5s+1)} \end{pmatrix} . \quad (3)
$$

The identified model (3) has a minimal realization with 4 states. Figure 3 compares the noise-free open-loop step responses of model (3) with those of the "true" model (1).

We see from the responses that we have obtained a reasonable identification of the individual SISO-transfer functions. Furthermore, we see from the identified model that we have been able to obtain reasonable estimates for the two poles $-1/\tau_1$ and $-1/\tau_2$. However, the multivariable interactions have not been captured as the model (3) has multivariable zeros at $-0.0217 = -1/46.1$ and $-0.426 = -1/2.35$ which do not cancel the poles. This also becomes clear if one considers the singular values of the true (1) and fitted (3) model respectively. The true model (1) has, as mentioned previously, a low-gain direction

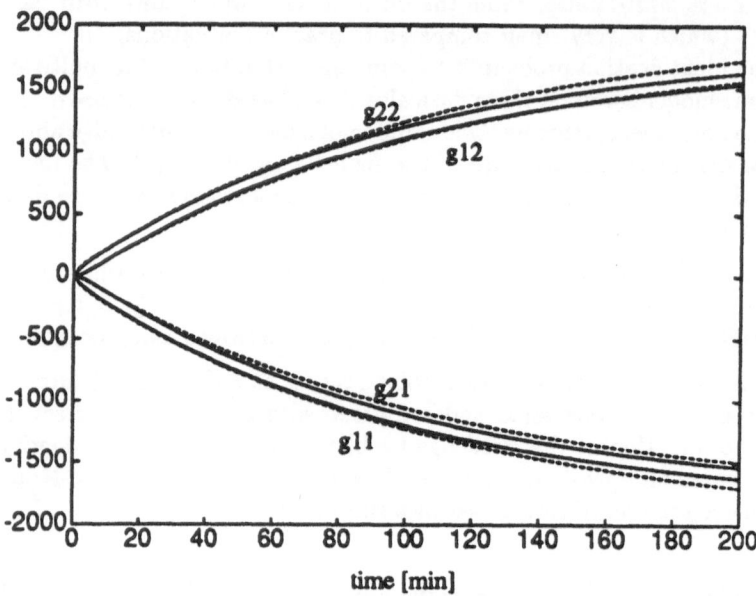

Fig. 3. Open-loop step responses of identified model (3) (dashed lines) and correct model (1) (solid lines). Labels g_{ij} denotes corresponding transfer-matrix element

with a single fast pole $-1/\tau_2$. However, the low-gain direction of the fitted model (3) has a significant part of its dynamics related to a slow pole around $-1/\tau_1$.

Figure 4 compares the closed-loop responses of the correct model (1) and the identified model (3) when output 1 is controlled with input 1 using the proportional feedback law $u_1 = K_c y_1$ with $K_c = 0.015$. We see that the identified model yields a good prediction for the controlled output y_1. However, for the uncontrolled output y_2 there is a large discrepancy between the process represented by (1) and the identified model (3). For the correct model the single slow pole is moved by the feedback controller and the response in the uncontrolled output y_2 is as fast as for y_1, while the identified model (3) contains an excessive slow pole which is left in the partially controlled model and results in a slow settling in output y_2.

4 Discussion

The noise levels of the data provided in this note are relatively small compared to what one should expect in a practical situation. Increasing the noise levels will mainly change the results obtained in a qualitative manner, that is, the excessive slow pole in the identified model will become even more marked.

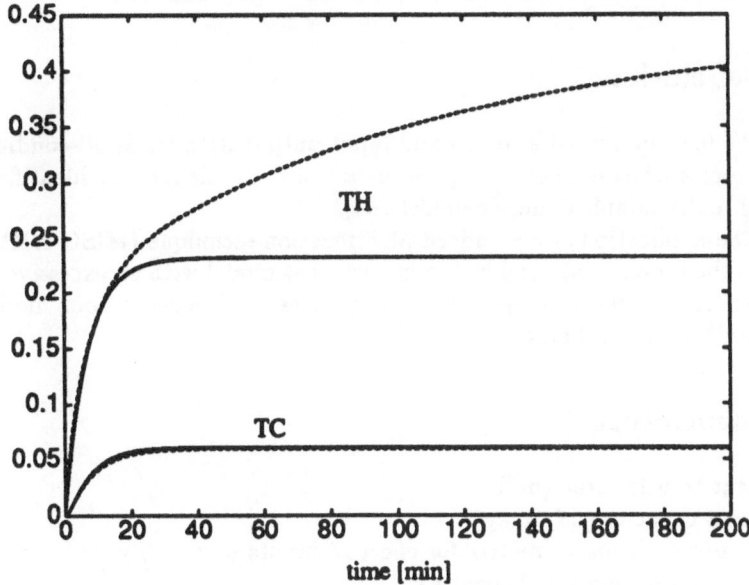

Fig. 4. Closed-loop responses of identified model (3) (dashed lines) and correct model (1) (solid lines) to step disturbance of magnitude 0.001 in the hot flow u_2. Output y_1 controlled by u_1 q_C using proportional controller with gain $K_c = 0.015$

An additional problem which may be encountered at higher noise levels is that of obtaining the correct sign of the determinant of the steady-state model. However, as mentioned in the introduction, this is usually a less crucial problem as the sign and approximate value of the determinant in many cases is known a priori.

The input sequence used to generate the "experimental" data in Appendix 2 are based on low-pass filtered PRBS signals with a minimum time between changes of 5 minutes, an experiment time of 100 minutes, and a sampling rate of 1 minute. Prolonging the time for the experiment with this set of input sequence does not seem to improve the identification.

It is worth noting that although we used a low-pass filtered input sequence, the main model error was at rather low frequencies, while the high frequency behavior of the process was reasonably well captured in the identified model. This may indicate that an input sequence with even more emphasis on low frequencies would yield better results. Indeed, with 500 minutes of experiments and a minimum time between changes of 25 minutes, we obtained better results as we have more information in the low-frequency region.

Even if the experiment time is fixed to 100 minutes, we do not rule out the possibility that a different input sequence may yield better results. The problem is how to determine the best possible input sequence when the process dynamics and multivariable interactions are largely unknown.

5 Conclusions

- We have presented a model and input-output data for an ill-conditioned process which we believe represents a "new" problem in the identification of multivariable dynamic models.
- The application of a standard identification technique (MISO ARMAX) to the process data yielded an inconsistent model with an excessive number of slow poles compared to the process, and hence a poor model for feedback control studies.

Nomenclature

A - heat transfer area (m^2)
c_P - heat capacity $(kJ/^{\circ}Ckg)$
$G(s)$ - process transfer-matrix for effect of inputs u
$g_{ij}(s)$ - transfer matrix element i,j
q_C - cold inlet flow (m^3/min)
q_H - hot inlet flow (m^3/min)
T_C - cold outlet temperature $(^{\circ}C)$
T_H - hot outlet temperature $(^{\circ}C)$
U - heat transfer coefficient $(kJ/m^2\ ^{\circ}Cmin)$
V_C - liquid volume cold side (m^3)
V_H - liquid volume hot side (m^3)

Greek symbols
τ_1 - dominant (largest) process time-constant (min.)
τ_2 - smaller process time-constant (min.)

Subscripts
s - setpoint change

References

1. Jacobsen, E.W., P. Lundström and S. Skogestad, "Modelling and Identification for Robust Control of Ill-Conditioned Plants - A Distillation Case Study", *Proc. 1991 American Control Conference*, Boston, 242-248, 1991.
2. Jacobsen, E. W. and S. Skogestad, "Inconsistencies in Dynamic Models for Ill-Conditioned Plants - with Application to Low-Order Models of Distillation Columns", Submitted to *Chem.Eng.Sci..*, 1992
3. Kapoor, N. and T.J. McAvoy, "An Analytical Approach to Approximate Dynamic Modelling of Distillation Towers", *Ind.Eng.Chem.Res.*, 26, 2473-2482, 1987.
4. Ljung, L., *System Identification Toolbox Version 3.0*, MathWorks Inc., 1991.

5. Skogestad, S. and M. Morari, "Understanding the Dynamic Behavior of Distillation Columns", *Ind. & Eng. Chem. Res*, **27**, 10, 1848-1862, 1988.
6. Skogestad, S., M. Morari and J.C. Doyle, "Robust Control of Ill-Conditioned Plants: High-Purity Distillation", *IEEE Trans. Autom. Control*, **33**, 12, 1092-1105, 1988.

Appendix 1. Simple model of heat exchanger

Consider a simplified heat exchanger with one mixing tank on each side as shown in Fig. 1. Assume constant volumes, V, on each side, and constant values of ρ and c_P. A heat balance for the cold and hot side then yields

$$\tau_C \frac{dT_C}{dt} = \frac{q_C}{q_C^*}(T_{Ci} - T_C) + \alpha_C(T_H - T_C) \tag{4}$$

$$\tau_H \frac{dT_H}{dt} = \frac{q_H}{q_H^*}(T_{Hi} - T_H) - \alpha_H(T_H - T_C) \tag{5}$$

where q^* denotes the nominal (steady-state) flow, and

$$\tau_C = \frac{V_C}{q_C^*}; \quad \alpha_C = \frac{UA}{\rho_C q_C^* c_{PC}} \tag{6}$$

$$\tau_H = \frac{V_H}{q_H^*}; \quad \alpha_H = \frac{UA}{\rho_H q_H^* c_{PH}} \tag{7}$$

Linearizing the model assuming UA and thus α constant (independent of flow and temperature), introducing deviation variables, and taking Laplace transforms yields

$$\tau_C s T_C(s) = T_{Ci}(s) - T_C(s) + (T_{Ci}^* - T_C^*)\frac{q_C(s)}{q_C^*} + \alpha_C(T_H(s) - T_C(s)) \tag{8}$$

$$\tau_H s T_H(s) = T_{Hi}(s) - T_H(s) + (T_{Hi}^* - T_H^*)\frac{q_H(s)}{q_H^*} - \alpha_H(T_H(s) - T_C(s)) \tag{9}$$

where the superscript $*$ denotes steady-state values. In the following we will assume $\tau_C = \tau_H = \tau = 100$ [min], $\alpha_C = \alpha_H = \alpha = 20$ and $q_C^* = q_H^* = q^* = 0.01$ [m^3/min] (see data in Table 1). Rearranging yields

$$\begin{pmatrix} T_C(s) \\ T_H(s) \end{pmatrix} = G(s) \begin{pmatrix} q_C(s) \\ q_H(s) \end{pmatrix} + G_d(s) \begin{pmatrix} T_{Ci}(s) \\ T_{Hi}(s) \end{pmatrix} \tag{10}$$

where

$$G_d(s) = \frac{1}{(\tau s + 1)(\tau s + 1 + \alpha)} \begin{pmatrix} \tau s + 1 + \alpha & \alpha \\ \alpha & \tau s + 1 + \alpha \end{pmatrix} \tag{11}$$

and

$$G(s) = G_d(s) \begin{pmatrix} (T_{Ci}^* - T_C^*)/q_C^* & 0 \\ 0 & (T_{Hi}^* - T_H^*)/q_H^* \end{pmatrix} \tag{12}$$

Inserting the numerical values finally yields

$$G_d(s) = \frac{0.02439}{(100s + 1)(2.439s + 1)} \begin{pmatrix} 21(1 + 4.76s) & 20 \\ 20 & 21(1 + 4.76s) \end{pmatrix} \qquad (13)$$

and $G(s) = G_d(s) \cdot \begin{pmatrix} -3659 & 0 \\ 0 & 3659 \end{pmatrix}$

$V_H = V_C$	$q_C = q_H$	T_{Ci}	T_{Hi}	T_C	T_H	UA	ρ	c_P
m^3	m^3/min	°C	°C	°C	°C	$kJ/°Cmin$	kg/m^3	$kJ/°Ckg$
1	0.01	25	100	61.59	63.41	300	500	3.0

c_P and ρ are equal for the hot and cold side.

Table 1. Steady-state data for heat-exchanger (see also Fig. 1)

Appendix 2. Matlab-file for generating input-output data of heat-exchanger

```
% This file generates inputs, u, and outputs, y,
% for heat exchanger identification problem:
rand('normal');
A=[-.21 .20;.20 -.21];B=[-36.5853 0;0 36.5853];C=eye(2);D=zeros(2);
%PRBS-signals (low-pass filtered):
q1=1.5e-3*[-1 -1 -1 1 1 1 -1 -1 1 1 1 1 1 1 1 1 1 1 -1 -1];
q2=3.5e-3*[-1 -1 -1 -1 -1 -1 1 -1 -1 -1 -1 -1 -1 1 1 -1 -1 -1 -1 -1];
% Inputs last for 5 minutes (sampling time 1 min.):
for i=1:length(q1),
u(1+5*(i-1):5*i,1)=q1(i)*ones(5,1); u(1+5*(i- 1):5*i,2)=q2(i)*ones(5,1);
end
%Noisy inputs for simulation:
usim(:,1)=u(:,1)+0.03*max(u(:,1))*rand(100,1);
usim(:,2)=u(:,2)+0.03*max(u(:,2))*rand(100,1);
% Obtain noise-free outputs:
t=1:100;
ysim=lsim(A,B,C,D,usim,t);
% Noise on outputs has variance 0.03 degrees centigrades:
y(:,1)=ysim(:,1)+0.03*rand(100,1);
y(:,2)=ysim(:,2)+0.03*rand(100,1);
```

Control Design and Implementation based on Experimental Wind Turbine Models*

Peter M.M. Bongers, Gregor E. van Baars

Mechanical Engineering Systems and Control Group,
Delft University of Technology, Mekelweg 2, 2628 CD Delft, The Netherlands.

1 Introduction

The recent years have shown an increasing interest in wind power plants as alternative schemes for electrical energy generation besides fossil or nuclear power plants. In times of growing environmental conscience a clean and renewable energy source deserves more attention.

However wind power plants are not economically attractive yet. The design of well controlled flexible wind turbines seems to be promising for commercial applications because lighter and less costly construction elements can be used over a longer lifetime.

It is straightforward that an accurate model describing the wind turbine system is necessary to achieve such a design of the wind turbine construction and accompanying control system. The available, first principles, wind turbine models [6] need to be verified in practice before they can be used as a control design tool. In this paper we will pursue the use of models estimated on measurements for control purposes.

A wind turbine is a non-linear system and the wind velocity, driving the turbine, has a stochastic nature. However the average wind velocity determines the operating condition of the turbine.

A problem arising in experimental modeling of wind turbine systems is the fact that the wind conditions are never the same. These effects are not negligible as for helicopters in hoover or flight, for which good models are available. Therefore experiments are not repeatable under the same operating conditions. Thus *the nominal model* describing the wind turbine is hard to find, but it is possible to derive models for several operating conditions. The same holds when a non-linear, first principles, model is approximated by a set of linear models. On the other hand for robust control design *one* low order nominal model is needed, while a model of the uncertainty is desirable.

---

* This research was supported by the CEC under grant JOUR-0110 and the Netherlands agency for energy and environment under grant 40.35-001.10

Iterative schemes for estimation and control design like [15, 19] are not applicable for these systems yet. A possible approach to experimental wind turbine control design is the following:

- Under more or less the same operating conditions a number of models are estimated. These models are allowed to be of different structure and of different order, denote the models by $\hat{P}_i$ for $i = 1, \cdots, N$.
- For the given set of models and a choosen uncertainty structure Δ, the nominal model $\hat{P}_o$ has to be obtained such that the uncertainty measure δ used to describe whole modelset is as small as possible. The set of all models to be controlled can be expressed by $\mathcal{S}_\Delta(\hat{P}_o, \delta)$
- Based on the nominal model $\hat{P}_o$ determine a controller C such that the closed loop system $(C, \hat{P}_o)$ will have the desired performance. Robust performance is achieved if the closed loop system remains stable with small performance degradation for all plants $\hat{P}_i \in \mathcal{S}_\Delta(\hat{P}_o, \delta)$. By the fact that a certain level of performance is achieved for the whole modelset, it is assumed that on the real plant the same level of performance is obtained.

For complicated uncertainty describtions the problem of estimating the nominal model is not solved yet. Therefore we take a practical approach: the nominal model is a member of the modelset. For a specific uncertainty structure Δ, this can be formalized as $\hat{P}_o = \arg\min_j \max_i d(\hat{P}_j, \hat{P}_i)$ for $i, j = 1, \cdots, N$. The uncertainty bound can be expressed by $\delta = \max_i d(\hat{P}_o, \hat{P}_i)$ for $i = 1, \cdots, N$, where $d(.,.)$ is a suitable distance measure. Both uncertainty structure and distance measure should be compatible with the robust control design. For these reasons normalized coprime factor perturbations will be used as uncertainty structure, with the infinity norm as distance measure.

It is not the aim to achieve the highest performance possible, but to achieve a reasonable increase of performance compared to an existing controller. For a more detailed discussion see [16].

The procedure mentioned above is applied to the UNIWEX test facility situated in Stuttgart, Germany. The UNIWEX set up offers the possibility to emulate a wide range of (flexible) wind turbine configurations without any hardware changes [14]. Unlike the electrical power generating wind turbines, at this remote test site the electrical conversion part is replaced by a hydraulic pump.

The layout of the paper is as follows: Section 2 considers the experimental wind turbine setup. In Sect. 3 identification experiments are designed such that all relevant input/output dynamics are excited, then these experiments are carried out on the UNIWEX wind turbine. Using standard prediction error methods experimental models having a Box-Jenkins structure are calculated. Different experimental models are estimated to account for uncertainty in the model description. Section 4 describes a controller design method and some robustness aspects. The nominal model is determined such that the uncertainty set containing all models is as small as possible. The controller will be designed on the nominal model such that it robustly stabilizes all models. The control computer calculates with integer arithmetics, hence the designed controller will be approximated by a controller using only integer arithmetics. Closed loop stability robustness for simultaneous controller and plant perturbaions will be investigated. Finally in Sect. 5 the performance of the implemented controller will be evaluated by means of measurements on the wind turbine.

2 Wind Turbine System Description

In order to describe the wind turbine dynamics and their interactions the following mathematical model has been developed [6]. The complete model consists of a description of each of the wind turbine parts and their mutual connections by interaction variables. Using this structure it is easy to describe different wind turbine configurations because different mathematical descriptions of each submodel are available [6, 8].

We will apply this approach to describe the dynamic behavior of the UNI-WEX turbine in particular. The mathematical model of this wind turbine

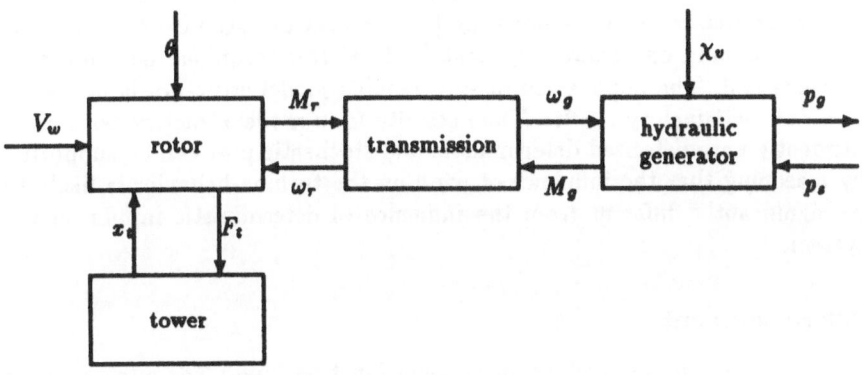

Fig. 1. Block scheme of UNIWEX turbine

consists of the interconnection of submodels having the following characteristics (more details can be found in [5, 14]): The rotor has two blades, without flexibility in the joints. Only pitch angle movements were allowed by the controller. In the aerodynamical model each blade is divided into 10 sections, each section has its own cord, mass, twist and profile. Based on the local wind speed and angle of attack of each section the aerodynamic loads are calculated. The local wind speed depends on wind-shear, tower shadow, dynamic inflow to account for wake effect and turbulence. The tower is considered to be rigid. The **transmission** is described by the first torsional mode of the rotor shaft. The **hydraulic generator** will be described by a spring characteristic.

The complete wind turbine system has the following input and output signals.

Input signals. As mentioned in Sect. 1 the wind speed V_w as felt by the rotor is considered to be a *stochastic* input. The counter torque M_s generated by the hydraulic generator can be adjusted by manipulating the valve position χ_v which is the first *deterministic* input. The blade pitch angle θ can be seen as the second *deterministic* input because it directly influences the angle of attack.

Output signals. In order to gain insight in the dynamic behavior of the wind turbine the following outputs are measured: the rotor shaft speed ω_r, the rotor shaft torque M_r and blade root bending moments which form an indication for the structural loads.

3 Experimental Modeling

3.1 Strategy and Technical Background

To estimate experimental models we apply system identification of the well known *prediction error method* type [11]. In previous studies on rigid wind turbines around one operating point [1, 5, 8] this technique has shown to be successful. The application of Box-Jenkins model structure is necessary to obtain satisfactory results. The necessity for a model structure with independently parameterized deterministic and stochastic part can be supported by reasoning that the influence of wind on the turbine behavior is likely to be significantly different from the influence of deterministic inputs on the system.

3.2 Experiments

We are interested in modeling the transfer function from blade pitch angle to rotor speed and rotor shaft torque, because this transfer function offers the most direct opportunity to design and implement a controller. Because of the ever present fluctuations in wind speed one can not expect the conditions to be constant for a long period. Therefore experiments have to be inspected in order to select intervals of relative constant operational conditions. During the experiments the blade pitch angle of both blades runs through a pseudo random sequence of stepwise changes between 2 and 7 degrees. An example of such a data set is given in Fig. 2. In this figure it can be seen that there is no clear impact on the rotor speed due to the applied pitch steps. Therefore it is possible to apply this procedure without aborting normal operation.

The measured signals are sampled with 50 Hz which implies that dynamics up to 25 Hz can be identified. This should be enough to cover the relevant

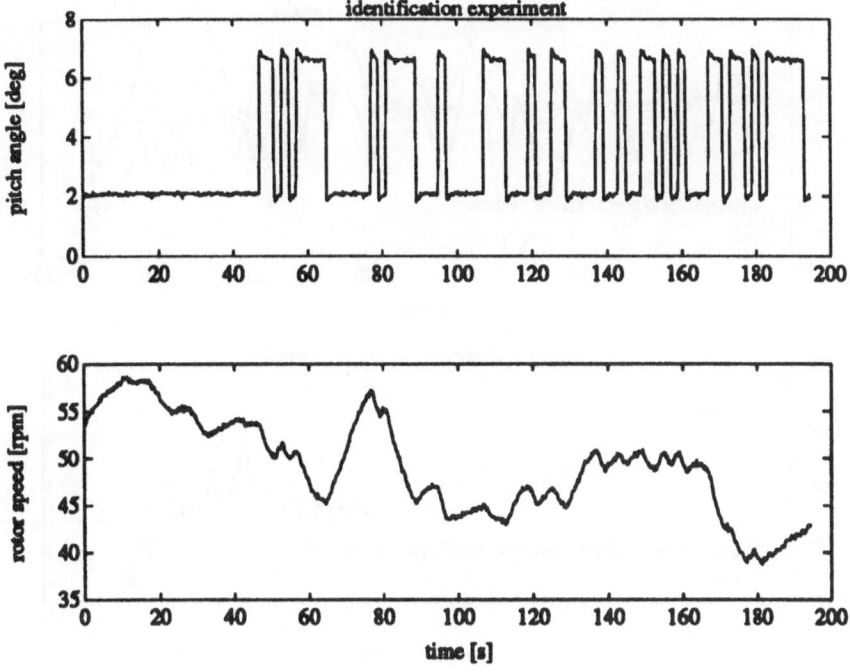

Fig. 2. Experimental data identification experiment

dynamics particularly because this transfer function covers aerodynamic features which are, by nature, rather slow.

3.3 Identification Results

Given the data the experimental transfer functions from blade pitch angle to rotor shaft speed, rotor shaft torque and flap moment are estimated. Different choices of model orders have been investigated, both for the deterministic part and the stochastic part. For deterministic orders > 3 and stochastic orders > 8 both the loss criterion and the step responses show no significant improvement, they only differ slightly. Hence we may assume that all significant linear relations in the data are explained by the model.

As the upper part of Fig. 3 shows the identification technique succeeded in finding a clear cut deterministic relation between pitch angle and rotor speed. The solid line represents the contribution of the deterministic part of the estimated model to the data. The dashed line presents the measured rotor speed which is result of both deterministic and stochastic phenomena. This is rather surprising because the identification data set did not show an evident correlation between pitch steps and rotor speed response. With aerodynamic behavior in mind the deterministic response to the pitch steps seems realistic.

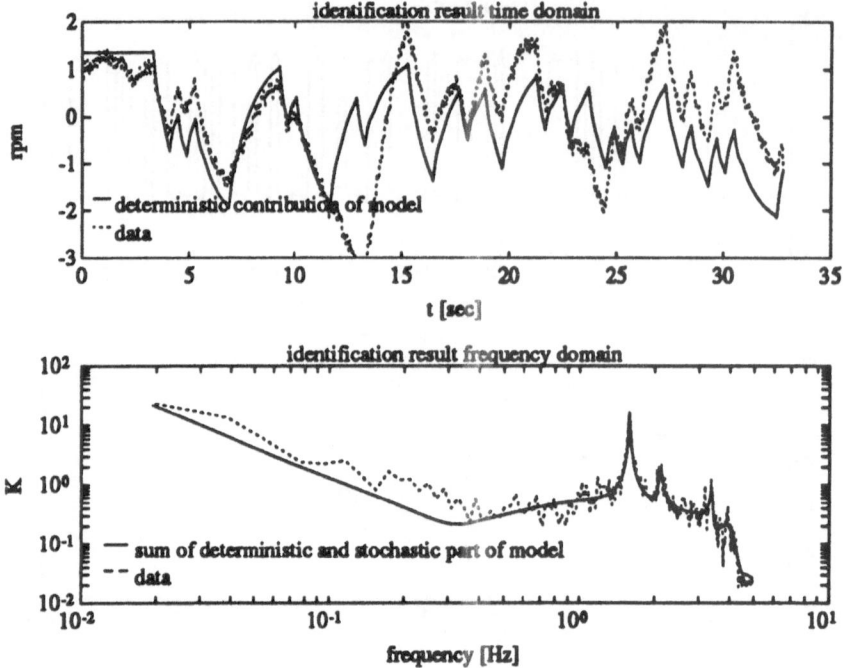

Fig. 3. Identification result time and frequency domain

The accuracy of the stochastic part of the identified models cannot be judged properly in time domain. In order to investigate the validity of the stochastic part we apply the following procedure. First an auto-spectrum of the rotor speed signal is produced. Next the frequency response of the deterministic and stochastic part of the estimated model is calculated. For the model to be reliable the sum of these spectra should correspond to the signal spectrum. The lower part of Fig. 3 shows the result. Clearly there is no perfect match but no essential dynamic behavior is missing in the experimental model so we can be quite confident about both the deterministic and stochastic part of the model.

4 Controller Design and Implementation

In this section the controller design and implementation will be discussed. A more thorough description of the applied control design method can be found in [3, 4].

The control design objective pursued in this paper is to maintain the rotor speed as constant as possible by means of blade pitch movements. Therefore the controller has to achieve to achieve a good servo–behavior. The wind turbine system will be approximated with a set of linear models. These models are estimated with the procedure described in Sect. 3 and the available theoretical model. The controller must therefore not only achieve the desired servo-behavior for the design model but also for the set of linear models. Even deviations of the designed controller have to be allowed in order to implement the controller successfully. For UNIWEX the main source of deviations in the controller is the limitations introduced by integer arithmetics of the control computer.

In this paper we will employ a controller design method based on coprime factorizations of the wind turbine model in order to achieve the desired objectives. First we need some definitions that will be used in the sequel.

Only finite dimensional linear time-invariant systems are considered. Let P be a multi-input multi-output transfer function.

Definition 1 (Factorizations [17]).

- A system P has a right (left) fractional representation if there exist stable $N, M(\tilde{N}, \tilde{M})$ such that:

$$P = NM^{-1} (= \tilde{M}^{-1}\tilde{N}) .$$

- The pair $M, N(\tilde{M}, \tilde{N})$ is right (left) coprime fractional representation (rcf or lcf) if it is a right (left) fractional representation and there exist stable $U, V(\tilde{U}, \tilde{V})$ such that:

$$UN + VM = I \ (\tilde{N}\tilde{U} + \tilde{M}\tilde{V} = I) .$$

- The pair $M, N(\tilde{M}, \tilde{N})$ is called normalized right (left) coprime fractional representation (nrcf or nlcf) if it is a coprime fractional representation and:

$$M^*M + N^*N = I \ (\tilde{M}\tilde{M}^* + \tilde{N}\tilde{N}^* = I)$$

with $M^* = M^T(-s)$.

According to [13] the graph Hankel singular values $\sigma_i^G(P)$ are defined as the Hankel singular values of the nrcf of P $\sigma_i^H(\begin{smallmatrix} M \\ N \end{smallmatrix})$

Definition 2 (Gap distance measure [9, 18]). Suppose P, P_Δ are two plants with nrcf $(N, M), (N_\Delta, M_\Delta)$ respectively.

The gap metric distance $\delta(P, P_\Delta)$ between the two plant is defined as

$$\delta(P, P_\Delta) = \max\{\delta(P, P_\Delta), \delta(P_\Delta, P)\}$$

$$\delta(P, P_\Delta) = \inf_{Q \text{ stable}} \left\| \begin{bmatrix} M \\ N \end{bmatrix} - \begin{bmatrix} M_\Delta \\ N_\Delta \end{bmatrix} Q \right\|_\infty$$

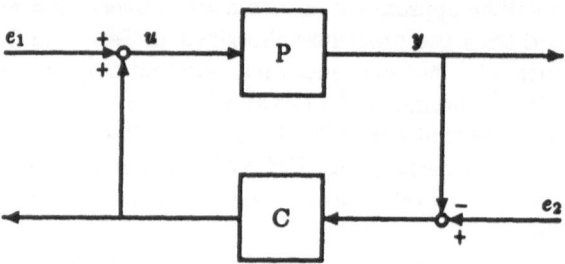

Fig. 4. Feedback configuration

The feedback system considered in the sequel is given in Fig. 4, where P is the wind turbine and C the controller. The closed loop transfer function $T(P, C)$ mapping the inputs (e_1, e_2) onto the outputs (u, y) is:

$$T(P, C) = \begin{bmatrix} I \\ P \end{bmatrix} (I + CP)^{-1} \begin{bmatrix} I & C \end{bmatrix} \tag{1}$$

The (2,2)-element of the $T(P, C)$ transfer function is precisely the servo-behavior, the (1,1)-element is the sensitivity function. Therefore the function $T(P, C)$ represents all closed loop properties to be considered.

In the next theorem a sufficient condition for stability of the closed loop is given when uncertainties in both plant and controller are present.

Theorem 3 (Robustness aspects [10]). *Suppose that $T(P, C)$ is stable, and that both plant and controller are perturbed to P_Δ, C_Δ respectively. Then for all pairs (P_Δ, C_Δ), $T(P_\Delta, C_\Delta)$ is stable provided*

$$\delta(P, P_\Delta) + \delta(C, C_\Delta) < \frac{1}{\|T(P, C)\|_\infty}$$

To obtain as much robustness as possible the controller C should induce the smallest $\|T(P, C)\|_\infty$. For the implementation in existing hardware the order of the controller should also be as low as possible.

Theorem 4 (Control synthesis [7]). *For a given plant P_n (with distinct $\sigma_i^G(P_n)$) of order n there exists a controller C_r of order r, with $r < n$, such that $T(P_n, C_r)$ is stable and the H_∞ bound on the closed loop transfer function is given by:*

$$\|T(P_n, C_r)\|_\infty \leq \frac{1}{\sqrt{1 - (\sigma_1^G + \sum_{i=r+2}^n \sigma_i^G)^2}}$$

Actual Design

On three independent datasets experimental models have been calculated. These models are denoted by s644, s668, s669, the theoretical model is also available as ST. In Fig. 5 the amplitude part of the frequency response of all these models is given. It can be seen that around the cross-over frequency these models are quite different. First a nominal model needs to be chosen.

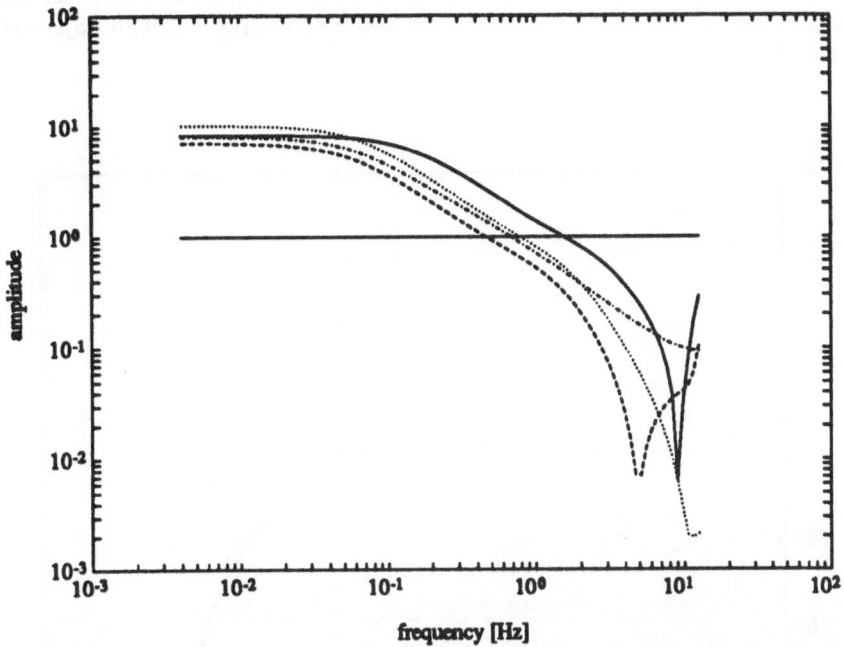

Fig. 5. Frequency response of the modelset. (—) s644, (- -) s668, (...) s669, (-.-) ST

According to Theorem 3 we choose as nominal model, that model which generates the smallest gap (Definition 2) with the other models within the model set. The model P :=s669 is selected to be the nominal model. The gaps between this nominal model end the other models in the model set are: δ(s669,s644) = 0.5512, δ(s669,s668) = 0.3309, δ(s669,ST) = 0.219. This means that if the controller induces $\|T(P,C)\|_\infty < 1.81$, all models are stabilized when there are no controller perturbations.

The design bandwidth of the servo–loop will be chosen to be 1Hz. If a bandwidth larger than 0.5 Hz is achieved, the designed controller will be an improvement of an existing controller. A PI pre–compensator is used to obtain a zero tracking error at low frequencies. This pre–compensator can be incorporated in the nominal model to act as a weighting function [12].

    Using the controller synthesis of Theorem 4 a first order controller for the weighted plant has been designed. The final controller C is composed by the controller of Theorem 4 and the pre–compensator. For the controller C and the nominal plant P, $\|T(P,C)\|_{\infty} = 1.5462$ and generates a robustness margin of 0.6467 which is large enough to incorporate the given perturbations. In Fig. 6 the amplitude part of the frequency response of the design model with PI-pre-compensator is given, together with the obtained sensitivity and complementary sensitivity. It can be seen that the achieved bandwidth for the design model is about 0.7 Hz, hence the nominal controller design objective is satisfied.

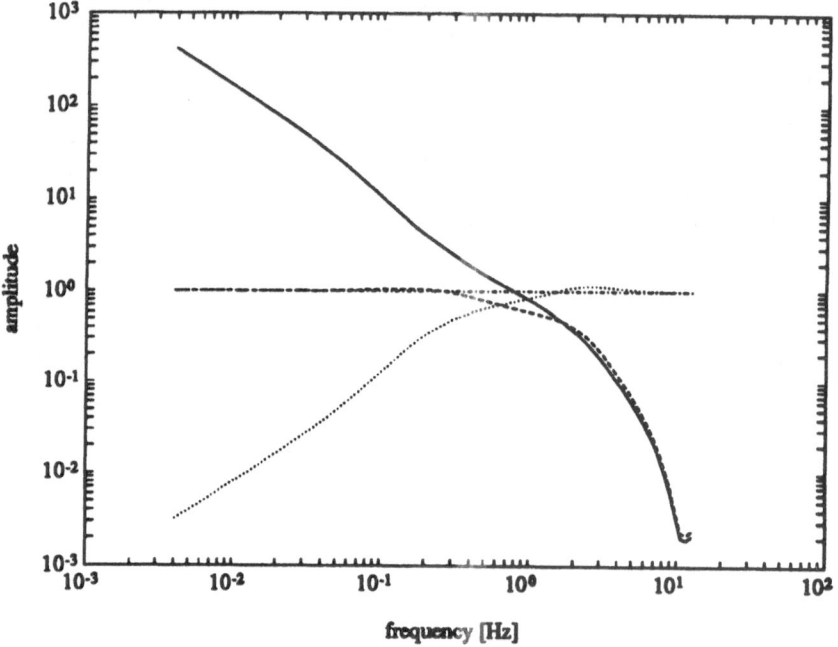

Fig. 6. frequency response closed loop system, (—) weighted s669, (- -) Sensitivity, (...) Complementary sensitivity

The theoretical controller is implemented in assembler code on a PDP 11 control computer. The available software constrains the implemented controller (CIMPL) coefficients to be integers. This implies that we have to approximate our theoretical controller and have to allow controller perturbations. The approximation consists of a sequence of binary shifts and converting floating point numbers to integers. The number of shifts is quite ad hoc.

In Fig. 7 the amplitude part of the frequency response of the theoretical controller, approximated controller and the additive error between the two controllers is given. It can be seen that the deviations are mainly at low frequencies, where they are of approximately the same amplitude as the controller. If the open loop response of both controllers are compared, the im-

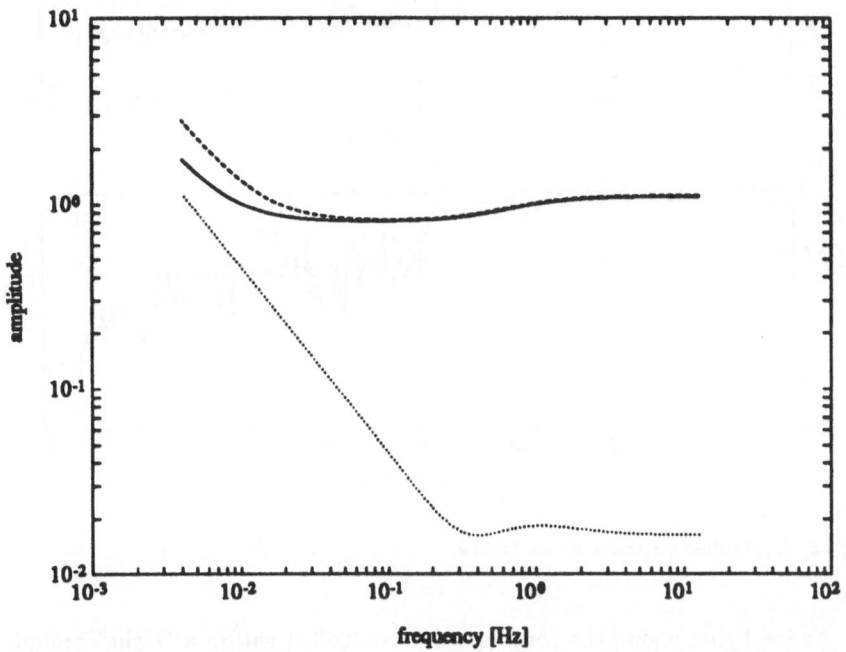

Fig. 7. Frequency response controller. (—) C, (- -) CIMPL, (...) C-CIMPL

plemented one can easily be rejected, see Fig. 7. The gap between the two controllers is $\delta(C, \text{CIMPL}) = 0.008$, which is surprisingly small. According to the robustness test of Theorem 3: $\max(\delta(P, P_i)) = 0.5512 + \delta(C, \text{CIMPL}) = 0.008$ is smaller than the allowed robustness margin of 0.6467, hence the whole model set is stabilized by CIMPL, therefore this controller will be implemented.

5 Experimental Results

After implementation of the controller some experiments were performed. The first experiment shows the difference between the controlled and uncontrolled

wind turbine behavior. In Fig. 8 a clear difference in behavior can be seen
before and after t=39s, the time instant the controller is switched on.

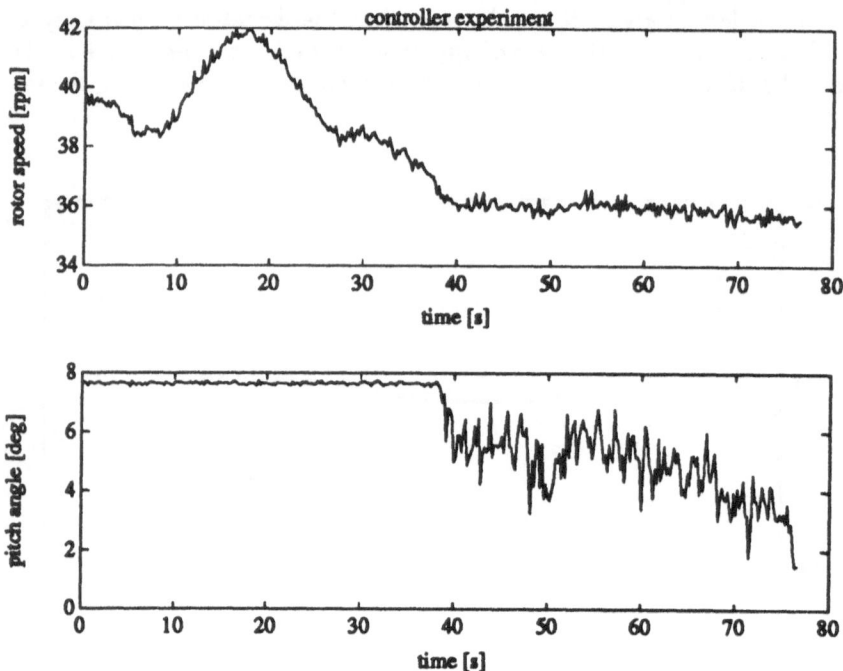

Fig. 8. Controller switches on at t=39s

As the figure shows the rotor speed is controlled within a "tight" bound.
The applied control action (the pitching of the blades) is not excessive and
remains in an acceptable range. Comparing the behavior of the rotor speed
at constant pitch (no control) and active pitch (control) it is clear that the
rotational speed variations due to variations of wind speed can be reduced
sufficiently by this controller.

The second controller experiment investigates the tracking performance.
Figure 9 shows that the controlled system responds accurately to stepwise
changes in the desired rotor speed.

Considering the large rotor inertia a time of about 2s, necessary to reach
the new set point, is quite an improvement compared to the existing controller. The time constant of the controlled wind turbine is approximately
1s, which implies that the designed bandwidth has been achieved. The control actions stay within a reasonable area.

A remark has to be made concerning a large stepwise increase of desired
rotor speed. This may implicate that the pitch angle is steered towards neg-

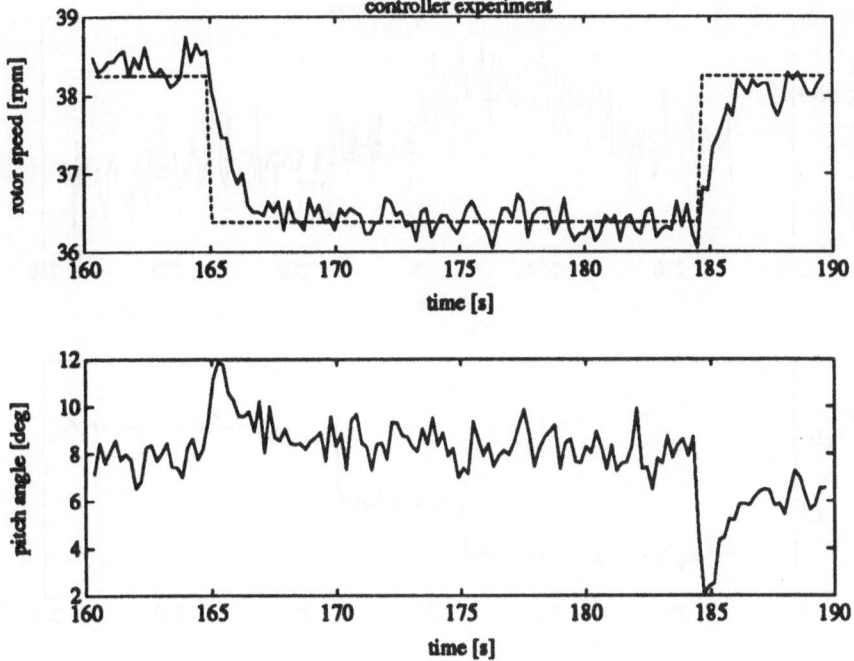

Fig. 9. Servo behavior

ative values. Physically this makes no sense from aerodynamic point of view and can lead to unstable behavior. In order to avoid this unstable behavior the pitch angle must be restricted to nonnegative values. The consequence of this is that such rigorous setpoint changes can not be followed as fast as desired.

The last controller experiment is the most drastic one. It evaluates the performance of the controller under stepwise changes in the absorbed torque by the generator system. This can be seen as abrupt disturbances that influence the system in addition to the fluctuating wind speed. In fact there are two kinds of disturbances acting on the system: the wind stochastics which are always present and the deterministic induced variations of the counter torque. As Fig. 10 shows the controller acts satisfactory particularly for these drastic disturbances (the torque is doubled). The speed remains in a range of approximately .5 rpm, which is slightly more than the inaccuracy of the rotor speed measurement.

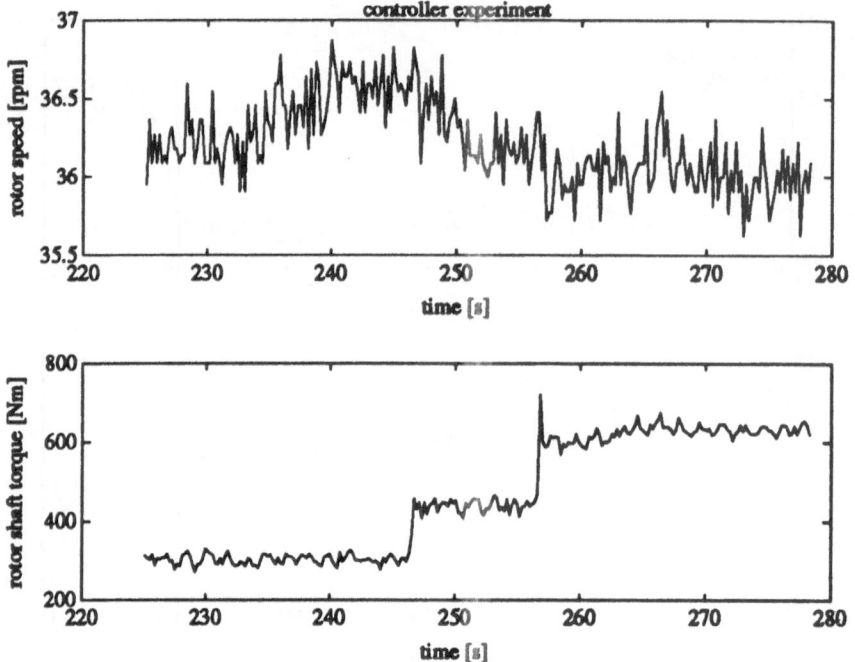

Fig. 10. Response to disturbances

6 Conclusions

A procedure consisting of identification, robust controller design and implementation has been successfully applied to a real life wind turbine system.

Since one can not expect the operational conditions during identification and controller experiments to be the same more than one experimental model has been estimated. A nominal model has been calculated such that it induces the smallest gap with the models in the set of estimated models.

A robust controller has been designed which stabilizes the set of all these experimental models. Implementation of the controller showed satisfactory behavior with respect to the design objective both for setpoint changes and stepwise disturbances of absorbed torque. Despite restrictions due to integer arithmetics in the control computer.

Further research will focus on application of this procedure to more flexible configurations.

References

1. Baars G.E. van (1992). *Experimental validation of the UNIWEX wind turbine.* Delft University of Technology, The Netherlands.

2. Bongers P.M.M. (1990). *DUWECS Reference Guide, Delft University Wind Energy Conversion Simulation Program*. Delft University of Technology, The Netherlands.

3. Bongers P.M.M. (1992) *Factorizational approach to robust control: application to a flexible wind turbine*. Dr. Dissertation, Delft University of Technology, The Netherlands, In preparation.

4. Bongers P.M.M. (1992) Robust control using coprime factorizations, application to a flexible wind turbine. *To appear in Proc.31th IEEE Conf. on Decision and Contr.*

5. Bongers P.M.M., G.E. van Baars (1991). Experimental validation of a flexible wind turbine model. *Proc. 30th IEEE Conf. on Decision and Control.*, Dec.11-13, 1991, Brighton, UK, pp. 1660-1661.

6. Bongers P.M.M., W.A.A.M. Bierbooms, Sj. Dijkstra, Th. van Holten (1990). *An Integrated Dynamic Model of a Flexible Wind Turbine*. Delft University of Technology, The Netherlands.

7. Bongers P.M.M., O.H. Bosgra (1990). Low order H_∞ controller synthesis. *Proc. 29th Conf. Decision and Control*. Hawaii, 1990, USA, pp. 194-199.

8. Bongers P.M.M., T.G. van Engelen, Sj. Dijkstra, Z.D.Q.P. Kock (1989). Optimal control of a wind turbine in full load. *Proc. European Wind Energy Conference*. 1989, Glasgow, UK.

9. El-Sakkary A.K. (1985). The gap metric: Robustness of Stabilization of Feedback Systems. *IEEE Trans. Automat. Contr.*, vol. AC-30, pp.240-247.

10. Georgiou T.T., M.M. Smith (1990). Optimal robustness in the Gap metric. *IEEE Trans. Automat. Contr.*, vol. AC-35, pp. 673-686.

11. Ljung L. (1987). *System Identification: theory for the user*. Prentice Hall, Inc. Englewood Cliffs, New jersey.

12. McFarlane D.C., K. Glover (1989). *Robust controller design using normalized coprime factor plant descriptions*. Lecture Notes in Control and Information Sciences, vol.138, Springer Verlag, Berlin, Germany.

13. Meyer D.G. (1988). A fractional approach to model reduction. *Proc. Amer. Control Conf.*, pp. 1041-1047.

14. Müller M. (1989). Experimental investigations with the universal test wind turbine. *Proc. European Wind energy Conf.*, 1989, Glasgow, Schotland, pp. 832-836.

15. R.J.P. Schrama (1992). *Approximate Identification and Control Design with Application to a Mechanical System*. PhD. Thesis, Delft University of Technology, May 1992.

16. Van den Hof, P.M.J., R.J.P. Schrama, P.M.M. Bongers (1992). On nominal models, model uncertainty and iterative methods in identification and control design. *this issue*

17. Vidyasagar M., H. Schneider, B.A. Francis (1982). Algebraic and topological aspects of feedback stabilization. *IEEE Trans. Automat. Contr.*, vol. AC-27, pp. 880-893.

18. Zames G., A.K. El-Sakkary (1980). Unstable systems and feedback: The Gap metric. *Proc. 18th Allerton Conf.*, pp. 380-385.

19. Zang Z., M.R. Gevers, R. Bitmead (1992). H_2 Iterative model refinement and control robustness enhancement.*IEEE Trans. Automat. Contr.*

Lecture Notes in Control and Information Sciences

Edited by M. Thoma

1989–1993 Published Titles:

Vol. 135: Nijmeijer, Hendrik; Schumacher, Johannes M. (Eds.)
Three Decades of Mathematical System Theory. A Collection of Surveys at the Occasion of the 50th Birthday of Jan C. Willems.
562 pp. 1989 [3-540-51605-0]

Vol. 136: Zabczyk, Jerzy W. (Ed.)
Stochastic Systems and Optimization.
Proceedings of the 6th IFIP WG 7.1 Working Conference, Warsaw, Poland, September 12-16, 1988.
374 pp. 1989 [3-540-51619-0]

Vol. 137: Shah, Sirish L.; Dumont, Guy (Eds.)
Adaptive Control Strategies for Industrial Use.
Proceedings of a Workshop held in Kananaskis, Canada, 1988.
360 pp. 1989 [3-540-51869-X]

Vol. 138: McFarlane, Duncan C.; Glover, Keith
Robust Controller Design Using Normalized Coprime Factor Plant Descriptions.
206 pp. 1990 [3-540-51851-7]

Vol. 139: Hayward, Vincent; Khatib, Oussama (Eds.)
Experimental Robotics I. The First International Symposium, Montreal, June 19-21, 1989.
613 pp. 1990 [3-540-52182-8]

Vol. 140: Gajic, Zoran; Petkovski, Djordjija; Shen, Xuemin (Eds.)
Singularly Perturbed and Weakly Coupled Linear Control Systems. A Recursive Approach.
202 pp. 1990 [3-540-52333-2]

Vol. 141: Gutman, Shaul
Root Clustering in Parameter Space.
153 pp. 1990 [3-540-52361-8]

Vol. 142: Gündes, A. Nazli; Desoer, Charles A.
Algebraic Theory of Linear Feedback Systems with Full and Decentralized Compensators.
176 pp. 1990 [3-540-52476-2]

Vol. 143: Sebastian, H.-J.; Tammer, K. (Eds.)
System Modelling and Optimizaton.
Proceedings of the 14th IFIP Conference, Leipzig, GDR, July 3-7, 1989.
960 pp. 1990 [3-540-52659-5]

Vol. 144: Bensoussan, A.; Lions, J.L. (Eds.)
Analysis and Optimization of Systems.
Proceedings of the 9th International Conference. Antibes, June 12-15, 1990.
992 pp. 1990 [3-540-52630-7]

Vol. 145: Subrahmanyam, M. Bala
Optimal Control with a Worst-Case Performance Criterion and Applications.
133 pp. 1990 [3-540-52822-9]

Vol. 146: Mustafa, Denis; Glover, Keith
Minimum Entropy H Control.
144 pp. 1990 [3-540-52947-0]

Vol. 147: Zolesio, J.P. (Ed.)
Stabilization of Flexible Structures. Third Working Conference, Montpellier, France, January 1989.
327 pp. 1991 [3-540-53161-0]

Vol. 148: Not published

Vol. 149: Hoffmann, Karl H; Krabs, Werner (Eds.)
Optimal Control of Partial Differential Equations. Proceedings of IFIP WG 7.2 - International Conference. Irsee, April, 9-12, 1990.
245 pp. 1991 [3-540-53591-8]

Vol. 150: Habets, Luc C.
Robust Stabilization in the Gap-topology.
126 pp. 1991 [3-540-53466-0]

Vol. 151: Skowronski, J.M.; Flashner, H.;
Guttalu, R.S. (Eds.)
Mechanics and Control. Proceedings of the 3rd
Workshop on Control Mechanics, in Honor of
the 65th Birthday of George Leitmann, January
22-24, 1990, University of Southern
California.
497 pp. 1991 [3-540-53517-9]

Vol. 152: Aplevich, J. Dwight
Implicit Linear Systems.
176 pp. 1991 [3-540-53537-3]

Vol. 153: Hajek, Otomar
Control Theory in the Plane.
269 pp. 1991 [3-540-53553-5]

Vol. 154: Kurzhanski, Alexander; Laseicka,
Irena (Eds.)
Modelling and Inverse Problems of Control for
Distributed Parameter Systems. Proceedings of
IFIP WG 7.2 - IIASA Conference, Laxenburg,
Austria, July 1989.
170 pp. 1991 [3-540-53583-7]

Vol. 155: Bouvet, Michel; Bienvenu, Georges
(Eds.)
High-Resolution Methods in Underwater
Acoustics.
244 pp. 1991 [3-540-53716-3]

Vol. 156: Hämäläinen, Raimo P.; Ehtamo, Harri
K. (Eds.)
Differential Games - Developments in
Modelling and Computation. Proceedings of
the Fourth International Symposium on
Differential Games and Applications, August
9-10, 1990, Helsinki University of Technology,
Finland.
292 pp. 1991 [3-540-53787-2]

Vol. 157: Hämäläinen, Raimo P.; Ehtamo, Harri
K. (Eds.)
Dynamic Games in Economic Analysis.
Proceedings of the Fourth International
Symposium on Differential Games and
Applications. August 9-10, 1990, Helsinki
University of Technology, Finland.
311 pp. 1991 [3-540-53785-6]

Vol. 158: Warwick, Kevin; Karny, Miroslav;
Halouskova, Alena (Eds.)
Advanced Methods in Adaptive Control for
Industrial Applications.
331 pp. 1991 [3-540-53835-6]

Vol. 159: Li, Xunjing; Yong, Jiongmin (Eds.)
Control Theory of Distributed Parameter
Systems and Applications. Proceedings of the
IFIP WG 7.2 Working Conference, Shanghai,
China, May 6-9, 1990.
219 pp. 1991 [3-540-53894-1]

Vol. 160: Kokotovic, Petar V. (Ed.)
Foundations of Adaptive Control.
525 pp. 1991 [3-540-54020-2]

Vol. 161: Gerencser, L.; Caines, P.E. (Eds.)
Topics in Stochastic Systems: Modelling,
Estimation and Adaptive Control.
1991 [3-540-54133-0]

Vol. 162: Canudas de Wit, C. (Ed.)
Advanced Robot Control. Proceedings of the
International Workshop on Nonlinear and
Adaptive Control: Issues in Robotics, Grenoble,
France, November 21-23, 1990.
Approx. 330 pp. 1991 [3-540-54169-1]

Vol. 163: Mehrmann, Volker L.
The Autonomous Linear Quadratic Control
Problem. Theory and Numerical Solution.
177 pp. 1991 [3-540-54170-5]

Vol. 164: Lasiecka, Irena; Triggiani, Roberto
Differential and Algebraic Riccati Equations
with Application to Boundary/Point Control
Problems: Continuous Theory and
Approximation Theory.
160 pp. 1991 [3-540-54339-2]

Vol. 165: Jacob, Gerard; Lamnabhi-Lagarrigue,
F. (Eds.)
Algebraic Computing in Control. Proceedings
of the First European Conference, Paris, March
13-15, 1991.
384 pp. 1991 [3-540-54408-9]

Vol. 166: Wegen, Leonardus L. van der
Local Disturbance Decoupling with Stability for
Nonlinear Systems.
135 pp. 1991 [3-540-54543-3]

Vol. 167: Rao, Ming
Integrated System for Intelligent Control.
133 pp. 1992 [3-540-54913-7]

Vol. 168: Dorato, Peter; Fortuna, Luigi;
Muscato, Giovanni
Robust Control for Unstructured Perturbations:
An Introduction.
118 pp. 1992 [3-540-54920-X]

Vol. 169: Kuntzevich, Vsevolod M.; Lychak,
Michael
Guaranteed Estimates, Adaptation and
Robustness in Control Systems.
209 pp. 1992 [3-540-54925-0]

Vol. 170: Skowronski, Janislaw M.; Flashner,
Henryk; Guttalu, Ramesh S. (Eds.)
Mechanics and Control. Proceedings of the 4th
Workshop on Control Mechanics, January
21-23, 1991, University of Southern
California, USA.
302 pp. 1992 [3-540-54954-4]

Vol. 171: Stefanidis, P.; Paplinski, A.P.;
Gibbard, M.J.
Numerical Operations with Polynomial
Matrices: Application to Multi-Variable
Dynamic Compensator Design.
206 pp. 1992 [3-540-54992-7]

Vol. 172: Tolle, H.; Ersü, E.
Neurocontrol: Learning Control Systems
Inspired by Neuronal Architectures and Human
Problem Solving Strategies.
220 pp. 1992 [3-540-55057-7]

Vol. 173: Krabs, W.
On Moment Theory and Controllability of
Non-Dimensional Vibrating Systems and
Heating Processes.
174 pp. 1992 [3-540-55102-6]

Vol. 174: Beulens, A.J. (Ed.)
Optimization-Based Computer-Aided Modelling
and Design. Proceedings of the First Working
Conference of the New IFIP TC 7.6 Working
Group, The Hague, The Netherlands, 1991.
268 pp. 1992 [3-540-55135-2]

Vol. 175: Rogers, E.T.A.; Owens, D.H.
Stability Analysis for Linear Repetitive
Processes.
197 pp. 1992 [3-540-55264-2]

Vol. 176: Rozovskii, B.L.; Sowers, R.B. (Eds.)
Stochastic Partial Differential Equations and
their Applications. Proceedings of IFIP WG 7.1
International Conference, June 6-8, 1991,
University of North Carolina at Charlotte, USA.
251 pp. 1992 [3-540-55292-8]

Vol. 177: Karatzas, I.; Ocone, D. (Eds.)
Applied Stochastic Analysis. Proceedings of a
US-French Workshop, Rutgers University, New
Brunswick, N.J., April 29-May 2, 1991.
317 pp. 1992 [3-540-55296-0]

Vol. 178: Zolésio, J.P. (Ed.)
Boundary Control and Boundary Variation.
Proceedings of IFIP WG 7.2 Conference,
Sophia- Antipolis,France, October 15-17,
1990.
392 pp. 1992 [3-540-55351-7]

Vol. 179: Jiang, Z.H.; Schaufelberger, W.
Block Pulse Functions and Their Applications in
Control Systems.
237 pp. 1992 [3-540-55369-X]

Vol. 180: Kall, P. (Ed.)
System Modelling and Optimization.
Proceedings of the 15th IFIP Conference,
Zurich, Switzerland, September 2-6, 1991.
969 pp. 1992 [3-540-55577-3]

Vol. 181: Drane, C.R.
Positioning Systems - A Unified Approach.
168 pp. 1992 [3-540-55850-0]

Vol. 182: Hagenauer, J. (Ed.)
Advanced Methods for Satellite and Deep
Space Communications. Proceedings of an
International Seminar Organized by Deutsche
Forschungsanstalt für Luft-und Raumfahrt
(DLR), Bonn, Germany, September 1992.
196 pp. 1992 [3-540-55851-9]

Vol. 183: Hosoe, S. (Ed.)
Robust Control. Proceedings of a Workshop
held in Tokyo, Japan, June 23-24, 1991.
225 pp. 1992 [3-540-55961-2]

Vol. 184: Duncan, T.E.; Pasik-Duncan, B.
(Eds.)
Stochastic Theory and Adaptive Control.
Proceedings of a Workshop held in Lawrence,
Kansas, September 26-28, 1991.
500 pages. 1992 [3-540-55962-0]

Vol. 185: Curtain, R.F. (Ed.); Bensoussan, A.; Lions, J.L.(Honorary Eds.)
Analysis and Optimization of Systems: State and Frequency Domain Approaches for Infinite-Dimensional Systems. Proceedings of the 10th International Conference, Sophia-Antipolis, France, June 9-12, 1992.
648 pp. 1993 [3-540-56155-2]

Vol. 186: Sreenath, N.
Systems Representation of Global Climate Change Models. Foundation for a Systems Science Approach.
288 pp. 1993 [3-540-19824-5]

Vol. 187: Morecki, A.; Bianchi, G.; Jaworeck, K. (Eds.)
RoManSy 9: Proceedings of the Ninth CISM-IFToMM Symposium on Theory and Practice of Robots and Manipulators.
476 pp. 1993 [3-540-19834-2]

Vol. 188: Naidu, D. Subbaram
Aeroassisted Orbital Transfer: Guidance and Control Strategies.
192 pp. 1993 [3-540-19819-9]

Vol. 189: Ilchmann, Achim
Non-Identifier-Based High-Gain Adaptive Control.
220 pp. 1993 [3-540-19845-8]

Vol. 190: Chatila, R; Hirzinger, G (Eds.)
Experimental Robotics II: The 2nd International Symposium, Toulouse, France, June 25-27 1991.
580 pp. 1993 [3-540-19851-2]

Vol. 191: Blondel, V.
Simultaneous Stabilization of Linear Systems.
212 pp. 1993 [3-540-19862-8]